Prospective Studies on Children With Sex Chromosome Aneuploidy

RECENT TITLES PUBLISHED BY ALAN R. LISS, INC.
FOR MARCH OF DIMES BIRTH DEFECTS FOUNDATION

Birth Defects Compendium, Second Edition, Daniel Bergsma, *Editor*

BIRTH DEFECTS: ORIGINAL ARTICLE SERIES

1982 — Volume 18

No. 1 **Craniofacial Anomalies: New Perspectives,** Carlos F. Salinas, *Editor*
No. 3A **Annual Review of Birth Defects, 1981, Prenatal Diagnosis and Mechanisms of Teratogenesis,** William L. Nyhan and Kenneth Lyons Jones, *Editors*
No. 3B **Annual Review of Birth Defects, 1981, Dysmorphology,** William L. Nyhan and Kenneth Lyons Jones, *Editors*
No. 4 **Children With Sex Chromosome Aneuploidy: Follow-Up Studies,** Donald A. Stewart, *Editor*
No. 5 **Cumulative Index: Birth Defects: Original Article Series Volumes I (1965)–XVI (1980),** Natalie W. Paul, *Editor*
No. 6 **Genetic Eye Diseases: Retinitis Pigmentosa and Other Inherited Eye Disorders,** Edward Cotlier, Irene H. Maumenee, and Elaine R. Berman, *Editors*
No. 7 **Thalassemia: Recent Advances in Detection and Treatment,** Antonio Cao, Ugo Carcassi, and Peter T. Rowley, *Editors*

1983 — Volume 19

No. 1 **Dentition: Genetic Effects,** Ronald J. Jorgenson, *Editor*
No. 3 **Primary Immunodeficiency Diseases,** Ralph J. Wedgwood, Fred S. Rosen, and Natalie W. Paul, *Editors*
No. 4 **Nervous System Regeneration,** Bernard Haber, J. Regino Perez-Polo, George A. Hashim, and Anna Maria Giuffrida Stella, *Editors*
No. 5 **Annual Review of Birth Defects, 1982, Birth Defects: Clinical and Ethical Considerations,** Sara C. Finley, Wayne H. Finley, and Charles E. Flowers, Jr., *Editors*

1984 — Volume 20

No. 1 **CMV: Pathogenesis and Prevention of Human Infection,** Stanley A. Plotkin, Susan Michelson, Joseph S. Pagano, and Fred Rapp, *Editors*
No. 3 **Craniofacial Mesenchyme in Morphogenesis and Malformation,** Kenneth S. Brown and Carlos F. Salinas, *Editors*

1985 — Volume 21

No. 2 **Diagnostic Accuracy: Effect on Treatment Planning,** Robert J. Shprintzen and Natalie W. Paul, *Editors*
No. 5 **The Patient Within the Patient: Problems in Perinatal Medicine,** Bruce K. Young, *Editor*

1986 — Volume 22

No. 1 **Bone Marrow Transplantation for Treatment of Lysosomal Storage Diseases,** William Krivit and Natalie W. Paul, *Editors*
No. 3 **Prospective Studies on Children With Sex Chromosome Aneuploidy,** Shirley G. Ratcliffe and Natalie Paul, *Editors*

Please contact the publisher for information about previous titles in this series published by Alan R. Liss, Inc.

March of Dimes Birth Defects Foundation
Birth Defects: Original Article Series, Volume 22, Number 3, 1986

Prospective Studies on Children With Sex Chromosome Aneuploidy

Editors

Shirley G. Ratcliffe
MRC Clinical and Population
Cytogenetics Unit
Western General Hospital
Edinburgh, Scotland, United Kingdom

Natalie Paul
March of Dimes Birth Defects Foundation
White Plains, New York

ALAN R. LISS, INC., NEW YORK

To enhance medical communication in the birth defects field, the March of Dimes Birth Defects Foundation publishes the *Birth Defects Compendium (2nd Edition)*, an *Original Article Series, a Reprint Series, Nursing Staff Development Modules, Genetics in Practice* (a quarterly newsletter, and provides a series of films and related brochures).

Further information can be obtained from:

Professional Education Department
March of Dimes Birth Defects Foundation
1275 Mamaroneck Avenue
White Plains, New York, 10605

Published by:
Alan R. Liss, Inc.
41 East 11th Street
New York, New York 10003

Library of Congress Cataloging-in-Publication Data
Prospective Studies on Children with Sex Chromosome Aneuploidy.

(Birth defects, original article series; v. 22, no.3)
Reports from a workshop sponsored by the March of Dimes Birth Defects Foundation, held in Edinburgh, June 1984.
Includes bibliographies and index.
1. Sex chromosome abnormalities in children—Longitudinal studies—Congresses. 2. Aneuploidy—Longitudinal studies—Congresses. I. Ratcliffe, Shirley G. II. Paul, Natalie W. III. March of Dimes Birth Defects Foundation. IV. Series: Birth defects original article series; v. 22, no. 3. [DNLM:
1. Aneuploidy—congresses. 2. Sex Chromosome Abnormalities—congresses. W1 BI966 v.22 no. 3 /
QS 677 P966 1984]
RG626.B63 vol. 22, no. 3 616′.043 s 86-21486
[RJ47.4] [618.92′0042]
ISBN 0-8451-1062-4

Second Printing, June 1987

The **March of Dimes Birth Defects Foundation** is dedicated to the goals of preventing birth defects and ameliorating their consequences for patients, families, and society.

As part of our efforts to achieve these goals, we sponsor, or participate in, a variety of scientific meetings where all questions relating to birth defects are freely discussed. Through our professional education program we speed the dissemination of information by publishing the proceedings of these and other meetings. From time to time, we also reprint pertinent journal articles to help achieve our goals. Now and then, in the course of these articles or discussions, individual viewpoints may be expressed which go beyond the purely scientific and into controversial matters. It should be noted, therefore, that personal viewpoints about such matters will not be censored but this does not constitute an endorsement of them by the **March of Dimes Birth Defects Foundation.**

Contents

Contributors

John D. Bailey, MD, FRCP (C), Division of Endocrinology, The Hospital for Sick Children, Toronto, Ontario M5G 1X8, Canada **[119]**

Anthony S. Bashir, PhD, Judge Baker Guidance Center, Boston, MA 02115 **[1]**

Bruce G. Bender, PhD, Department of Psychiatry, University of Colorado School of Medicine, National Jewish Center for Immunology and Respiratory Medicine, Denver, CO 80206 **[23]**

Joyce B. Borelli, MD, Department of Psychiatry, University of Colorado School of Medicine, National Jewish Center for Immunology and Respiratory Medicine, Denver, CO 80206 **[23]**

Michael F. DeNapoli, BA, Judge Baker Guidance Center, Boston, MA 02115 **[1]**

R. de von Flindt, PhD, Departments of Pediatrics and Psychiatry, University of Manitoba and Children's Hospital, Winnipeg, Manitoba R3E 0W3, Canada **[183]**

Shinzo Egi, MD, Department of Pediatrics, University of Tokyo, Tokyo, Japan **[251]**

Gunnar Eriksen, Gynecological-Obstetric Department Y, Århus Municipal Hospital, Århus, Denmark **[209]**

J.A. Evans, PhD, FCCMG, Department of Human Genetics, University of Manitoba and Children's Hospital, Winnipeg, Manitoba, Canada R3E 0W3 **[183]**

Jens Faaborg-Andersen, Gynecological-Obstetric Department Y, Århus Municipal Hospital, Århus, Denmark **[209]**

John M. Graham, Jr., MD, ScD, Judge Baker Guidance Center, Boston, MA 02115 **[1]**

C. Greenberg, MD, FCCP, FCCMG, Department of Human Genetics, University of Manitoba and Children's Hospital, Winnipeg, Manitoba R3E 0W3, Canada **[183]**

J.L. Hamerton, DSc, FCCMG, Department of Human Genetics, University of Manitoba and Children's Hospital, Winnipeg, Manitoba R3E 0W3, Canada **[183]**

Karl Brogaard Hansen, Gynecological-Obstetric Department Y, Århus Municipal Hospital, Århus, Denmark **[209]**

Makoto Higurashi, MD, Department of Health Sciences, Yamanashi Medical College, Yamanashi, Japan **[251]**

The numbers in brackets are the opening page numbers of the contributors' articles.

Lone Hvidman, Gynecological-Obstetric Department Y, Århus Municipal Hospital, Århus, Denmark **[209]**

Kumiko Iijima, Department of Maternal and Child Health, University of Tokyo, Tokyo, Japan **[251]**

Yukie Ikeda, DHS, Department of Education for Handicapped Children, Tsukuba University, Ibaragi, Japan **[251]**

Shigehiko Kamashita, Department of Pediatrics, University of Tokyo, Tokyo, Japan **[251]**

Hisashi Kawai, Department of Psychology, Psychiatric Research Institute of Tokyo, Tokyo, Japan **[251]**

Bente Krag-Olsen, Gynecological-Obstetric Department Y, Århus Municipal Hospital, Århus, Denmark **[209]**

Nicholas T. Lange, MA, Judge Baker Guidance Center, Boston, MA 02115 **[1]**

Martha F. Leonard, MD, Yale Child Study Center and Department of Pediatrics, Yale University Medical School, Yale University, New Haven, CT 06510 **[221]**

Ib Moulvad, Gynecological-Obstetric Department Y, Århus Municipal Hospital, Århus, Denmark **[209]**

Lynne Murray, MA, PhD, MRC Clinical and Population Cytogenetics Unit, Western General Hospital, Edinburgh EH4 2XU, Scotland, United Kingdom **[73]**

Charles T. Netley, PhD, Department of Psychology, The Hospital for Sick Children, Toronto, Ontario M5G 1X8, Canada **[119, 293]**

Johannes Nielsen, Cytogenetic Laboratory, Århus Psychiatric Hospital, Risskov, Denmark **[209]**

Elizabeth Park, PhD, Research Institute of The Hospital for Sick Children, Toronto, Ontario M5G 1X8, Canada **[119]**

Mary H. Puck, MA, Department of Psychiatry, University of Colorado School of Medicine, National Jewish Center for Immunology and Respiratory Medicine, Denver, CO 80206 **[23]**

Theodore T. Puck, PhD, Eleanor Roosevelt Institute for Cancer Research, Departments of Biochemistry, Biophysics, Genetics, and Medicine, University of Colorado Health Sciences Center, Denver, CO 80262 **[307]**

S. Ramsay, RN, Department of Human Genetics, University of Manitoba and Children's Hospital, Winnipeg, Manitoba R3E 0W3, Canada **[183]**

Shirley G. Ratcliffe, MB, BS, FRCP, MRC Clinical and Population Cytogenetics Unit, Western General Hospital, Edinburgh EH4 2XU, Scotland, United Kingdom **[xiii, 73]**

Julius B. Richmond, MD, Judge Baker Guidance Center, Boston, MA 02115 **[1]**

Arthur Robinson, MD, Departments of Pediatrics, Biochemistry, Biophysics, and Genetics, University of Colorado School of Medicine, National Jewish Center for Immunology and Respiratory Medicine, Denver, CO 80206 **[23]**

Joanne Rovet, PhD, Department of Psychology, The Hospital for Sick Children, Toronto, Ontario M5G 1X8, Canada **[119]**

James A. Salbenblatt, MD, Department of Pediatrics, University of Colorado School of Medicine, National Jewish Center for Immunology and Respiratory Medicine, Denver, CO 80206 **[23]**

Masaya Segawa, Segawa Neurological Clinic for Children, Tokyo, Japan **[251]**

Annette R. Silbert, PhD, Judge Baker Guidance Center, Boston, MA 02115 **[1]**

Sara Sparrow, PhD, Yale Child Study Center (Psychology) and Department of Psychology, Yale University, New Haven, CT 06510 **[221]**

Donald A. Stewart, MD, FRCP (C), Outpatient Department, The Hospital for Sick Children, Toronto, Ontario M5G 1X8, Canada **[119]**

Fumihiko Tanaka, Department of Pathology, University of Tokyo, Tokyo, Japan **[251]**

Peter Teague, BSc, MSc, MRC Clinical and Population Cytogenetics Unit, Western General Hospital, Edinburgh EH4 2XU, Scotland, United Kingdom **[73]**

Alice Theilgaard, DMSc, Department of Psychiatry, University Hospital, Copenhagen, Denmark **[277]**

Poul Videbech, Gynecological-Obstetric Department Y, Århus Municipal Hospital, Århus, Denmark **[209]**

Stanley Walzer, MD, Judge Baker Guidance Center, Boston, MA 02115 **[1]**

Jeremy S.D. Winter, MD, Departments of Paediatrics and Medicine, University of Manitoba, Winnipeg, Manitoba, Canada R3E 0W3 **[23]**

Mogens Wohlert, Gynecological-Obstetric Department Y, Århus Municipal Hospital, Århus, Denmark **[209]**

Introduction

The March of Dimes Birth Defects Foundation continued its support of research into birth defects by sponsoring the fourth workshop for investigators carrying out prospective longitudinal studies on children with sex chromosome abnormalities (SCA). This took place in Edinburgh in June 1984; reports from that meeting are gathered together in this monograph, which also includes some data collected in the interval between the workshop and the publication of this volume.

Newborn cytogenetic surveys commenced in the mid-1960s and continued over the next decade — at the present time such studies are continuing only in Denmark.

307 children with SCA have been identified in seven studies of consecutive livebirths covering 199,898 infants from Denver, Colorado; Boston, Massachussetts; New Haven, Connecticut; Toronto and Winnipeg, Canada; Århus, Denmark and Edinburgh, Scotland.

In addition, we have included a report of the study from Japan (Page 251) although here the children with SCA were identified by a combination of sex chromatin screening and examination for congenital malformations.

In order to extend the age range into the young adult period, Dr. Alice Theilgaard from Copenhagen joined the workshop for the first time and gave a detailed description (Page 277) of the psychological findings in young men with SCA identified in the Witkin study of young military recruits. These men were selected from a complete birth cohort from 1944 to 1947 with the restriction that they should be over 184 cm tall. Each study has its own characteristics and it may be helpful to the reader to reiterate these briefly.

The Denver Study (Page 23) was carried out from 1964 to 1974 using mainly X chromatin examination of amnion cells from the placenta. 40,371 consecutive liveborn male and female infants were tested and this resulted in the identification of 69 infants with SCA. Eight of these died in the first month of life. The prospective study has covered physical and psychological development, and has used siblings as controls, at times for group comparisons, less often as matched pairs. These investigators have made a particular feature of anticipatory guidance to mitigate the anxiety that knowledge of the sex chromosome abnormality may produce in the parents during the child's maturation.

The Boston Study (Page 1) took place from 1970 to 1974 and covered male infants only, in total 13,751, and using heel prick blood, resulted in the ascertainment of 23 boys with SCA. Here a different group of controls was taken, balanced familial autosomal abnormalities, again with the aim of controlling for parental information of the sex chromosome abnormality. The current report discusses the interaction of behavioural style and of intelligence on the educational achievements and difficulties of the XXY boys.

The largest newborn population studied was in Toronto (Page 119), where 71,004 male and female infants were screened for X chromatin abnormalities in amnion cells from 1969 to 1971. In the last year of the survey Y chromatin was also analysed in 5,395 males. A total of 74 children with SCA were identified (51 males and 23 females). The present report covers 39 males and nine females, with detailed psychological, endocrinological and anthropometric examinations using siblings as control groups. There are two points of particular interest in this report; first, the description in the XXY group of the comparison of nine boys with, and twelve boys without supplementary testosterone therapy over a two year period from the age of 13. The second area to be treated in depth is more speculative but of great importance and involves an attempt to describe the mechanism by which additional sex chromosomes disturb normal development. Testing of the hypothesis can now be extended to the other newborn studies.

The Winnipeg Study (Page 183) describes chromosome analysis on 14,069 newborn infants from 1970 to 1973, using cord blood, which revealed 20 with SCA. Follow-up studies are described on 13 children. In this group, the controls were selected from the newborn population, together with siblings of cases, and siblings of controls. This study differs from the others in that it examined a population of marked ethnic variation, with Canadian Indians, and immigrants from Eastern and Western Europe as well as the United Kingdom. A unique aspect of this study is the evaluation of dysmorphic features in the children with SCA.

The New Haven Study (Page 221) covered 4,500 infants born over one year from October 1967 in one hospital where 18% of the population were Afro-Caribbean and the remainder Caucasian – mainly Italian. Controls were selected from the same newborn population of the same sex, race, and approximate birthweight, born just before or after the infant with SCA. In the case of a twin, the chromosomally normal co-twin was used. Here follow-up evaluations have taken place with intervals of several years.

The Danish Study (Page 209) has to date screened 20,222 newborns using cord blood; initially it extended over the years 1969 to 1974 then restarted in 1980 with the aim of gaining further knowledge of the effects of environmental conditions on the development of the fetus and of de novo chromosome abnormalities. 58 children with SCAs have been ascertained. Information on antenatal diagnosis of SCA and the decisions of parents concerning termination of the pregnancy are included in this chapter and demonstrate that up to the end of 1982, 78% of such pregnancies were aborted medically. With greater knowledge of the minor nature of the alterations produced by additional sex chromosomes the hope is expressed that this high level of termination may fall.

The Edinburgh Study (Page 73) extended from 1967 to 1979 and used, at times, full chromosome analysis on heel prick blood, and at other times sex chromatin screening for both X and Y numerical abnormalites. A total of 34,380 infants were screened and this resulted in the diagnosis of 70 infants with SCA. A longitudinal prospective study of these children and a large control group from the same newborn population have been evaluated on a regular basis with detailed growth and psychological studies. 93% of the children with SCA and 87% of the controls continue to attend for longitudinal study.

While the various investigators used different methods of assessment and different controls, an attempt has been made to summate the psychological findings in the penultimate chapter, and we hope this will be of use to counsellors when considering the crucial question of intellectual potential. The final chapter by Dr. Theodore Puck discusses future use of the new techniques available and their potential for understanding the fundamental processes of human genetic-biochemical regulation.

For those working in cytogenetic laboratories, the urgent need for accurate information is reflected in the steadily increasing number being diagnosed. On notification of an abnormality, prospective parents are faced with the agonising decision of whether to continue the pregnancy or seek termination. The necessity to define the extent of that abnormality puts a heavy burden on the counsellor, and highlights the need for up-to-date information which has been obtained from a source that does not introduce its own bias. In addition, paediatricians are becoming more aware of the sex chromosome abnormalities and their diagnosis, while still unusual, is increasing, and requires knowledge of the natural history of these conditions.

It is the hope of those contributing to this volume that the information obtained from the newborn cytogenetic surveys will promote a fuller understanding of both the effects and ultimately the mechanisms of actions of alterations in the sex chromosome constitution.

Shirley G. Ratcliffe

Behavioral Development of Boys With X Chromosome Aneuploidy: Impact of Reactive Style on the Educational Intervention for Learning Deficits

Stanley Walzer, MD, Anthony S. Bashir, PhD, John M. Graham, Jr., MD, ScD, Annette R. Silbert, PhD, Nicholas T. Lange, MA, Michael F. DeNapoli, BA, and Julius B. Richmond, MD

Judge Baker Guidance Center, Boston, Massachusetts 02115

This longitudinal prospective study of boys with sex chromosome aneuploidy (XXY, XYY) is investigating the relationships that exist between the sex chromosome complement and specific aspects of personality and functional development. Male infants with sex chromosome variations and a control group were located through a chromosome screening of neonates. The children are being followed in a long-term study with the purpose of identifying the reactive (temperamental) styles and developmental patterns that are related to variations in sex chromosome number.

Since this study is concerned with group-related commonalities in reactive style and in cognitive development of boys with sex chromosome aneuploidy, we can examine how these temperamental and cognitive factors interact. Furthermore, since the method employed in this investigation provides extensive information about the social and psychologic development of the children, we can also study how the temperamental and cognitive variables interact with environmental factors to influence developmental outcome.

METHODS

Infants with chromosome variations were located through a chromosome screening of newborn males [1,2]. In all cases, the parents were informed

Birth Defects: Original Article Series, Volume 22, Number 3, pages 1–21

about the chromosome abnormality and the known developmental implications. Parents of infants with the XXY karyotype were told that the phenotypic consequences of this chromosome variation would include testicular changes and, possibly, delayed puberty. The limited knowledge about the behavioral implications of this genotype was explained. Emphasis was placed on the need for longitudinal prospective studies of children with this chromosome abnormality to determine whether affected individuals were at risk for future developmental or educational problems.

The need to share this information with the parents determined the nature of the control sample. This was comprised of neonates having a chromosome abnormality that did not affect the phenotype adversely (ie, infants with familial balanced autosomal rearrangements). These infants were selected as controls because their parents also had to be informed about the existence of a variation.

It was possible to give the parents of the control infants considerable reassurance about the insignificance of the chromosome variation for future development by indicating that one of the parents carried the same variation and that it had not affected their development adversely. Thus the information given the families about the chromosome variation and its possible developmental significance was different for the aneuploid and control groups; however, this was the most adequate control possible for the requirement to inform parents about the variation. Since over 30 families with balanced autosomal rearrangements were available, we had a variety of rearrangements represented in our control sample and were also able to match the aneuploid and control groups for socioeconomic status and for race.

The data, obtained through semistructured parental interviews recorded verbatim and through direct observations of the child, were reduced to operationally defined categories of behavior that represented the child's development and functioning in the most relevant aspects of psychosocial adaptation [3]. The method we employed was a modification of that reported by Thomas et al [4] in their longitudinal study of the development of temperamental organization in infants and children.

If the parents consented to participate in the longitudinal study, they were interviewed at home twice yearly. During the first 4 years of life, the children were also observed directly in their homes by a second member of the study staff within 2 weeks of each parental interview.

Parental Interviews

Using a court stenograph machine, the interviewer recorded verbatim the parent's historical account of the child's development according to a semistructured interview schedule. Since the data sought from the parents were the descriptive facts about the behavioral items selected for study, we requested specific, concrete descriptions of behavior whenever the parents gave subjective interpretations of the child's behavior. For example, if the parents stated that the infant "did not like the food," upon request for concrete examples of behavior, the parent might report: "I know he did not like that food because he spit it out, turned his head, pushed my hand away and screamed loudly when I tried to feed more of it to him."

Data Reduction

The typed protocols were reduced to operationally defined categories of behavior or development by using a rigorous content analysis that emphasized specific criteria for the selection of scoreable items. Although 12 categories were employed in the data reduction, only the following six are relevant for this report.

1) *Activity level.* This category described the extent to which motoric behavior characterized the child's functioning. In scoring this category, we used all the data in the protocol concerned with motility.

2) *Pliancy.* This category reflected the disposition of the child either to yield to and depend on another individual or to function independently or impose his will on another. Most behaviors that we coded in this category had to do with dependence/autonomy, passivity/self-assertiveness, and obedience/disobedience.

3) *Intensity of responsiveness.* Parental statements referring to specific concrete behaviors were scored in one of three categories of "energy content": intense (robust, ebullient), moderate, mild (lethargic, weak).

4) *Approach-withdrawal.* This category referred to the child's responses to novel events, such as new foods, toys, strangers, places, and situations. Taking introductory steps to become more fully acquainted with the novel experience was scored as *approach.* Turning away from or holding back involvement with the new experience was scored as *withdrawal.*

5) *Adaptability.* This category was concerned with the sequential course that was followed as the child adapted to altered situations. In contrast to the *approach-withdrawal* category, it was not concerned with the initial response, but rather with the ease or difficulty with which the child attempted to adapt to the change over time.

6) *Capacity to relate.* Relatedness referred to the nature or the quality of the child's interaction with people.

The typed parental interview was subdivided into consecutively numbered samples whose sizes were limited by the changing content of the narrative.

Each sample was scored for the categories that were appropriate to its content. All categories had three conditions (ie, measures) from which the most appropriate condition could be selected and scored. For most categories the conditions of positive, variable, and negative were used; with the categories of *activity level* and *intensity of responsiveness* high, moderate, and low were employed.

The data were then computer-processed to yield a single number (the preponderance rating) for each behavioral and developmental category. The rating was on a scale from 0 to 200, with a score from 99 down to 0 representing an increasingly lower category value (ie, toward lower activity levels, withdrawal, nonadaptability, etc) and a score from 100 to 200 representing an increasingly higher value (ie, toward higher activity levels, approach, adaptability, etc).

The mathematic derivation of the preponderance rating has been described previously [3]. For each category, the rating represented the relative proportion of the interview that was scored in a positive or high direction as compared to the portion scored in a negative or low direction. The preponderance ratings derived from these data defined a child's reactive style or developmental pattern for each category. Thus, a child can be described as highly active, moderately intense in the energy content of his responsiveness, withdrawn in his response to new experiences, etc.

Direct Observations of the Child

During the child's first 4 years of life, a second investigator visited the home within 2 weeks of the home visit at which the interview was done. A 2 hour observation of the child's behavior was systematically recorded in 30 second intervals, both in a free-field setting and during one feeding.

Each 30 second sample of the observer's protocol was scored for the categories appropriate to its content. The categories applied to the observational data included: *activity level, pliancy, intensity of responsiveness, approach-withdrawal, mood,* and *capacity to relate*. A preponderance rating for each category was calculated [3].

As previously reported, we achieved good agreement between the data obtained through direct observations and those obtained from the parental interviews [3]. The parental reports appeared to be valid descriptions of the children's behaviors.

Statistical Analysis

The preponderance ratings for aneuploid and control children were combined and ranked for each category in 6 month intervals from birth to age 5 years. The category preponderance ratings for the two 6 month intervals in the sixth and seventh years of life were averaged to yield a mean yearly preponderance rating for each category.

The Kruskal-Wallis one-way analysis of variance was employed, and, in all instances when the results were significant, Dunn's procedure for multiple comparisons was used to compare each aneuploid group (XXY, XYY) to the control group.

To maximize the chance of detecting differences between the aneuploid and control groups, we employed a type I error rate of $\alpha = 10\%$ in the multiple comparisons. A more stringent criterion for incorrect rejection of the null hypothesis of equality between groups might have caused us to overlook potentially important differences. The 10% α rate assured us of a small β error (ie, the probability of not detecting differences when they exist).

Isolated "significant" cells were given little credibility in interpreting the results in the cross-section analysis. Instead, patterns over time (sequences of "significant" cells) were viewed as being more meaningful in terms of identifying important differences between the study groups. This reduced the possibility that random "significant" differences would unduly affect the overall conclusions of the analyses.

Other Evaluations and Assessments

Preliminary conclusions about developmental variations based on this systematic sampling method were tested independently by formal psychologic and neuropsychologic assessments, speech/language evaluations, and reading/learning assessments at predetermined times throughout the child's life. Because of the severe language-processing difficulties that were unexpectedly evidenced within the first 5 years of life in the XXY boys, we also included in our study protocol a systematic investigation of communication skills.

RESULTS

This report summarizes the results on 13 subjects with an XXY complement and nine controls with balanced autosomal rearrangements. All of the aneuploid and control boys in the study had passed their eighth birthday at the time of this report.

Reactive Style Categories

The results obtained from the interviews will be compared for those behavioral categories that are relevant for this report. The results for the reactive style categories are given in Table I.

TABLE I. P Values (Kruskal-Wallis one-way analysis of variance)* of Comparisons (Dunn's procedure) and the Reactive Style Categories for 13 XXY and 9 Control Children†

	Age (years)											
	½	1	1½	2	2½	3	3½	4	4½	5	6	7
Activity level	.000‡	.000 Low	.000‡	.000 Low	.000 Low	.000 Low	.000 Low	.000 Low	.000 Low	.000‡	.000 Low	.000 Low
Intensity of responsiveness	.000‡	.000‡	.000 Low	.000 Low	.000 Low	.000 Low	.000 Low	.000 Low	.000 Low	.000 Low	.000 Low	.000 Low
Pliancy	.000‡	.000 Pliant	.000‡	.000‡	.001‡	.000‡	.000 Pliant	.000 Pliant	.002 Pliant	.000 Pliant	.000 Pliant	.000 Pliant
Approach-withdrawal	—	.021‡	.033‡	.001‡	.002 Withdr.	.003 Withdr.	.009 Withdr.	.001 Withdr.	.001 Withdr.	.000 Withdr.	.001 Withdr.	.001 Withdr.
Adaptability	—	—	—	—	—	—	—	—	—	—	—	—
Relatedness	—	.019‡	—	—	—	.018 Neg.	.002 ***	—	.033 ***	.051 ***	—	—

*The Kruskal-Wallis one-way analysis of variance was employed on the reactive style preponderance ratings for 13 XXY, and nine control children.
†Key to each cell: P value; direction of the significant difference, XXYs as compared to the controls.
‡The Kruskal-Wallis procedure achieved a level of significance; however, Dunn's procedure for multiple comparisons employed for the 13 XXY and nine control children did not achieve the assigned level of significance.

1) *Activity level.* During ten of the 12 periods covered by this report, the XXY children showed lower levels of activity than did the control children. At 13–18 months, when the children in both groups showed significant progress in walking, group differences were not apparent.

2) *Intensity of responsiveness.* From 13 months on, the parental descriptions of the behavior for XXY children were scored more frequently than the controls as being low in the intensity of energy content.

3) *Pliancy.* Group differences between the XXY and control children reached significance at 7–10 months and after 3 years of age, with the pliancy scores being higher for the XXY children. When the children were young (under 4 years), these high scores reflected the parents' perceptions that the children were easy to manage, accepted parental restrictions willingly, and adjusted their own schedules and activities to the requirements imposed by the parents. From 5 years on, however, the higher pliancy scores for the XXY children tended, in part, to reflect the parents' perceptions that the children lacked the level of self-assertiveness that they expected in their encounters with adults and other children and in the early learning situation at school. Between 30 and 36 months, the parents of both XXY and control children reported the negativism and self-assertiveness associated with the increasing autonomy expected at this age; therefore, group differences were not evident in this time period.

4) *Approach-withdrawal.* The XXY children were scored more frequently as withdrawing from new and novel events than were the controls; the difference became significant after 2 years of age. The parents of XXY children reported hesitancy and caution as the usual response of their offspring to new situations, people, food, or experiences.

5) *Adaptability.* This category did not discriminate between groups throughout the time period covered by this report. Although XXY children were frequently reported to respond to novelty by initial withdrawal, their ability to adapt to change *over time* was as adequate for them as it was for the control children.

6) *Capacity to relate.* This category did not discriminate the XXY from the control children in any consistent way during the time period covered by this report. Although they tended to be shy and to avoid strangers, their social relationships with the significant familiar persons in their everyday lives did not differ from those reported for the control children.

Other Developmental Variables

Early speech/language development. As we reported previously, by 38 months of age, nine of the 13 children with the XXY chromosome complement manifested delayed speech and language development, whereas similar delays were seen in only one of the nine controls [3,5]. For the nine XXY children, the parents reported difficulties with pronunciation or articulation, difficulties in finding specific words to express their thoughts, and problems with sentence formation. The parents of one control child reported difficulties with pronunciation.

Clinical speech and language assessments of the XXY children revealed that receptive (understanding) language skills were age-appropriate. However, reductions were present in immediate auditory memory skills. In addition, the children manifested deficits in preserving the serial order of auditorily presented material. Problems in expressive language skills were evident and included difficulties with syntax (sentence building skills), with word-finding abilities (dysnomia), and with the production of oral narrative.

Cognitive development. The age-appropriate Wechsler Scale of Intelligence was administered to the 13 XXY boys and nine controls at ages 5–7, and 9–11 years. Prior to age 6 years 9 months the Wechsler Preschool and Primary Scale of Intelligence (WPPSI) was used. Until 1974, we employed the Wechsler Intelligence Scale for Children (WISC) with the boys aged 6 years or older. Since 1974, when the revised edition of the WISC (WISC-R) became available, we have used this instrument with the children over 6 years 9 months of age.

On the 5 to 7-year-old testing, the mean Full Scale IQ of the XXY boys and that of their controls were not significantly different (Table II). Whereas the mean Performance IQ of the XXY children was higher than that of the controls, the difference between the mean scores was not large enough to achieve significance. However, the mean Verbal IQ of the XXY group was significantly lower than the mean of the controls (P ~ .02).

A comparison of the Verbal and Performance Scale scores for the 13 XXY boys revealed significant disparities between the subscale scores for nine children (Table III). In these nine boys (69% of the XXY children being reported), the Performance Scale scores were consistently higher than the

TABLE II. Mean IQ Scores on the WPPSI, WISC, WISC-R for XXY and Control Children at Ages 5–7 Years

		Score		
	N	Full Scale	Verbal	Performance
XXY	13	110.2	101.46*	117.5
Controls	9	114.77	113	113.55

*P ~ .0175 (Student t test).

TABLE III. Disparity Between Verbal and Performance Scale Scores on the WPPSI/WISC/WISC-R for XXY Children at Ages 5–7 Years

Child	Age (years: months)	Test form	Full Scale IQ	Verbal IQ	Performance IQ	Disparity	Significance level*
1	7:4	WISC	115	105	122	+17	<.05
2	7:4	WISC	99	89	111	+22	<.01
3	6:2	WPPSI	93	86	101	+15	<.01
4	7:7	WISC-R	128	121	129	+8	
5	5:6	WPPSI	115	101	127	+26	<.001
6	6:3	WPPSI	111	100	120	+20	<.001
7	6:4	WPPSI	111	104	118	+14	<.01
8	5:0	WPPSI	128	116	135	+19	<.01
9	5:7	WPPSI	118	102	131	+29	<.001
10	7:0	WISC	101	97	106	+9	
11	6:8	WISC	76	79	79	0	
12	6:5	WISC-R	105	94	117	+23	<.001
13	6:2	WISC-R	133	125	132	+7	

*Sattler [8], p 450, Table C-18, gives the probability of obtaining the designated differences between the Verbal and Performance IQs on the WPPSI and the WISC. For the WISC-R, the probabilities (to the .05 level of significance) were obtained from Wechsler ["Manual for the Wechsler Intelligence Scale for Children—Revised." New York: Psychological Corporation, 1974]. The formula in Sattler [p 445] was used to compute the criteria for the disparities at the .01 and .001 significance levels on the WISC-R.

Verbal Scale scores. One boy (child 11, Table III) was in the borderline range of intelligence in the Verbal, Performance, and Full Scale scores.

Among the nine control boys, only one (child 3, Table IV) had a Verbal Scale score that was significantly higher than the Performance score. The reverse was noted in the scores of a second boy (child 7, Table IV).

The XXY and the control groups were compared with respect to the direction of the disparities between the individual boy's Verbal and Performance Scale scores. There was a significant association between group membership and the direction of the disparity (Table V).

The Wechsler intelligence scores for the 13 XXY boys, who were assessed between ages 9 and 11 years, are presented in Table VI. The XXY children continued to demonstrate greater proficiency in nonverbal skills than in verbal skills. Ten of the 13 boys (76.9% of the XXY boys being reported) had a Performance IQ greater than Verbal IQ at levels of statistical significance that ranged from .06 to <.001. In two other boys (children 4 and 8, Table VI), the Verbal score was lower than the Performance score, but the disparity was not significant. Only one XXY boy (child 13, Table VI) showed a Verbal score that was higher than the Performance score, but the two-point

TABLE IV. Disparity Between Verbal and Performance Scale Scores on the WPPSI/ WISC/WISC-R for Control Children at Ages 5–7 Years

Child	Age (years: months)	Test form	Full Scale IQ	Verbal IQ	Performance IQ	Disparity	Significance level*
1	6:9	WISC	111	108	112	+4	
2	5:8	WPPSI	108	110	104	−6	
3	6:8	WISC-R	100	108	90	−18	<.01
4	5:10	WPPSI	134	130	131	+1	
5	5:4	WPPSI	116	117	112	−5	
6	7:1	WISC	120	120	115	−5	
7	6:11	WISC-R	122	107	133	+26	<.001
8	5:6	WPPSI	115	114	114	0	
9	5:8	WPPSI	107	103	111	+8	

*Sattler [8], p 450, Table C-18, gives the probability of obtaining the designated differences between the Verbal and Performance IQs on the WPPSI and the WISC. For the WISC-R, the probabilities (to the .05 level of significance) were obtained from Wechsler ["Manual for the Wechsler Intelligence Scale for Children—Revised." New York: Psychological Corporation, 1974]. The formula in Sattler [p 445] was used to compute the criteria for the disparities at the .01 and .001 significance levels on the WISC-R.

TABLE V. Direction of Disparity Between Verbal and Performance Scale Scores on the WPPSI/ WISC/WISC-R for XXY and Control Children at Age 5–7 Years

	No. of children	
	V < P	V ≥ P
XXY	9	4
Controls	1	8

$\chi^2 = 7.364$, df 1, $P < .01$.

difference was insignificant. We will not report the results of the 9- to 11-year-old IQ testing for the control group, since three of the nine control children have not yet been tested.

Academic performance. All the 13 XXY boys were beyond the fourth grade at the time of this report. Eleven of the 13 boys were referred independently by the teachers for learning disability evaluations because of persistent difficulties with reading and spelling. Subsequently, ten of the eleven were enrolled in special education programs for children with learning disabilities. One child is receiving private tutorial assistance in reading. Only two of the 13 XXY boys (children 4 and 13, Tables III and IV) have done consistently well in all areas of academic achievement.

TABLE VI. Disparity Between Verbal and Performance Scale Scores on the WISC/ WISC-R for XXY Children at Ages 9–11 Years

Child	Age (years: months)	Test form	Full Scale IQ	Verbal IQ	Performance IQ	Disparity	Significance level*
1	11:10	WISC-R	110	105	115	+10	.06
2	9:11	WISC-R	98	91	106	+15	.01
3	11:9	WISC-R	92	87	100	+13	<.05
4	10:9	WISC-R	120	115	120	+5	
5	10:7	WISC-R	107	98	117	+19	<.01
6	11:11	WISC-R	108	97	120	+23	<.001
7	10:3	WISC-R	99	94	105	+11	.06
8	10:5	WISC-R	99	97	101	+4	
9	10:9	WISC-R	118	101	132	+31	<.001
10	10:11	WISC	88	82	97	+15	.01
11	11:1	WISC-R	69	55	86	+31	<.001
12	10:0	WISC-R	96	85	111	+26	<.001
13	9:11	WISC-R	133	131	129	−2	

*Sattler [8], p 450, Table C-18, gives the probability of obtaining the designated differences between the Verbal and Performance IQs on the WISC. For the WISC-R, the probabilities (to the .05 level of significance) were obtained from Wechsler ["Manual for the Wechsler Intelligence Scale for Children—Revised." New York: Psychological Corporation, 1974]. The formula in Sattler [p 445] was used to compute the criteria for the disparities at the .01 and .001 significance levels on the WISC-R.

All nine control children were beyond the third grade at the time of this report. One control boy (child 7, Table IV) has manifested a persistent learning disability; eight of them are doing well academically.

Communication study. The specific aspects of the communication disorder in the XXY boys were confirmed in a study by Graham and his colleagues [6,7]. They compared the communication skills of a group of 14 unselected XXY boys (13 from this study, one from another study) with a group of 15 control boys matched for race, age, educational level, and socioeconomic status.

Graham et al [6,7] confirmed the presence of language impairment that affected language production more than comprehension in this group of unselected XXY boys. Significant difficulties with immediate auditory memory abilities and with certain aspects of auditory-processing capabilities were identified. The results suggest that the language abilities of XXY boys are characterized by a pattern of deficits that includes problems in rate and order processing of auditory stimuli, problems in the understanding of complex grammatic constructions, and difficulties with oral language production that include deficits in morphology, word-retrieval abilities, and oral narrative construction.

Specific problems in reading and spelling were noted in the XXY boys who had also demonstrated specific language disorders. This suggested the existence of a continuum of language learning disabilities seen during the preschool years as deficits in language processing and production, and associated in the school years with reading and spelling difficulties.

DISCUSSION

Continuity Between Three Wechsler IQ Tests

In this long-term, longitudinal follow-up study we have used three Wechsler test instruments (WPPSI, WISC, WISC-R) to obtain IQ scores at different ages. Therefore, we must examine the continuity between the three test forms if we are to interpret a child's later test performance in the light of the IQ score earned in earlier years. It is important to define 1) the correlation between the different test forms and 2) the congruence or closeness of the relationship between the abilities or constructs being measured by the respective test forms.

Numerous studies have reported the correlations between the IQ scores obtained by the three Wechsler test forms, and the coefficients between the comparable scales have been high. Comparisons of the WPPSI and the WISC have shown correlations ranging from .57 to .91 for the Verbal Scale, from .43 to .82 for the Performance Scale, and from .65 to .90 for the full scale [8, p 210].

Sattler [8, p 514] has reported a comparison between the WPPSI and WISC-R in 50 children who were given the two test forms at age 6. The correlations between the three scales were .80 for the Verbal and Performance Scales and .82 for the Full Scale.

Several studies have presented correlations between the IQ scores obtained by the WISC and its successor the WISC-R [9–11]. High coefficients (ie, in the .80s) between comparable scales were reported. From these many reports we can assume that the three Wechsler test forms are related to a significant degree.

Factor analytic investigations of the Wechlser instruments have shown that the specific subtests are more reliable estimates of specific abilities when they are combined with other subtests to yield pooled factors. The factors can be used to infer a child's level of functioning in separate skill areas. Factor analytic studies of the WISC and WISC-R have yielded three major ability-related pooled factors (verbal comprehension, perceptual organization, and freedom of distractability) that were consistent in children from ages 6 to 16 [12, 13]. A similar study of the WPPSI also yielded three factors, two of which (verbal comprehension and visual organization) were similar to those found in the WISC and WISC-R [14].

Swerdlik and Schweitzer [15] compared the factorial composition of two Wechsler instruments (WISC, WISC-R) that were administered to 164 subjects and found a high degree of congruence (.98) between the verbal comprehension and perceptual organization factor scores obtained from each test form. It appears, therefore, that there is a high level of congruence between the abilities measured by the WISC and WISC-R.

The three Wechsler forms used in this study contain the subtests in the verbal battery required to compute the verbal comprehension factor (ie, the information, similarities, vocabulary, and comprehension subtests). Therefore, we may compute this factor for each child's assessments and compare the earlier (ages 5–7) and later (ages 9–11) test results in terms of the two factor scores. However, the WPPSI does not include the subtests required to compute a perceptual organization factor (ie, the picture completion, picture arrangement, block design, object assembly, and mazes subtests). Therefore, we cannot compare the earlier and later IQ assessments in terms of this factor score.

A comparison of each XXY boy's verbal comprehension factor scores from the earlier and later testings shows a decline of 0.5–1.50 points in eight children, a decline of 2.5–4.25 points in three boys, and a gain of 1.25 points in two other boys. It is of interest that the three boys with the greatest decline (ie, 2.5–4.25 points) also manifested the greatest difficulties with academic achievement.

The X Chromosome and Cognition

Careful studies of developing children, or of adults, with X chromosome abnormalities have demonstrated that specific intellectual deficits appear to be associated with specific X chromosome complements. Deficits in language development have been reported in approximately one half of XXY boys being followed in prospective studies around the world [16]. Auditory rate processing difficulties and deficits in short-term auditory memory have been noted [5,6]. Although several investigators have reported greater impairments on measures of expressive rather than receptive language [3,5,6,17,18], this finding has not been consistent [19].

The severe language-processing difficulties have resulted in deficiencies in reading and spelling skills [5–7, 20–22]. Graham et al [6,7] found a high degree of correlation between language test performance, reading, and spelling proficiency. They found that abilities in auditory processing, auditory memory, language, reading, and spelling were intercorrelated. They suggested that many of the relationships depend on basic linguistic knowledge and performance.

Several prospective longitudinal studies of XXY boys, ascertained at birth and followed developmentally, have reported significantly higher Performance Scale IQ scores than Verbal Scale scores [3,5,19,21,23,24], although this has not always been a consistent finding [22]. In one study, the XXY boys showed a significant Verbal/Performance IQ Scale score disparity (with a lower Verbal score) when the boys were aged 4–5 years, but the disparity disappeared when they were retested at 8–10 years of age [21,25]. In our study, the Verbal/Performance IQ Scale score disparity was seen in both the 5–7 and 9–11-year-old assessments.

It is not surprising that, in some instances, there may be a decline in verbal subtest scores as the XXY children get older. Good performance in these subtests depends on those areas of functioning that are most impaired in our XXY boys. Thus the severe reading and spelling disabilities, the difficulties with abstract and analytic forms of verbal reasoning, the problems with verbal expression, and the deficits in short-term memory and auditory rate processing take their toll on the abilities of XXY boys to achieve well in tasks requiring verbal skills.

The mild neuromotor dysfunctions often found in the XXY boys may interfere with psychomotor efficiency on the nonverbal items in the WISC-R and, therefore, affect their Performance subtest scores. These subtests are timed, and the various norms become more rigorous as the boys advance in prepuberty and into the adolescent epoch. Therefore, the XXY boys may earn fewer time bonuses because of their neuromotor difficulties.

In certain XXY boys, there may be an uneven decline over time in the Verbal and Performance subtest scores. As a result, the disparity between the Verbal and Performance IQ Scale scores may drop below a level of statistical significance, or it may disappear entirely. However, the specific cognitive disabilities remain, and they must be considered in the prescriptive, academic curriculum planning for each learning-disabled XXY boy.

Since an extra X chromosome appears to be associated with specific language and reading disorders, it is not surprising that studies of XXX girls have demonstrated disabilities in verbal areas, with fewer difficulties in nonverbal areas. The 47,XXX girls have demonstrated deficiencies in auditory processing and problems in recalling serially ordered verbal information from memory [26]. Although deficits have been demonstrated in both receptive and expressive language skills, the latter appear to be more impaired

[22,25]. The language-processing deficits have been associated with poor academic performance [21,22,25,26].

Children with extra X chromosome material have cognitive problems that are the reverse of the deficits seen in girls lacking an X chromosome. Girls with Turner syndrome have been shown to suffer a specific deficit in nonverbal cognitive functioning. Although language-processing difficulties are not characteristically seen in these girls, they do demonstrate deficits in the spatial aspects of cognitive processing [21,25,27–32]. The implications of these studies is that a quantitative imbalance in the number of X chromosomes influences cognitive development and functioning.

The cognitive deficits seen in children with extra X chromosomes are also observed in reading-disabled children with normal sex chromosome complements. Since 1937, when Orton [33] suggested that children with reading disorders had concomitant perceptual impairments, many clinical reports have supported the presence of associated auditory and visual processing difficulties. Other authors have emphasized that these children manifest difficulties with the perception of serial or temporal order [34,35].

Children with extra X chromosome material have been shown to present with auditory rate processing and auditory memory deficits. They also manifest expressive language, reading, and spelling difficulties. The careful examination of the cluster of cognitive deficiencies seen in these individuals may help to explain the interactive nature of similar deficits in chromosomally normal children who have learning disabilities.

Reactive Style and Cognitive Deficits: Interactional Implications

In this study, certain of the reactive style variables impacted significantly on the learning disability of the XXY children and also tended to influence the special needs educators in a way that was counterproductive for the learning-disabled XXY boy to realize his full academic potential. To clarify this point, we will first describe the legislatively mandated programs for children in need of special education that have emerged over the last decade in the United States.

Legislative Mandates and the Special Needs Child

A major federal statute, the Education for All Handicapped Children Act of 1975 (P.L. 94-142), and other statutes in many of the 50 states require that each community provide for the special educational needs for all children having developmental, cognitive, or psychologic problems that impair the ability of the children to learn. The cornerstone of these legislative acts has been the notion of the "least restrictive environment"; this requires that the schools provide educational services in settings that maximize normal social experiences.

Many statutes mandate the specific levels of intervention that the communities must provide. The intervention may simply be some individualized help within the regular classroom. Alternatively, it may be required that the child spend a part of each day in a special needs classroom while spending the remainder of his school day in a regular classroom setting. A more impaired child might require day-long experience in a special needs program. Related services such as physical therapy, psychiatric treatment, and speech/language therapy are to be available as required.

The aim of this legislation has been to guarantee that each child receives a free and appropriate public education experience at no cost to the family. Furthermore, there is heavy emphasis on the requirements for careful diagnosis and assessment of the children as a prerequisite to the provision of the special services. Only then can the school provide an individualized plan tailored to meet each child's unique needs.

The Educational System and the XXY Boys

In this study, the reactive styles of the learning-disabled X aneuploid children affected their classroom behavior and, thereby, influenced the decision about the special educational supports provided them. It was not that the children showed a homogeneous personality pattern, since personality development is determined by many interacting biopsychosocial factors. However, the XXY boys had significantly lower scores than the controls for the categories of *activity level* and *intensity of responsiveness* (Table I). As such, in the classroom, the XXY boys were likely to be low-key and quiet and were not prone to the hyperactivity or the overt behavioral difficulties seen frequently in children diagnosed as having an attention deficit disorder. Furthermore, their high *pliancy* scores (Table I) reflected, in part, a tendency to "go along with" the behavioral expectations of the teachers. Thus their teachers often saw these children as being warm and likeable, anxious to please, and helpful.

The X aneuploid group also demonstrated a significant tendency to withdraw (*approach-withdrawal*, Table I) from new and novel situations and experiences. The XXY boys tended, therefore, to avoid challenging situations by quiet withdrawal. The teachers reported that the boys would quietly "daydream" in the class when they were supposed to be working on new material. These children did not accept the challenge to try actively to master material that was particularly difficult for them, considering their language-processing problems.

This withdrawal was often viewed by the teachers from a judgmentally negative (ie, a blaming) vantage point. Thus at the educational planning conferences we frequently heard statements about the children such as: "they are lazy," "they could do the work if they wanted to," "if they only tried to do the work, they would be doing well," or "they have good potential, but they are not using it because they lack motivation or interest."

Because of these biases on the part of the classroom teachers and the special educators, the children were often provided an inappropriate level and/or style of intervention. Although manifesting serious language-processing difficulties, the boys were usually well liked and presented no classroom behavior problems. Since the learning difficulty was viewed as being easily overcome if the child only worked harder, the educators often wished to retain them primarily in the mainstream classes while providing inadequate special needs enrichment. Even though the children tended not to make adequate academic progress with the shallow attempts at therapeutic intervention, the educators often resisted our efforts to press for a more intensive program. "Social promotions" were not unusual. We were frequently frustrated in our attempts to emphasize that these were the very children who could best be helped to overcome their disabilities by an intensive commitment of time and resources to a therapeutic program tailored to the specific elements of their learning deficits. Although we suggested specific modes of intervention for remediating the deficits in auditory rate processing, immediate auditory recall, and expressive language functions, our suggestions were not implemented sufficiently.

The teachers frequently emphasized that they had many children who were far more handicapped, and, therefore, required their limited resources. They referred to children who were not necessarily more handicapped in their learning but were more active, disruptive and behaviorally difficult or suffered from serious thought disorders.

Implications for Management

We are concerned ultimately with the child's functional learning capacity—his ability to realize his academic potential. A variety of biologic, sociocultural, and psychologic factors and forces contribute to, and shape, the child's capacity to learn. In this report, we are focusing on the impact of the specific cognitive deficits, the reactive style, and the availability of individualized educational enrichment on the functional learning capacity of XXY boys (Fig. 1). The cognitive impairments and the reactive style, alone and in interaction with each other, impair somewhat the ability of the boys to master certain learning tasks. Each child's successful adaptation to the learn-

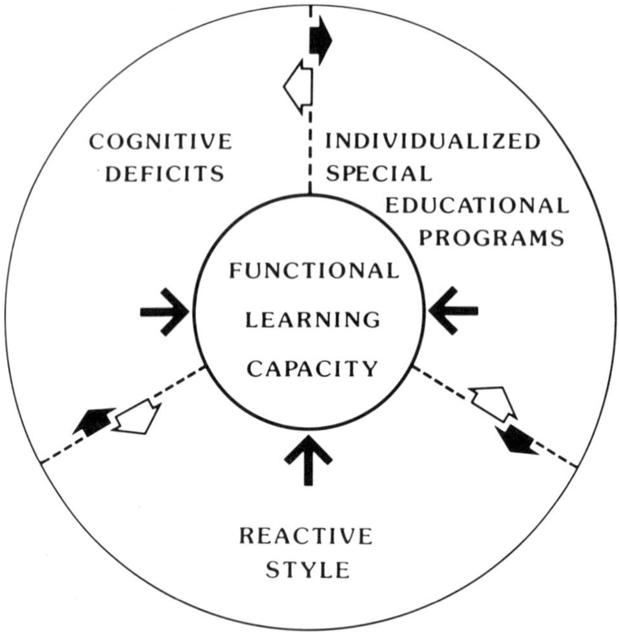

Fig. 1. The impact of three interacting factors on the functional learning capacity of the XXY boys.

ing situation depends on the implementation of an individualized special educational experience that is responsive to the unique elements of his cognitive deficit and to his reactive style. Future progress in the management of all children with learning difficulties depends on our ability to define specific functional subgroup profiles. We are moving toward the development of a better taxonomy.

These are not intellectual exercises without practical implications. Rather, as we define more precisely the biologic, psychologic, and social dynamics, we can design more specific and effective intervention strategies. Thus the significance of this aspect of our longitudinal study is that of demonstrating the specific functional profile of the learning problems that these children face. As a consequence, special educators should find it possible to tailor better their efforts to fit each child's unique needs. There should be a resulting decrease in the frustrations borne of inadequate diagnostic formulations and inappropriate methods for intervention. Teachers, children, and families should benefit from such investigations.

ACKNOWLEDGMENT

This research was supported by NIH grant MH38502.

REFERENCES

1. Walzer S, Breau G, Gerald PS: A chromosome survey of 2,400 normal newborn infants. J Pediatr 74:438–448, 1969.
2. Walzer S, Gerald PS: A chromosome survey of 13,751 male neonates. In Hook E, Porter IH (eds): "Population Cytogenetics." New York: Academic Press, 1977, pp 45–61.
3. Walzer S, Wolff PH, Bowen D, Silbert AR, Bashir AS, Gerald PS, Richmond JB: A method for the longitudinal study of behavioral development in infants and children: The early development of XXY children. J Child Psychol Psychiatr 19:213–229, 1978.
4. Thomas A, Chess S, Birch HG, Hertzig ME, Korn S: "Behavioral Individuality in Early Childhood." New York: New York University Press, 1963.
5. Walzer S, Graham JM, Bashir AS, Silbert AR: Preliminary observations on language and learning in XXY boys. In Stewart DA (ed): "Sex Chromosome Aneuploidy: Follow-up Studies." New York: Alan R. Liss, Inc., for the March of Dimes Birth Defects Foundation. BD:OAS 18(4):185–192, 1982.
6. Graham JM, Bashir AS, Walzer S, Stark RE, Gerald PS: Communication skills among unselected XXY boys. Pediatr Res 15:562, 1981.
7. Graham JM, Bashir AS, Stark RE, Walzer S, Silbert AR: Reading and spelling abilities of XXY boys: A prospective longitudinal study. Pediatrics (In press)
8. Sattler JM: "Assessment of Children's Intelligence," Rev Ed Philadelphia: WB Saunders, 1974.
9. Hamm H, Wheeler J, McCallum S, Herrin J, Hunter D, Catoe C: A comparison between the WISC and WISC-R among educable mentally retarded students. Psychol School 13:4–8, 1976.
10. Larrabee GJ, Holroyd RG: Comparison of WISC and WISC-R using a sample of highly intelligent children. Psychol Rep 38:1077–1080, 1976.
11. Davis EE: Matched pair comparison of WISC and WISC-R scores. Psychol Schools 14:161–166, 1977.
12. Cohen J: The factorial structure of the WISC at ages 7-6, 10–6, and 13–6. J Consul Psychol 23:285–299, 1959.
13. Kaufman AS: Factor analysis of the WISC-R at eleven age levels between 6 ½ and 16 ½ years. J Consul Clin Psychol 43:135–147, 1975.
14. Dokecki PR, Frede MC, Gautney DB: Criterion, construction and predictive validities of the Wechsler Preschool and Primary Scale of Intelligence. Proc Am Psychol Assoc 4:505–506, 1969.
15. Swerdlik ME, Schweitzer J: A comparison of factor structures of the WISC and WISC-R. Psychol Schools 15:166–172, 1978.
16. Robinson A, Lubs HA, Nielsen J, Sørensen K: Summary of clinical findings: Profiles of children with 47,XXY, 47,XXX and 47,XYY karyotypes. In Robinson A, Lubs HA, Bergsma D (eds): "Sex Chromosome Aneuploidy: Prospective Studies on Children." New York: Alan R. Liss, Inc., for The National Foundation—March of Dimes. BD:OAS XV(1):261–266, 1979.
17. Nielsen J, Sillesen I, Sørensen AM, and Sørensen K: Follow-up until age 4 to 8 of 25 unselected children with sex chromosome abnormalities, compared with sibs and controls.

In Robinson A, Lubs HA, Bergsma D (eds): "Sex Chromosome Aneuploidy: Prospective Studies on Children." New York: Alan R. Liss, Inc., for the March of Dimes Birth Defects Foundation. BD:OAS XV(1):15–73, 1979.

18. Robinson A, Puck M, Pennington B, Borelli J, Hudson M: Abnormalities of the sex chromosomes: A prospective study on randomly identified newborns. In Robinson A, Lubs HA, Bergsma D (eds): "Sex Chromosome Aneuploidy: Prospective Studies on Children." New York: Alan R. Liss, Inc., for the March of Dimes—Birth Defects Foundation. BD:OAS XV(1):203–241, 1979.
19. Netley C, Rovet J: Verbal deficits in children with 47,XXY and 47,XXX karyotypes. A descriptive and experimental study. Brain Language 17:58–72, 1982.
20. Pennington B, Puck M, Robinson A: Language and cognitive development in 47,XXX females followed since birth. Behav Genet 10:31–41, 1980.
21. Pennington B, Bender B, Puck M, Salbenblatt J, Robinson A: Learning disabilities in children with sex chromosome anomalies. Child Dev 53:1182–1192, 1982.
22. Ratcliffe SG, Tierney I, Nshaho J, Smith L, Springbett A, Callan S: The Edinburgh study of growth and development of children with sex chromosome abnormalities. In Stewart DA (ed): "Children with Sex Chromosome Aneuploidy: Follow-Up Studies." New York: Alan R. Liss, Inc., for the March of Dimes Birth Defects Foundation. BD:OAS 18(4):41–60, 1982.
23. Evans JA, de von Flindt R, Greenberg C, Ramsay S, Hamerton JL: A cytogenetic survey of 14,069 newborn infants: IV. Further follow-up on the children with sex chromosome anomalies. In Stewart DA (ed): "Sex Chromosome Aneuploidy: Follow-up Studies." New York: Alan R. Liss, Inc., for the March of Dimes Birth Defects Foundation.BD:OAS 18(4):169–184, 1982.
24. Ratcliffe SG, Bancroft J, Axworthy D, McLaren W: Klinefelter's syndrome in adolescence. Arch Dis Child 57:6–12, 1982.
25. Robinson A, Bender B, Borelli J, Puck M, Salbenblatt J, Webber ML: Sex chromosomal abnormalities (SCA): A prospective and longitudinal study of newborns identified in an unbiased manner. In Stewart DA (ed): "Sex Chromosome Aneuploidy: Follow-Up Studies." New York: Alan R. Liss, Inc., for the March of Dimes Birth Defects Foundation. BD:OAS 18(4):7–39, 1982.
26. Rovet J, Netley C: The triple X chromosome syndrome in childhood: Recent empirical findings. Child Dev 54:831–845, 1983.
27. Shaffer JW: A specific cognitive defect observed in gonadal aplasia (Turner's syndrome). J Clin Psychol 18:403–406, 1962.
28. Alexander D, Walker HT, Money J: Studies in direction sense. Arch Gen Psychiatr 10:337–339, 1964.
29. Alexander D, Ehrhardt AA, Money J: Defective figure drawing geometric and human in Turner's syndrome. J Nerv Ment Dis 142:161–167, 1966.
30. Garron D: Intelligence among persons with Turner's syndrome. Behav Genet 7:105–127, 1977.
31. Silbert A, Wolff PH, Lilienthal J: Spatial temporal processing in patients with Turner syndrome. Behav Genet 7:11–21, 1977.
32. Rovet J, Netley C: Processing deficits in Turner's syndrome. Dev Psychol 18:77–94, 1982.
33. Orton ST: "Reading, Writing and Speech Problems in Children." London: Chapman and Hall, 1937.

34. Bakker DJ: "Temporal Order in Disturbed Reading; Developmental and Neuropsychological Aspects in Normal and Reading Retarded Chldren." Rotterdam: Rotterdam University Press, 1972.
35. Corkin S: Serial-ordering deficits in inferior readers. Neuropsychologia 12:347–354, 1974.

Sex Chromosomal Aneuploidy: Prospective and Longitudinal Studies

Arthur Robinson, MD, Bruce G. Bender, PhD, Joyce B. Borelli, MD, Mary H. Puck, MA, James A. Salbenblatt, MD, and Jeremy S.D. Winter, MD

Departments of Pediatrics (A.R., J.A.S.), Psychiatry (B.G.B., J.B.B., M.H.P.), and Biochemistry, Biophysics and Genetics (A.R.), University of Colorado School of Medicine, National Jewish Center for Immunology and Respiratory Medicine, Denver, Colorado 80206; Departments of Paediatrics and Medicine, University of Manitoba, Winnipeg, Manitoba, Canada (J.S.D.W.)

INTRODUCTION

The Denver Family Development Study concerns itself primarily with individuals with sex chromosome anomalies identified during the screening of newborns between 1964 and 1974 [1]. Since then, we have added to the study group some who have been diagnosed serendipitously in utero and whose parents have chosen to continue the pregnancy. In addition, a small number of referred cases are included. This latter "biased" group, identified by virtue of problems, is utilized to demonstrate some of the more extreme aspects of the phenotype. Table I lists all children studied. Previous reviews of this continuing study have been published [2,3]. This chapter concentrates on the period since 1981.

One factor in planning this as a "family development study" has been the opportunity to use sibs of propositi as controls. Because sex- and age-matched sib controls were not available from within every propositus's family, sibs from various families were selected to form male and female control groups with distributions of age and socioeconomic status similar to those of the propositi groups. Except for the discussion of psychosocial adaptation, in which comparisons of each propositus were made with his or her own sibs, the term *control* refers to these pooled sib groups.

Birth Defects: Original Article Series, Volume 22, Number 3, pages 23-71

TABLE I. Case Load

Karyotype	Newborn series				Referral series				Prenatal diagnostic series			
	No. of propositi	Age range (years)	No. of sib controls	Age range (years)	No. of propositi	Age range (years)	No. of sib controls	Age range (years)	No. of propositi	Age range (years)	No. of sib controls	Age range (years)
47,XXY	15	10–19	38	3–34	3	11–16	7	10–23	4	1–5	3	7–24
47,XXX	11	12–19	23	9–35	0	—	0	—	2	1	3	5–12
45,X	6	11–19	16	3–25	2	1–15	2	9–17	0	—	0	—
46,XXq−	2	12–17	2	13–18	0	—	0	—	0	—	0	—
45,X/46,X,r(X)	1	12	1	15	0	—	0	—	0	—	0	—
47,XYY	4	10–13	5	4–15	3	1–17	4	11–19	2	1–5	1	2
46,XY/47,XXY	2	11–18	5	2–24	1	19	2	13–15	1	1	0	—
45,X/46,XX/47,XXX	1	18	1	15	0	—	0	—	0	—	0	—
45,X/46,XX	3	17	6	12–20	0	—	0	—	0	—	0	—
46,XX/47,XXX	2	11–19	6	4–21	0	—	0	—	0	—	0	—
45,X/47,XXX	1	13	2	16–22	0	—	0	—	1	1	2	3–6
48,XXYY	0	—	0	—	1	18	3	19–29	0	—	0	—
48,XXXY	0	—	0	—	1	8	1	6	0	—	0	—
49,XXXXY	0	—	0	—	2	11–18	2	13–24	0	—	0	—
Totals	48	10–19	105	2–35	13	1–19	21	6–29	10	1–5	9	2–24

The team consists of a geneticist, a developmental pediatrician, a psychiatric social worker, a child psychiatrist, a pediatric psychologist, a physical therapist, and a speech therapist. Propositi are checked as closely as possible by annual examinations. Families are given support for physical and emotional problems, and close telephone contact is maintained throughout the year.

The propositi and their sibs continue to be seen annually. Measurements include height and weight [4], head circumference [5], and blood pressure [6]. Sexual maturation is rated according to Marshall and Tanner [7,8]. Penile stretched length is compared to the norms of Schonfeld [9], and testicular volume is measured by palpable comparison using a Prader orchidometer [10]. Serum concentrations of follicle-stimulating hormone (FSH) [11], luteinizing hormone (LH) [12], testosterone (T) [13], and estradiol (E2) [13] are determined by radioimmunoassay.

47,XXY

Physical Assessment

The age range of the 15 boys currently being followed is 9–17 years. They are normal in general appearance. Their head circumferences cluster below the mean but are within the normal range (Fig. 1, Appendix A). Presently, all have a height at or greater than the 45th percentile (range 45–>95%; Fig. 2). Their height has previously been documented to be significantly greater than that of their sibs and population norms [3]. There is no genital ambiguity, and penile size is within the normal range. The timing of onset of clinical puberty is consistent with control sibs, and all have virilized normally. Minor genital anomalies include one subject with undescended testes and another with inguinal hernias. The prepubertal mean testicular volume is 1 ml, with an increase in size coinciding with the onset of puberty. Following the initial increase in volume, testicular growth ceases. This is followed by a subsequent decrease in size, and by midpuberty the boys have a mean testicular volume of 3 ml (range 1–6 ml).

Neuromuscular Assessment

Clinical impressions, in addition to parent and teacher reports, indicated that neuromaturational lags that were present in early childhood had not completely resolved. Parents confirmed our impression that these delays adversely affect school performance in writing skills, speed on timed activities, and athletic activities with peers. To document these delays more precisely, the Bruininks-Oseretsky test of motor proficiency [14] was administered to 14 unselected 47,XXY subjects and a matched group of control sibs by a physical therapist who was blind to the diagnosis.

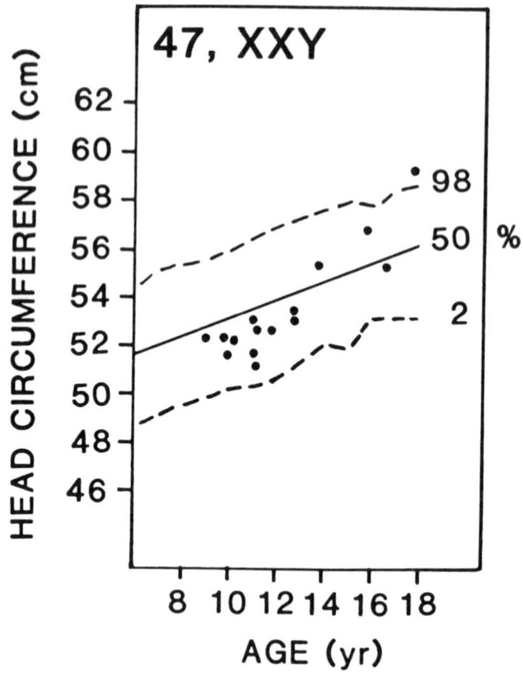

Fig. 1. Percentiles for head circumference of 47,XXY probands [5].

On the basis of blind evaluations, the physical therapist correctly identified 11 of 14 (70%) of the aneuploid boys. Six of the 14 (43%) 47,XXY boys have a test battery composite score within the lowest three stanines, which is at or below the 22nd percentile. There is only one in the upper three stanines. This is in contrast to the control group, who are evenly distributed. The 47,XXY boys achieve lower mean scores than the controls on tasks involving both gross and fine motor skills, coordination, speed and dexterity, and strength. These results correlated with clinical findings, including the persistence of primitive reflexes, joint stability, sensory-motor integration, and visual-perceptual-motor integration.

Endocrine Studies

In a continuing effort to document pituitary-gonadal function prior to and throughout puberty, basal serum hormone concentrations and acute responses to stimulation with gonadotropin-releasing hormone (GnRH) and human chorionic gonadotropin (hCG) are studied. These longitudinal data have been analyzed and published [15]. Forty subjects with a 47,XXY karyotype have been tested. Twenty-seven (ages 1–16 years) are unselected, having been identified through neonatal screening programs, whereas 13 (ages 7–19 years) were referred because of behavioral problems, gynecomastia, small testes, or micropenis and were later found to have sex chromosome aneuploidy. The timing of the onset of clinical puberty was normal, and no subject received androgen supplementation for at least 4 months prior to being studied. A second group of 22 chromatin-negative male control sibs from the Denver Family Development Study were available as a control group for comparison with previously published age-specific normal hormone values [16–18].

Serum concentrations of FSH, LH, T, and E2 were determined from venous blood. When possible, data collection began before the onset of puberty, and there are from one to seven specimens per subject.

During childhood and even into early puberty, pituitary-gonadal function in 47,XXY subjects is normal. FSH and most LH values were normal before age 12, but by age 14 the values for both were uniformly elevated. The magnitude and timing of the initial adolescent rise in serum T were normal,

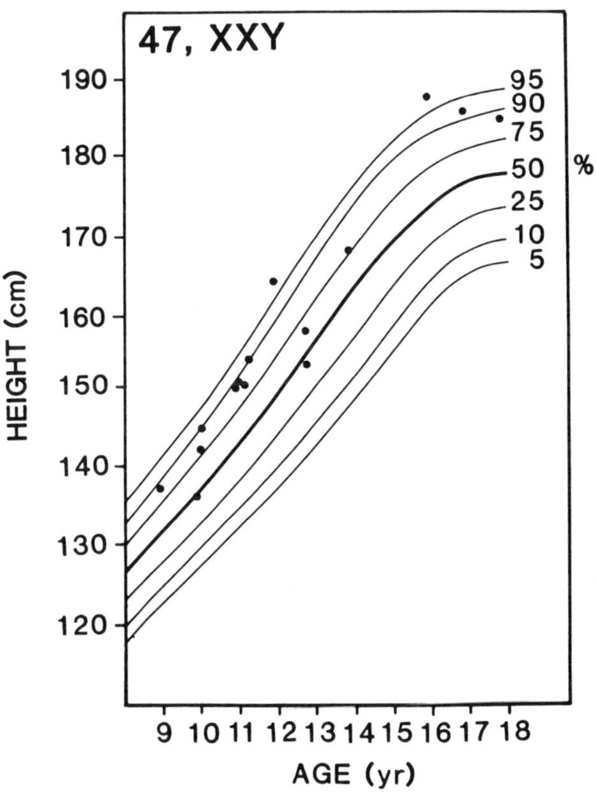

Fig. 2. Percentiles for height of 47,XXY probands [4].

but T levels plateaued after age 14, and individual values subsequently failed to rise above the middle of the normal range. Serum FSH, LH, and T values for the control sibs were within the normal range. Six of the aneuploid subjects had slight elevations of E2 (up to 2 ng/dl) before the clinical onset of puberty and before any rise in testosterone. A similar unexplained slight elevation was observed in one of the sibs, although E2 concentrations at this age (8–10 years) were not significantly different between the two groups. Serum T and E2 levels in six boys at the time of diagnosis of gynecomastia were not significantly different from those in age-matched 47,XXY subjects without gynecomastia.

At age 8–10 years, there were no significant differences between median serum hormone levels in the 47,XXY boys and control sibs. At 11–12 years, T and E2 levels were higher in the 47,XXY boys, a possible indication that they had entered puberty slightly earlier than the sibs. By 13–14 years, their sex steroid concentrations were within the normal range, but the 47,XXY subjects are clearly hypergonadotropic.

In 14 47,XXY individuals, ages 6–14 years, more detailed testing of pituitary-gonadal function was carried out with sequential GnRH and hCG stimulation tests. In response to hCG stimulation, all 14 boys increased their serum testosterone level. In all but two, the peak T value was at least as high as in normal boys at the same level of development [16] even though several already had elevated basal FSH and/or LH concentrations. The median post-hCG peak serum E2 concentration was within the normal range in the prepubertal and early pubertal subjects [16]. Similarly, hCG administration did not induce any significant abnormality in the E2/T ratio at this stage of puberty.

Exaggerated responses to a single intravenous dose of GnRH were observed in 47,XXY boys who already had elevated basal serum gonadotropin values. However, acute GnRH stimulation did not evoke abnormal responses in subjects with normal basal FSH and LH levels.

Cognitive Development

Intelligence. Fourteen 47,XXY boys have now been administered the Wechsler Intelligence Scale for Children—Revised at 8–10 years of age. Results indicated that the full-scale, verbal, and performance IQs of this group appear to be slightly diminished but not significantly so relative to 14 male controls selected for this investigation (Table II). Although the WISC-R verbal IQ is lower than the performance IQ, it is lower by the same degree in the control group. Verbal IQ, therefore, does not appear to capture the cognitive difference between 47,XXY males and 46,XY males.

The IQ range of 71–122 in the 47,XXY group reflects considerable variability and does not allow drawing of a single profile of the intellectual

TABLE II. IQ Scores (WISC-R) of SCA and Control Children

Karyotype (N)		VIQ	PIQ	FSIQ
47,XXY (14)	$\bar{X}$	94.5	99.0	97.1
	SD	15.5	11.3	15.3
	P*	.081	.136	.063
47,XYY (4)	$\bar{X}$	94.0	111.5	101.8
	SD	2.45	11.9	6.70
	P	.05	.463	.373
46,XY controls (14)	$\bar{X}$	105.2	106.0	105.9
	SD	9.36	12.7	10.0
47,XXX (11)	$\bar{X}$	83.3	88.8	84.6
	SD	20.4	14.5	18.2
	P	.013	.004	.006
45,X and variants (9)	$\bar{X}$	90.6	85.8	87.3
	SD	21.7	18.7	20.7
	P	.144	.016	.040
Mosaic (8)	$\bar{X}$	104.8	105.8	106.5
	SD	14.7	15.5	14.9
	P	.938	.883	.910
46,XX controls (15)	$\bar{X}$	104.0	105.8	105.2
	SD	16.9	9.7	13.5

*P values reflect statistical comparisons between each propositi group and the control group of the same sex.

ability of 47,XXY boys. Four scores were below a statistically defined average of 85–115 (one standard deviation above and below the mean), nine were in the average range, and one was above average (122). In the control group, with a range of 88–131, no scores were below average, 11 were average, and three were above average.

Perceptual-motor. Mean scores from the developmental test of visual-motor integration [19], a test of ability to copy designs, did not show significant differences between the two groups (Table III). However, the trend is similar to that observed in the IQ scores and suggests that perceptual-motor skills of the 47,XXY boys are slightly but not significantly lower than those of the controls.

Memory. The 47,XXY boys performed significantly lower than controls on the digit span test [20]. This striking deficit in short-term verbal memory cannot be attributed solely to the slight reduction in general intelligence among propositi, nor is it, by our clinical observation, a secondary effect of inadequate attention (Table III).

Language. Three tests of language skills were selected for this investigation, measuring auditory perception (Goldman-Fristoe-Woodcock test of auditory discrimination) [21], receptive language (token test) [22], and expressive language (Boston naming test) [23]. Mean scores were not signifi-

TABLE III. Cognitive Test Scores of Male Propositi and Controls*

Karyotype (N)		Perception (visual-motor integration)	Memory (digit span memory)	Language: Auditory discrimination	Language: Token test	Language: Boston naming test	Achievement: Reading	Achievement: Writing
47,XXY (14)	$\overline{X}$	87.1	89.9	89.9	90.5	72.9	91.2	90.5
	SD	18.4	14.1	16.2	19.1	14.2	16.0	15.5
	P†	.179	.003	.217	.182	.013	.038	.028
47,XYY (4)	$\overline{X}$	84.3	90.5	91.8	88.8	69.5	96.3	96.0
	SD	11.8	11.0	15.8	6.29	12.3	4.99	8.76
	P	.323	.062	.740	.674	.059	.467	.410
46,XY controls (14)	$\overline{X}$	97.4	111.1	94.4	92.9	86.1	100.2	101.6
	SD	24.5	19.4	13.1	18.8	14.8	10.2	13.4

*All scores are quotient scores with mean of 100 and SD of 15.

†P values reflect paired statistical comparisons between each propositi group and the control group.

cantly different from controls for the tests of auditory perception and receptive language. The 47,XXY boys were significantly surpassed by the controls on the Boston naming test, a word-retrieval test requiring the child to name pictures of decreasingly familiar objects. The mild dysnomia that characterized the 47,XXY group represents another deficit in verbal memory, in this case inability to retrieve information from long-term storage (Table III).

Achievement. Two sections of the Woodcock-Johnson psychoeducational battery [24] were given, examining ability to read individual words and to read and comprehend short passages (reading cluster) and to use spelling and grammar (written language cluster). The 47,XXY boys achieved significantly lower mean scores on both tests relative to the controls (Table III).

Educational progress. Twelve of 14 47,XXY boys have been identified by school personnel (unaware of their karyotype) as having learning difficulties requiring special education (Table IV). Eleven of fourteen boys required help for poor reading skills; three of them were also weak in other academic subjects. Ten of the children received part-time special education help while otherwise attending regular classes; two were placed full time in special education classrooms. In the control group, only one of 15 boys was assigned to special education, in this case part-time help for weak reading skills. The incidence of educational intervention in the control group is currently less than when the children were younger [25]. In contrast, the propositi continue to come to attention because of their ongoing struggle with academic material.

Psychosocial Adaptation and Adolescence

Seven of the 15 47,XXY boys have fairly dependable adaptive skills as observed in the context of their families, schools, and peers (Appendix B). Emotional stability for the remaining eight boys has been more difficult. Diminished self-esteem usually began to appear around age 7 with awareness of school problems. The same impaired language and motor functions that identified them as candidates for special education programs also interfered with normal adaptive processes, resulting in diffuse anxiety and immature behavior. These obstacles, when combined with a nonsupportive or even hostile social environment, increased the incidence of prolonged maladjustment. Six of these eight boys have a high risk for continued unsatisfactory psychosocial adaptation because of unstable family relationships and their own limited resources for coping with stress.

In the nine dysfunctional families studied, six of the 47,XXY boys have been more vulnerable than their sibs who shared the same family setting. For example, one propositus at age 17 was admitted to a private psychiatric hospital for treatment of aggressive defiance of his father, whereas his two older sibs had chosen more conventional shortcuts to emancipation through college and early marriage. Maladaptive behavior—stealing, running away

TABLE IV. Current Incidence of Specific Learning Disorders Requiring Educational Intervention in Propositi and Controls as Reported by Schools

Karyotype (N)	No problem	Reading	Reading and language	Math	Math and language	All school subjects	Mental retardation
47,XXY (14)	2	6	3	0	0	3	0
47,XYY (4)	0	2	2	0	0	0	0
Male controls (16)	15	1	0	0	0	0	0
47,XXX (11)	2	1	3	0	3	1	1
45,X and variants (9)	3	1	0	0	0	4	1
Female mosaics (8)	8	0	0	0	0	0	0
Female controls (16)	14	0	0	0	1	1	0

from home, aggression at school, social withdrawal, and encopresis—was also noted in five of the affected prepubertal boys whose sibs retained relative equanimity. Conversely, the youngest boy in the affected group is at age 10 apparently free from serious disturbance despite incidents like a suicide attempt in one of his five older sibs.

Five boys are currently of high school age. Two have the necessary academic credentials and plan to enter college; one of these anticipates an athletic scholarship and a career in professional golf. One boy who was asked to leave a small church school when he displayed a can of beer on his 18th birthday now works as a cook in a fast-food shop. He and the other two boys have no specific vocational goals beyond a vague desire for technical training in the Army or Air Force. The attitudes of these five adolescents toward eventual marriage and family are still too incomplete to assess. However, all have masculine identity and a heterosexual orientation.

Summary

The 47,XXY boys have height above the 45th percentile, are normal in physical appearance, and have small testicular size. Onset of clinical puberty has been consistent with the controls. After age 14, testosterone levels plateau and the propositi are clearly hypergonadotropic.

Neuromaturational lags have been identified as reduced gross and fine motor skills, coordination, speed and dexterity, and strength. Half of the 47,XXY boys have a specific dyslexia. They have intelligence in the average range, exhibit impaired language and verbal memory, are slow in cognitive processing (ie, even when they can produce a correct answer they require more time to do so), and have a history of reading difficulty in school requiring special education. Seven of the boys have demonstrated adequate psychosocial adaptation, but eight have had difficulty in their adjustment related to diminished self-esteem, anxiety, and immature behavior.

47,XXX

Physical Assessment

The age range of the 11 girls currently being followed is 12–19 years (Appendix C). They are normal in general appearance, and their head circumferences remain within the normal range (2–80%; Fig. 3). Presently, ten have a height at or greater than the 75th percentile (range 40–>95%; Fig. 4). The present mean height is at the 87th percentile, and five have a height greater than the 95th percentile. Their height has previously been documented to be significantly greater than that of their sibs and population norms [2]. Two girls had true precocious puberty, with breast development beginning at age 7. The remaining nine girls had the onset of Tanner stage 2

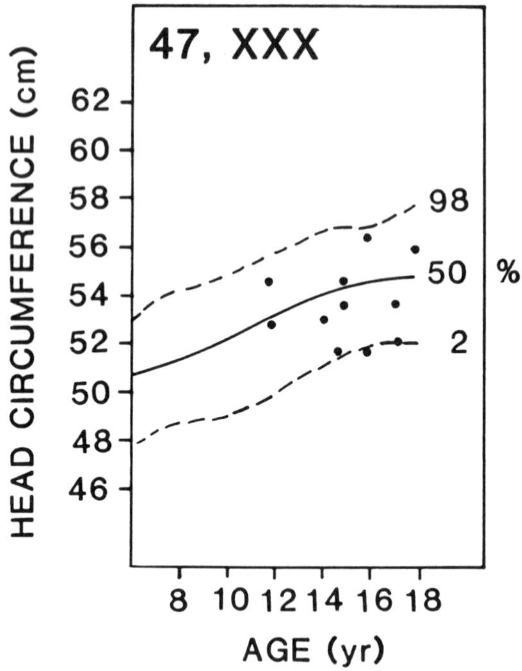

Fig. 3. Percentiles for head circumference of 47,XXX probands [5].

breast development between 9 and 12 years of age, which is consistent with population norms [26]. The mean age of menarche is 12.5 years. One subject at the age of 16 had a son who was reportedly developing normally until he died of sudden infant death syndrome at 3 months.

Medically, five subjects have experienced allergies including upper respiratory symptoms, eczema, or asthma. Five have experienced some degree of scoliosis, which has responded to medical management without serious progression. Somatization is present in seven girls, five of whom have a history of nonspecific abdominal pain. One girl developed trichotillomania. They are generally immature, which is in contrast to their tall stature, resulting in a discrepancy between expectation and performance.

Neuromuscular Assessment

We have previously reported the problems with early motor dysfunction in 47,XXX girls [27,28]. Gross and fine motor development continues to be below that of siblings, particularly involving balance, equilibrium, and visual-motor and sensory-perceptual integrative dysfunction. Therefore, the Bruininks-Oseretsky test of motor proficiency was administered by a physical therapist who was blind to the diagnosis of ten unselected 47,XXX girls and an equal number of control sibs.

Strikingly, nine of the aneuploid girls have a test battery composite score in the lowest two stanines, which places them at or below the 10th percentile. The single subject who is functioning at the fifth stanine is generally the most competent. In contrast to the aneuploid girls, the female controls have a normal distribution of motor function with a test battery composite stanine range of 2–7.

Endocrine Studies

Pituitary-gonadal function is studied prior to and throughout puberty by determining random serum hormone concentrations of FSH, LH, E2, and T. Ten of our 11 propositae with the 47,XXX karyotype have been tested (Appendix D).

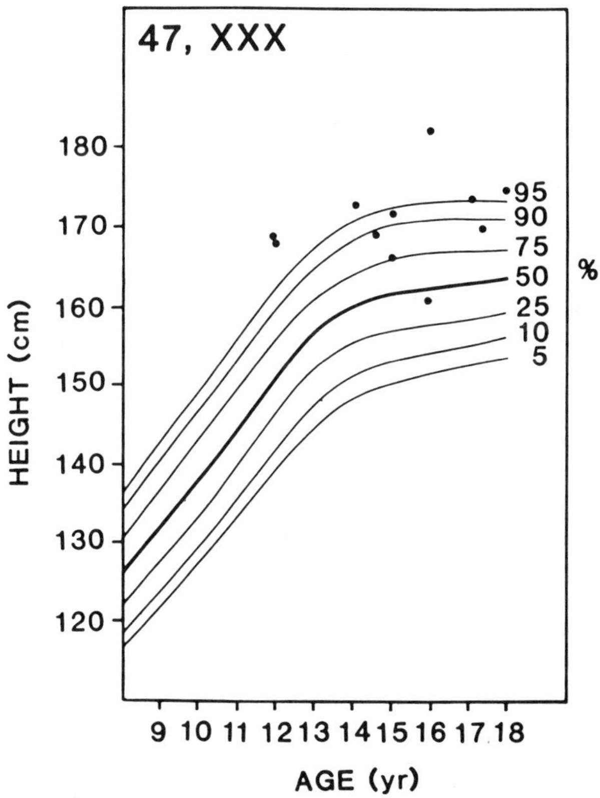

Fig. 4. Percentiles for height of 47,XXX probands [4].

The interrelationship of the hormone values is normal in specimens analyzed between ages 7 and 16 years. Mean values at 2 year age increments are consistent with previously published norms [18] for estradiol, luteinizing hormone, and testosterone. In contrast, FSH is consistently generous, with many values lying outside the previously reported range. Although the female hormones are variable and cyclic, the persistently high FSH levels remain unexplained. The two individuals with precocious puberty had pubertal hormone values at the time of their early breast development. For these two girls, menarche followed 2 and 4 years after the initial physical changes.

Cognitive Development

Intelligence. The cognitive development of 47,XXX girls has been poorer than that of any other group of propositi in this study. The IQ range (53–112) is larger than that of the group of 15 female controls (83–128), and the mean verbal IQ, performance IQ, and full-scale IQ of the propositae are all about 20 points lower than (and significantly different from) those of the controls (Table II). Five girls have full-scale IQs in the average range (85–115). Six have IQs below this range, with five in the borderline range (70–79). Four propositae with histories of poor emotional development have been difficult to test, and, perhaps because of this, have shown 14–26 point differences between the preschool (WPPSI) and school-age (WISC-R) IQ scores. Three of these were lower at the older test age. The girl with lowest IQ score (53) from our test results was reported by her school to have an IQ of 75. The latter result is closer to our estimates of her potential. One control girl has an IQ below average, and three are above average.

Perceptual, memory, and language test scores. Mean scores from tests of perceptual-motor, memory, and language abilities are reported in Table V. A selected battery of cognitive tests was administered to nine available 47,XXX girls, three of their female sibs and six female sibs of other propositi. On all eight tests, scores of propositae were significantly lower than those of controls. As evidenced also in their IQ scores and multiple special educational needs, the 47,XXX girls do not demonstrate a deficit specific to any one area of cognition but are characterized by a more generalized reduction in abilities and slow cognitive processing. Poor language skills are most apparent because difficulty in communication is observed quickly in any academic or social setting.

Education progress. Nine of 11 47,XXX girls have received special educational assistance from their schools, in contrast to two of 16 controls (Table IV). Three have been placed in classes for severely learning-disabled and emotionally disturbed children. Special educational programs for the remaining six have consisted of part-time classroom placements. The academic skills targeted for intervention (Table IV) frequently included math,

TABLE V. Cognitive Test Scores of Female Propositae and Sib Controls*

Karyotype (N)		Perceptual		Memory (digit span memory)	Language			Token test
		Spatial relations	Visual-motor integration		Auditory association	Grammatic closure	Picture vocabulary	
47,XXX (9)	$\overline{X}$	85.4	66.3	97.3	70.8	74.8	85.0	73.4
	SD	16.2	15.1	24.1	31.3	42.5	16.9	21.5
	P†	.019	.001	.017	.042	.048	.048	.011
45,X and variants (9)	$\overline{X}$	87.2	71.8	99.8	94.5	99.3	103.9	79.3
	SD	18.0	16.6	28.1	23.9	32.8	18.5	21.4
	P	.020	.002	.069	.410	.240	.300	.046
Female mosaics (6)	$\overline{X}$	98.6	98.3	107.5	113.3	113.8	105.2	101.0
	SD	14.5	11.9	16.7	17.6	4.49	15.1	21.3
	P	.313	.898	.088	.265	.463	.936	.857
46,XX controls (9)	$\overline{X}$	109.4	97.31	122.4	99.9	106.1	104.3	101.2
	SD	22.0	16.0	14.5	24.2	29.6	21.6	18.1

*All scores are quotient scores with mean of 100 and SD of 15.

†P values reflect statistical comparisons between each propositi group and the control group.

language, and reading. Seven propositae required assistance in more than one academic skill. Two of 16 control girls have been placed in part-time learning disability classrooms. The two propositae with the highest IQs (109 and 112) have had no need for special education. One has completed high school with a B average, is working at a fast-food restaurant, and is attempting to enter a community college. The other is currently in high school, maintaining an A–B average, and plans to enter a university after graduation.

Psychosocial Adaptation and Adolescence

Appendix E lists the 11 propositae in rank order of overall adaptation and summarizes the parameters of growth and development on which this evaluation is based. Two girls are reasonably well adjusted in all respects; six girls have more difficulty than peers or sibs; three girls have inadequacies that jeopardize or preclude normal independent functioning as adults. In this admittedly small sample, the high proportion of developmental delays and the relative severity of their limitations cause us greater concern for the prognosis of 47,XXX girls than for any of the other sex chromosome aneuploidies represented in the Denver study. At the same time, standards of child care for this group as a whole are noticeably lower, and social stability is more rare. A child's successful adaptation to the presence of functional deficits or unusual patterns of physical development (as described above) depends strongly on consistent efforts by the family to provide supportive encouragement. For various reasons, conditions for optimal success have been unavailable to most of the 47,XXX girls.

Major stressful events with long-lasting consequences occurred in eight of the 11 families. Two were investigated by community welfare agencies for allegations of child abuse. Four fathers are alcoholic, and there were six divorces. Except for two of these families (those of the most successful girls), held together by strong, resourceful mothers, the affected chidlren have suffered from ambiguous relationships at home. Another kind of ordeal for one girl was the crib death at age 3 months of her baby, the only one born so far to a 47,XXX girl in the study. Her depression is responding to family and community support.

Most of the nine girls with a history of poor school adjustment are aware of their sibs' and peers' superior academic achievement and feel incompetent. Six of them developed psychosomatic complaints, including enuresis and trichotillomania for one girl at age 12. School problems also interfere with setting goals for the future. Early dependent marriage relationships are the first choice for the future for six of the affected girls. Their parents also accept marriage as a satisfactory basic attainment for their daughters although not necessarily for their sons. Three girls expect to attend college, and at least one will probably succeed.

The three girls listed at the bottom of Appendix E as having seriously impaired overall functioning have never achieved satisfactory family or peer relationships. In the sensitive period of adolescence, their marked isolation and withdrawn behavior may be the most suggestive precursors of adult psychopathology. Reliable cognitive skills, proficiency in language, motor coordination, and perhaps above all unequivocal family support would appear to constitute the most important elements predictive of success in the developmental history of these 11 47,XXX girls.

Summary

Two of the 11 47,XXX girls function normally in all respects. Eight girls, although exhibiting varying degrees of developmental difficulty, are not perceived as conspicuously different from other family members. Only one girl contrasts sharply with her sibs, who display markedly greater social and academic achievement.

General appearance, birth weight, and head circumference of the entire group are within normal limits. However, five girls are below average in intellectual ability, one is the borderline range, six have severe language impairment, eight have significant motor dysfunction, and three have been unable to achieve satisfactory psychosocial adjustment. Tall stature is a unique physical characteristic—ten probands are above the 75th percentile (including five above the 95th percentile). Two girls became pubertal at age 7.

It is concluded that the 47,XXX condition is associated with increased risk for developmental deficits and small physical deviations from the norm. Family counseling to ameliorate the effects of these disadvantages on the child's psychosocial adaptation is recommended.

MOSAICS

Physical Assessment

Appendix F summarizes the current anthropometric data on females with sex chromosome mosaicism, including the initial chromosome ratio and timing of the onset of puberty. Hormone data are summarized in Appendix G.

Of the three girls with 45,X/46,XX karyotype being followed only No. 21 has a height below the 5th percentile, and she also has a broad chest, triangular facies, and low posterior hairline. Pubertal changes occurred spontaneously and menstrual periods are regular, although she experiences primary dysmenorrhea and premenstrual tension syndrome. The remaining two subjects have no physical stigmata and also had spontaneous onset of puberty.

Subject No. 35 with a 45,X/47,XXX karyotype is presently 13 years old and has a height below the 5th percentile and deep-set nails. Breast development began spontaneously, but she has not yet had a menstrual period. After the neonatal period, the 45,X cell line has been greater than 50% in blood leukocytes and skin fibroblasts.

The two subjects with 46,XX/47,XXX karyotypes are normal on physical examination and entered puberty spontaneously. Subject No. 3, at the age of 16 years, had a son with a 46,XY karyotype.

One subject with 45,X/46,XX/47,XXX mosaicism is being followed, and her physical examination is normal.

Two boys with 46,XY/47,XXY mosaicism are being followed. The percentages of aneuploid cells are 2% and 12%. At age 17 years, one has short stature, which appears to be familial, normal physical and secondary sexual development, and a mean testicular volume of 20 ml. Hormone levels reveal normal pubertal values. The second subject is presently 11 years of age and is normal on physical examination, with early pubertal development and a mean testicular volume of 4 ml (Appendix H).

Cognitive Development

Intelligence. No significant differences were found between the female mosaic and female control groups in verbal, performance, or full-scale IQ. As can be seen in Table II, the mean IQ scores for the mosaic group were very similar to those of the control group. The full-scale IQ range for the mosaic children was 81–129. Two propositi had above average IQs, five were average, and one was below average.

Perceptual, memory, and language test scores. Mean scores for the female mosaic group on seven tests measuring perceptual, memory, and language abilities did not differ from those of the controls (Table V). As on all other measure of developmental skills, the mosaics are more similar to their controls than any other group of propositi.

Educational progress. In contrast to all other groups of propositi, none of the female mosaic girls has required special help in school since our last report (Table IV). Of the two oldest mosaic propositi, one found school difficult, received poor grades, and left to get married before completing 4 years of high school; her three sibs also found school difficult. The other entered college with a scholastic scholarship but lost the scholarship because of her inability to maintain a satisfactory grade point average. Two of the remaining propositi showing competent academic achievement have sibs who have had marked learning difficulties.

Psychosocial Adaptation and Adolescence

The emotional development of nine chidlren with mosaicism has progressed according to family patterns without obvious interference by physical or intellectual disadvantages (Appendix I). Six are well adjusted and secure in school and social situations, either responding to favorable influences at home or, in two instances, even surpassing other members in effective use of adaptive skills. Three of these are adolescent girls.

At opposite ends of the socioeconomic scale are two additional adolescent girls who have no serious emotional disturbance but who nevertheless have been consistently unhappy and dissatisfied and lack a sense of direction with respect to the future. Both belong to families accustomed to friction in personal relationships, and both have sibs with similar adjustment difficulties. At age 18, the one with social advantages was a National Merit Scholar but failed to make passing grades in the first year of college. The other, at age 19, is the mother of a 3-year-old son, and her husband is physically disabled and unemployed.

The most severely impaired child in the mosaic group, and the study's only antisocial offender, has a history of early truancy and delinquency associated with his father's pornographic activities. When his mother died during his father's second prison term, his custody was assigned to an older sister later convicted of prostitution. At age 18, he lost his legal protection of juvenile status, and reportedly has been jailed once since that time on charges of sexual assault.

It is concluded that no significant differences in psychosocial adaptation between this group of mosaic children and their sibs can be identified from this study.

Summary

One of two mosaic girls with a 45,X cell line and height below the 5th percentile (subjects 21 and 35) has primary dysmenorrhea and marked physical stigmata suggestive of Turner syndrome. The remaining five have normal physical examinations. None has required special education, and their cognitive test scores do not differ from those of female controls. Emotional development has progressed at least as well as that of sibs.

47,XYY

Physical Assessment

Four boys with an age range of 10–12 years are being followed (Appendix J). Physical appearance is normal, with no consistent dysmorphic features. One (No. 41) had a congenital dislocated hip, bilateral inguinal hernias, hydrocele, and marked neonatal facial acne. One has chronic asthma, as does the sib of another. Tall stature, from the 50th to 95th percentiles, is a consistent physical finding. The height-weight ratio is within the normal range. Three remain prepubertal, and the oldest at age 13 is in early puberty. Penile length and testicular volume are age-appropriate.

Neuromuscular Assessment

Hypotonia and oral-motor dysfunction suggesting neuromuscular problems were found on clinical examination in all four boys. The Bruininks-Oseretsky test of motor proficiency [14], administered by a physical therapist blind to the diagnosis, documented them to be functioning in the lower stanines for age. They were all identified as having sensory-motor integration dysfunction. Two sibs are functioning in the middle stanines. A third sibling, in a family in which all three children have delayed motor development, is less affected than the propositus.

Endocrine Studies

Random hormone studies, including FSH, LH, T, and E2, have been completed on two subjects and their two sibs, and the values are within the normal range with no evidence of pituitary or gonadal dysfunction.

Cognitive Development

Intelligence. The small number of subjects in this group makes meaningful statistical analysis impossible. Nonetheless, paired comparisons were completed for the 47,XYY group in order to have data analyses similar to those of the other propositi groups. The mean full-scale and performance IQs of the 47,XYY boys did not differ significantly from those of male controls. All full-scale IQs were in the average range. The performance IQs of two 47,XYY boys were 26 points higher than their verbal IQs. In contrast, the mean verbal IQ for the propositus group was significantly lower than that of the control group, indicating relatively weaker language and verbal-conceptual skills (Table II).

Perceptual-motor. The 13 point discrepancy between the 47,XYY and control groups did not result in a significant difference, although it suggested that propositi, as indicated in the neuromotor scores, have difficulty with visual-motor integration (Table III).

Memory. Mean scores on the digit-span memory subtests approached significance and suggested impaired auditory short-term memory (Table III).

Language. Mean scores were not significantly different from those of controls on two language tests, auditory discrimination and token test. However, mean scores on the Boston naming test approached significance and suggested a specific impairment in expressive language among propositi (Table III).

Educational achievement. The 47,XYY boys achieved mean scores on the reading and writing subtests of the Woodcock-Johnson psychoeducational battery that did not differ significantly from those of controls (Table III).

Educational progress. Despite their average achievement test scores, all four 47,XYY boys have difficulty mastering school curricula and have received part-time special education in reading, language, and writing skills. In tasks of decoding words or comprehending sentences, slowness rather than lack of ability has been noted by teachers. This decreased speed in processing information appears to be similar to that of the 47,XXY boys (Table IV).

Psychosocial Adaptation and Adolescence

In this sample of four 47,XYY boys, all show mild learning disorders and impaired athletic skills on the playground, and all have been subjected to academic pressure to compensate for these difficulties (Appendix K). Nevertheless, two of the boys who have reasonably supportive families display satisfactory psychosocial adaptation with respect to family and peer relationships and to age-appropriate activities. The other two boys were born into families with stressful situations, have unstable relationships with at least one parent, and display behavior patterns of moderate insecurity, isolation, and lack of initiative. In all cases, the quality of emotional adjustment is indistinguishable from that of chromosomally normal sibs or peers with similar cognitive deficits. The small number of subjects and their preadolescent stage of development preclude any sweeping generalizations, but the data suggest the importance of family support in coping with the mild learning disabilities accompanying this condition.

Summary

Common characteristics shared by the four 47,XYY boys are 1) normal intellectual and physical development, 2) moderately impaired sensory-motor integration and language skills, and 3) increased height. Their psychosocial adaptation is comparable to that of sibs or peers with similar mild learning disorders.

45,X AND PARTIAL X MONOSOMY

Physical Assessment

45,X. Six girls with this karyotype are being followed, and at the time of their last examination the age range was 11–18 years (Appendix L). Head circumferences are within normal limits, although two are just below the 2nd percentile (Fig. 5). Their height is significantly below that of their sibs and population norms, with the exception of one girl having few stigmata. Her height is presently at the 25th percentile, and her family has tall stature (Fig. 6). The shortest individual's family has low-normal stature. The degree of facial dysmorphism is variable, and the number of physical anomalies is greatest in the 45,X group (Appendix M).

Medical concerns include the short stature, difficulty with weight control, elevation of the diastolic blood pressure, coarctation of the aorta, and

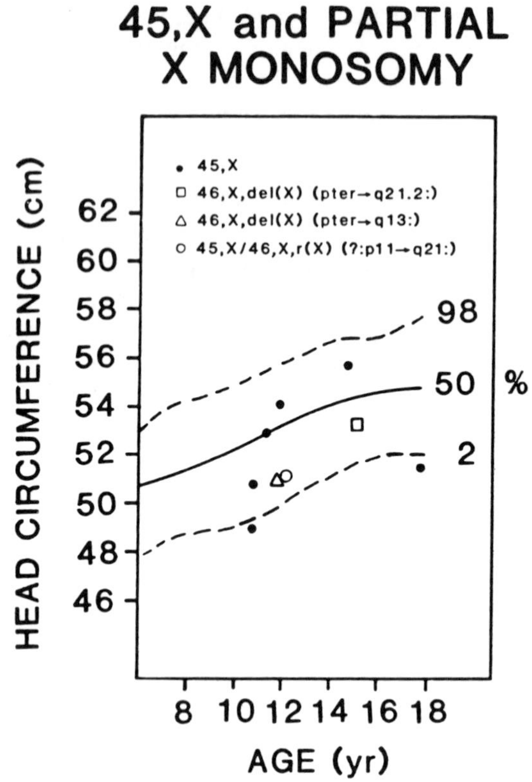

Fig. 5. Percentiles for head circumference of 45,X and partial monosomy X probands [5].

recurrent serous otitis media. There is no evidence of hypothyroidism or growth hormone deficiency in any of the girls. The pituitary and gonadal hormone values are summarized in Appendix N.

Partial X monosomy. Two girls with a partial deletion of the X chromosome are being followed (Appendix L). The oldest was last seen at age 15, has no minor congenital anomalies, was normal on neurological examination, and had breast development at Tanner stage IV. Her karyotype is 46,X,del(X)(pter→q21.2:). Following the onset of spontaneous puberty, her FSH levels have steadily increased with a low estradiol level (Appendix N). Her menstrual periods have been variably irregular.

In contrast, the younger girl with karyotype 46,X,del(X)(pter→q13:) at age 12 has a high arched palate and short broad hands and feet and is impulsive and distractible with a short attention span. Her weight is elevated. Her estradiol is low with elevated FSH levels. One girl with a 45,X/46,X,r(X)(?:p11→q21:) karyotype has had serous otitis media and evidence of palatal dysfunction.

During the early school years, the short stature begins to have a psychological impact on the subjects. Our present protocol is to offer a 2 year course of oxandrolone (Anavar) 2.5 mg per day at approximately 11 years of age. Three girls (Nos. 8, 49, and 117) have completed the treatment course, and No. 39 has completed 1 year of therapy. Subject No. 7 is androgen-untreated and began estrogen replacement at age 12. Subject No. 66 is receiving oxandrolone and growth hormone.

The pretreatment growth velocity was determined over a period of approximately 1 year. Height measurement was by stadiometer (Holtain). Hand and wrist radiographs were obtained before, during, and after treatment for determination of skeletal maturation. Pretreatment growth velocities range from 2.6 to 3.7 cm/year. Oxandrolone treatment resulted in an increase in growth velocity for each subject (Fig. 7). During the first year of therapy,

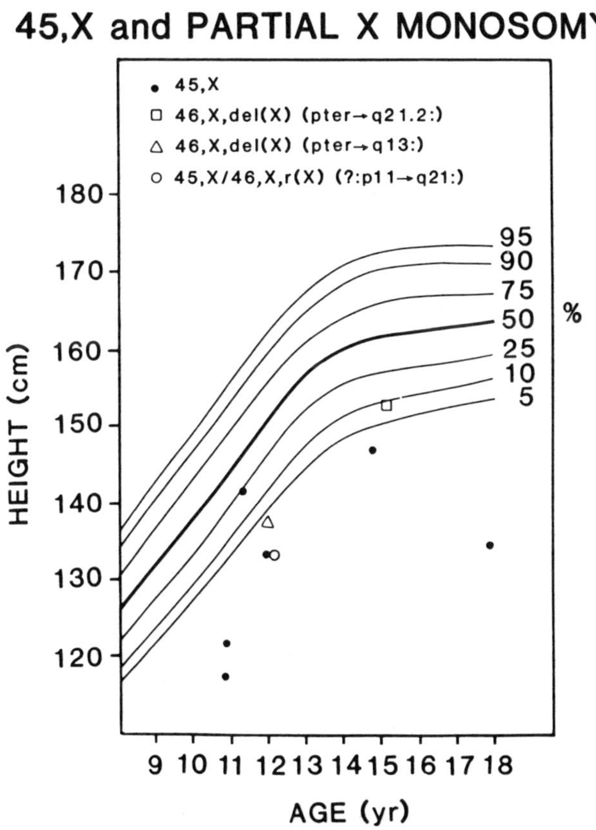

Fig. 6. Percentiles for height of 45,X and partial monosomy X probands [4].

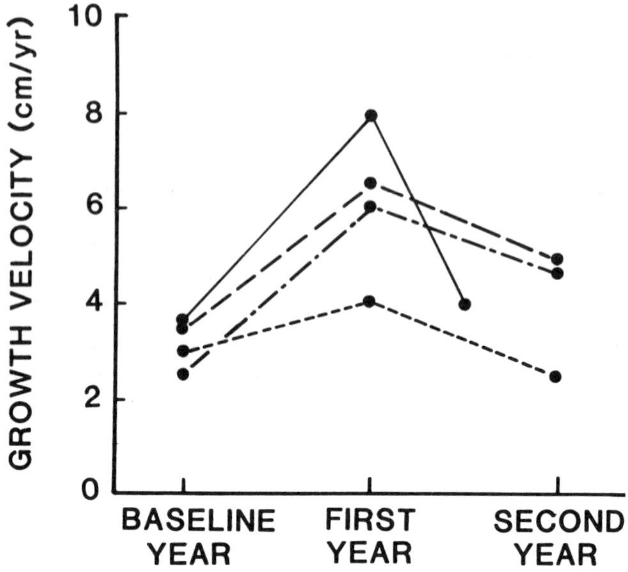

Fig. 7. Growth velocity prior to and during oxandrolone treatment—45,X and partial monosomy X probands.

the growth rate essentially doubled in three subjects. Subject No. 8 with the smallest response began treatment at age 14 and is the shortest. During the second year of treatment, growth velocity slowed for all probands. The only documented side effect was the development of pubic hair. Although there is an immediate psychologic benefit from an increase in growth rate, the long-term effect on final adult height remains unclear.

Neuromuscular Assessment

The Bruininks-Oseretsky test of motor proficiency [14] has been administered to five girls with a pure karyotype and subject No. 39 with a ring X chromosome. Subject No. 56, who does not have short stature and has few stigmata, has a battery composite score in the fourth stanine. The other five girls have significant sensory-motor integration dysfunction, with a battery composite at the first stanine. There is decreased perceptual awareness of the body in space, with some girls exhibiting impulsivity and distractibility. This suggests that neurologic impairment is more extensive than previously recognized in the literature.

Cognitive Development

Intelligence. The mean performance and full-scale IQ scores for the group of nine girls with 45,X and partial X monosomy were significantly lower than those of the control girls. The data also suggest a lower mean verbal IQ, but this difference was not significant (Table II).

One 45,X girl with an IQ of 50 is moderately retarded, and two are functioning in the borderline retarded range with IQs of 72 and 73. A fourth girl had a low-average IQ of 84. Four had average IQs ranging from 92 to 109, and one girl had an above average IQ of 118.

Perception. The results of the group of 45,X and partial X monosomy girls were significantly lower on two tests of perceptual skill (Table V), one measuring ability to mentally rotate geometric shapes (spatial relations test) [29] and the other measuring ability to copy geometric drawings (developmental test of visual-motor integration), thus demonstrating a striking deficit in perceptual skills in contrast to the control group.

Memory. The 22 point spread on scores for the digit-span memory test [20] between the X monosomy and control groups does not constitute a significant difference but approaches significance ($P = .069$; Table 5).

Language. The 45,X and partial X monosomy group achieved mean scores comparable to those of the control group on auditory association, grammatic closure [19], and picture vocabulary [30] (Table 5), indicating adequate skills in both receptive and expressive language. The propositae performed significantly below the control group on the token test, which measures ability to follow a series of directions.

Attention deficit. In accordance with the diagnostic manual of the American Psychiatric Association (DSM III) [31], five of the nine 45,X and partial X monosomy girls were diagnosed as having an attention disorder characterized by inability to remain focused and attend to test material. Two of these were also hyperactive, as observed in their difficulty with motor activity. Distractability and poor concentration are believed to affect the cognitive abilities of the propositae, as is the case with eukaryotic children who have attention deficits. In support of this finding, the mean IQ for the propositae with attention deficit was 75.6, and for those without the deficit it was 100.8. We have observed that two tests, auditory memory and token test, are particularly sensitive to variations in ability to attend. Consequently, the reduced performance of the 45,X and partial X monosomy group on these two tests must be viewed as tentative before any conclusion is made that these propositae have impaired auditory short-term memory or receptive language.

Educational progress. Six of the nine 45,X and partial X monosomy girls have learning difficulties requiring special education (Table IV). The nature of these learning disorders is nonspecific. Four propositae required help in all school subjects because of generalized poor achievement. Three of them attend full-time special educational classes, one for mental retardation and two for severe learning disabilities. Three of the nine girls have been noted by parents and schools to have poor organizational skills, immature handwriting, and some confusion in left-right directions; two of these have been placed in part-time special education classes, and one has received no special education. The oldest proposita is attending college.

Psychosocial Adaptation and Adolescence

Physical limitations associated with X chromosome monosomy have a pervasive influence on psychosocial adaptation (Appendix O). One of the nine girls in this group has no abnormalities in physical appearance or serious health problems and is the only one with a completely satisfactory perspective on herself and her world. In contrast, extremely short stature, below the 5th percentile, is characteristic of seven girls, creating for them the difficult task of maintaining age-appropriate behavior while being treated as much younger than their chronologic years. Six of these girls have the additional burden of chronic health problems requiring special medical care (coarctation of the aorta, hypertension, otitis media and hearing loss, back pain, and obesity), and four have heavy stigmata of Turner syndrome.

Three girls are now in late adolescence. At age 19, one of them is enrolled in a 2 year junior college program, studying radio communication and writing a psychology term paper on Turner syndrome. As a younger teenager, she was socially apprehensive and overly concerned with somatic complaints, but she was also able to accept psychiatric counseling as an aid to establishing her individual identity. A second adolescent girl at age 17 has similar difficulties. So far, her adaptation has been hampered by increasing social isolation and irrational fears. She describes herself as feeling like a freak. The third adolescent girl in this group is the study's only clearly retarded child, with significant intellectual as well as physical limitations. At age 19, she is eligible for a sheltered vocational workshop and Social Security disability benefits for the rest of her life. She and her mother have a mutually supportive relationship. It is of interest to note that her six chromosomally normal sibs also rely on public funds for support—three sisters receive aid for their dependent children, and three brothers are in prison.

Wide variations in cognitive, behavioral, and physical development are evident in this group of nine girls. Their psychosocial difficulty depends in part on the extent of physical deviation. Supportive family relationships appear to be even more essential in establishing the basis for favorable emotional growth and self-esteem in these cases than they are for normal children.

Summary

Physical and cognitive development of six girls with total X monosomy is variable, although most show at least some impairment in both categories. In the small group of three girls with partial X monosomy, physical anomalies are less frequent, but patterns of cognitive function are virtually indistinguishable from those in the 45,X condition. Mean test scores for the combined group reveal adequate expressive and receptive language, slightly impaired memory function, and striking deficits in perceptual and motor skills. With one exception, full-scale and performance IQ scores were lower than those of sibs. Children with pronounced physical stigmata may possess a normal IQ.

Medical concerns include hypertension, recurring otitis media, difficulty with weight control, coarctation of the aorta, and short stature. Five girls responded to administration of oxandrolone by increased height velocity during the initial period of treatment, although its effect on ultimate height is not yet known. As with other sex chromosome aneuploidies, there is evidence that appropriate parenting can ameliorate the associated developmental risks and promote adequate psychosocial adaptation.

PSYCHIATRIC ASSESSMENTS OF SCA CHILDREN AND THEIR SIBS

Two Denver child psychiatrists have developed a method for the assessment of emotional adjustment in adolescence. They conduct a blind, semi-structured, 1 hour interview with propositi and siblings. It is video taped for later review to ensure consistent, standardized scoring. The purpose of this psychiatric interview is explained to each child and his/her parents, and consent forms are signed by both parents and child.

The interviews focus on four specific areas of inquiry. The first is an assessment of the quality of the relationship of the adolescent with family members and the extent to which he/she has begun to separate from the family of origin. The second assessment is the child's sense of self-esteem as revealed by response to direct inquiry about personal strengths and concerns and plans for the future. The third is a determination of sexual identity and preference. The fourth is the assessment of responses to life stress and a rating of the child's capacity to respond appropriately throughout the interview. Specific projective questions concerning dreams and fantasies are included at the end of the interview. A DSM III diagnosis [31] is made if appropriate. As an estimate of overall functioning and adaptation to adolescence, each child is then given a numerical rating based on the children's Global Assessment Scale [32] developed at the Columbia University School of Medicine (see Appendix P).

About half the total number of available propositi and controls have been interviewed. Until the project is completed, the children's identity will remain concealed from the interviewing psychiatrists, and no attempt will be made to analyze the data. However, it is of interest to note that results of the 43 interviews so far are in general agreement with conclusions drawn from many years of recorded history in each child's chart. Median scores of the sib and mosaic groups are higher than those of the nonmosaic groups; the range of scores for propositi is 35–87 and for sibs 52–87.

MEANING OF EARLY KNOWLEDGE OF A CHILD'S INFERTILITY IN FAMILIES WITH 47,XXY AND 45,X CHILDREN

In the course of taking a closer look at the impact of the study on participating families [33], we realized that the parents of 47,XXY boys and 45,X girls had, in addition to their worry about how to tell their children about the chromosomal diagnosis, the burden of knowing that their son or daughter would probably be infertile and of wondering how and when to communicate this to their children. This prompted us to explore the nature of these feelings and their effect on the emotional climate surrounding the child.

Successful adoption, should building a family by adoption later be the choice, cannot take place unless there has been an adaptive resolution of infertility issues [34]. The pain and sadness parents feel in the knowledge that the choice to become a parent is lost to their child must be appropriately shared in the family so that the loss does not come to be viewed as a "defect," but rather as something unusual that has happened to the child that can be integrated into a positive self-concept. By the time children are 4 years old, for example, they understand that making a family, where parents stand by their child and help that child grow up, is not the same as simply making a baby.

The parents of 15 47,XXY boys and 9 45,X and partial monosomy X girls (now 8–18 years old) were interviewed. Fathers' expressed concerns were more directly focused on sons, whereas mothers expressed concerns more directly about daughters. This reflects what is known about a parent's particular identification with a child of the same sex. The fathers' expressed concerns were divided between anxiety about possible homosexuality (every father interviewed expressed concern about this) and regret about infertility. The mothers' expressed concerns were focused more on infertility. Infertility, however, was the source of shared grief and sense of loss in both parents.

The profound feelings of loss experienced by childless couples wishing to establish a family and the accompanying "crisis of infertility" have recently been recognized and described in the literature [35]. The questions asked by the parents communicated vicarious and anticipatory grief, as if they were going through a "crisis of infertility" involving themselves and their child. Parents who otherwise had little difficulty in talking to their children reported feeling inarticulate because of sadness while watching them "play house," casting each other as mother and father. Clearly, intellectual recognition of adoption as an alternative path to parenthood was not enough to ease their pain.

The parents of potentially infertile children in our study have expressed feelings related to the elements associated with the "crisis of infertility": guilt, mourning, depression, loss of health, loss of self-esteem, and frustration. Fathers were concerned that their sons might be less than healthy, expressing anxiety about developing gender identity, future sexual orientation, and function. Mothers were anxious about fertility and about maternal role fulfillment for their daughters. Mothers' communications were most often colored by depressed affect and grief, whereas anxiety and depressed affect most often were expressed by fathers. Both mothers and fathers spoke of guilt, not necessarily about the chromosomal diagnosis but, as one couple put it, "she's here because we're here." Children readily pick up such feelings and anxieties in the family. Lest these become the source of confused worries about invisible defects that may negatively affect the child's self-image, they need to be identified and appropriately shared with the child.

We were alerted anew to the importance of fertility as part of the family identity when some parents of 47,XXX girls asked if their daughters could expect to have children. Parents, in recognizing an abnormality of a sex chromosome, readily worry about reproduction. These parents' continued need for clarification reemphasized for us not only the importance of the issue itself but also the importance of repeatedly exploring with participating families what their understanding has been of issues discussed at various times. In four families, there was a dread that, despite evidence to the contrary and in the face of reassuring communication from the research staff, a silent and fatal illness would carry off their child in early adolescence. It is not yet clear whether this is related to fears of infertility and therefore of genetic death.

The profoundly unsettling effects of infertility on couples have been described in the literature [35], and it is now known that men can suffer as deeply as women when they lose the choice to become biologic parents [36,37]. Parental grief and frustration when they see the choice to become biologic parents taken away from their children is therefore understandable. That these feelings should emerge as early in the life of an infertile child as they did (in one case as early as 2.5 years) was surprising. Parents spoke of sadness and regret about their child's anticipated loss and about their own loss of grandchildren and expressed concern about their children's wholeness and, by extension, their own.

Studies of infertile adults show that the capacity to find adaptive ways to deal with grief, anger, and sadness is associated with a positive integrated sense of self [38]. For children, the most effective impetus toward development of a positive self-image comes from parents. Openness between parents and child is therefore encouraged, since children need to know the facts about family feelings concerning them so that they do not feel isolated and alone. There is natural pain in sharing feelings and facts about infertility, but children stand to gain because they find that they can trust their parents' honesty as well as their own perceptions. Early factual knowledge also gives them the time to adapt gradually, in steps appropriate to their maturity and age, to a reality that might otherwise be experienced as a crisis.

Questions about how to talk to their child about the chromosomal variation and its possible future effects are asked regularly. We believe that as soon as a child learns to talk and ask questions, parents can respond in developmentally appropriate terms; for example, in the context of 4 year olds characteristically fantasizing about future roles, about how families become families and about what makes a parent truly a parent. The importance of fostering a sense of personal value in their children apart from fertility can be emphasized to parents so that as the family develops, infertility will be seen in perspective. We encourage parents openly to discuss issues of parenthood, sexuality, and fertility at a level appropriate to the child's development. When the child has reached an age at which an intellectual understanding of infertility can be achieved, the following facts are helpful in increasing parents' confidence in their ability to talk to their child. 1) One in six couples wishing to have a child is infertile [38]. Our families are therefore not isolated in confronting this problem. 2) Infertility is not a sexual disorder and has nothing to do with the ability to engage in sexual relations. Infertile people have the same emotional and physical feelings in sexual relations as other people do. 3) Infertility is not a punishment for mistakes in either the recent or the distant past.

INTRAUTERINE DIAGNOSIS OF SCA

Diagnosis of SCA in utero has become quite common with the increasing use of genetic amniocentesis. In deciding whether to continue such pregnancies, involved couples need to know the prognosis for these conditions. The following in essence is what our group has been telling them and their referring physicians.

All these fetuses represent a group at increased risk of developmental problems, with the amount varying with the specific karyotype. The severity of developmental delay varies also within a specific karyotype. In general, mental retardation is not a major problem, although the probands usually do not do as well as their sibs. We stress the positive effect of a good supportive family environment and of "anticipatory guidance" by the physician. Since neuromuscular development is often delayed, this should be checked early in life and when indicated an infant stimulation program started. Language development may be slow and there may be learning disabilities evident when school starts. With the exception of the 45,X group, we stress that they are usually phenotypically normal. We then give any information unique to a given karyotype.

CONCLUSIONS

This study has been in progress for 20 years. The importance of determining the prognosis of these conditions has been emphasized by the frequency with which these karyotypes are diagnosed in utero. We continue to be impressed by the variability of the expression of the individual karyotypes and by the influence of environmental factors on their expression. Continuing support of the families and propositi is a great help in management. It is our continuing belief that these "experiments of nature" will illuminate normal processes of physical, emotional, and cognitive development.

ACKNOWLEDGMENTS

This study was supported in part by grant 5R01-HD10032 from the USPHS; grant RR-69 from the General Clinical Research Centers Program of the Division of Research Resources, NIH; and The Genetic Foundation.

REFERENCES

1. Robinson A, Puck T: Studies on chromosomal nondisjunction in man II. Am J Hum Genet 19:112–129, 1967.
2. Robinson A, Puck M, Pennington B, Borelli J, Hudson M: Abnormalities of the sex chromosomes: A prospective study on randomly identified newborns. Birth Defects 15(1):203–241, 1979.
3. Robinson A, Bender B, Borelli J, Puck M, Salbenblatt J, Webber ML: Sex chromosomal abnormalities (SCA): A prospective and longitudinal study of newborns identified in an unbiased manner. Birth Defects 18(4):7-39, 1982.
4. National Center for Health Statistics: NCHS growth charts. Monthly Vital Statistics Report 25(3)[Suppl HRA]:76–1120, 1976.
5. Nellhaus G: Head circumference from birth to eighteen years. Pediatrics 41:106–114, 1968.
6. Nadas AS, Fyler DC: "Pediatric Cardiology." Philadelphia: WB Saunders, 1972.
7. Marshall WA, Tanner JM: Variations in the pattern of pubertal changes in boys. Arch Dis Child 45:13–23, 1970.
8. Marshall WA, Tanner JM: Variations in the pattern of pubertal changes in girls. Arch Dis Child 44:291–303, 1969.
9. Schonfeld WA: Primary and secondary sexual characteristics: Study of their development in males from birth through maturity, with biometric study of penis and testes. Am J Dis Child 65:535–549, 1943.
10. Zachmann M, Prader A, Kind HP, Hafliger H, Budliger H: Testicular volume during adolescence: Cross-sectional and longitudinal studies. Helv Paediatr Acta 29:61–72, 1974.

11. Faiman C, Ryan RJ: Radioimmunoassay for human follicle stimulating hormone. J Clin Endocrinol Metab 27:444–447, 1967.
12. Faiman C, Ryan RJ: Radioimmunoassay for human luteinizing hormone. Proc Soc Exp Biol Med 125:1130, 1967.
13. Winter JSD, Hughes IA, Reyes FI, Faiman C: Pituitary-gonadal relations in infancy: 2. Patterns of serum gonadal steroid concentrations in man from birth to two years of age. J Clin Endocrinol Metab 42:679, 1976.
14. Bruininks RH: "Bruininks-Oseretsky Test of Motor Proficiency Examiner's Manual." Circle Pines, MN: American Guidance Service, 1978.
15. Salbenblatt, JA, Bender BG, Puck MH, Robinson A, Faiman C, Winter JSD: Pituitary-gonadal function in Klinefelter syndrome before and during puberty. Pediatr Res 19:82–86, 1985.
16. Winter JSD, Taraska S, Faiman C: The hormonal response to hCG stimulation in male children and adolescents. J Clin Endocrinol Metab 34:348, 1972.
17. Winter JSD, Faiman C: Pituitary-gonadal relations in male children and adolescents. Pediatr Res 6:126–135, 1972.
18. Winter JSD: Prepubertal and pubertal endocrinology. In Falkner F, Tanner JM (eds): "Human Growth, Vol 2. Postnatal Growth." New York: Plenum Press, 1978, p 196.
19. Beery KE: "Developmental Test of Visual-Motor Integration." Chicago: Follet Publishing Co., 1967.
20. Kirk SA, McCarthy JJ, Kirk W: "Illinois Test of Psycholinguistic Abilities." Urbana, IL: University of Illinois Press, 1968.
21. Goldman R, Fristoe R, Woodcock R: "Goldman-Fristoe-Woodcock Test of Auditory Discrimination." Circle Pines, MN: American Guidance Service, 1970.
22. DeRenzi E, Vignolo LA: The token test: A sensitive test to detect receptive disturbances in aphasics. Brain 85:665–678, 1962.
23. Goodglass H, Kaplan E: "The Assessment of Aphasia and Related Disorders." Philadelphia: Lea and Febiger, 1972.
24. Woodcock RW, Johnson MB: "Woodcock-Johnson Psycho-Educational Battery." Hingham, MA: Teaching Resources Corporation, 1977.
25. Pennington, BF, Bender B, Puck M, Salbenblatt J, Robinson A: Learning disabilities in children with sex chromosome anomalies. Child Dev 53:1182–1192, 1982.
26. Harlan WR, Harlan EA, Grillo GP: Secondary sex characteristics of girls 12 to 17 years of age: The U.S. Health Examination Survey. J Pediatr 96:1074–1078, 1980.
27. Pennington B, Puck M, Robinson A: Language and cognitive development in 47,XXX females followed since birth. Behav Genet 10:31–41, 1980.
28. Tennes K, Puck M, Orfanakis D, Robinson A: The early childhood development of 17 boys with sex chromosome anomalies: A prospective study. Pediatrics 59:574–583, 1977.
29. Thurstone TG: "Primary Mental Abilities Test." Chicago: Science Research Associates, 1962.
30. Dunn LM: "Peabody Picture Vocabulary Test." Circle Pines, MN: American Guidance Service, 1965.
31. "Diagnostic and Statistical Manual of Mental Disorders (DSM III), ed 3." Washington, DC: American Psychiatric Association, 1980.
32. Shaffer D, Gould M, Brasic J, Ambrosini P, Fisher P, Bird H, Aluwahlia S: A children's global assessment scale (CGAS). Arch Gen Psychiatry 40:1228–1231, 1983.
33. Puck MH, Bender BG, Borelli JB, Salbenblatt JA, Robinson A: Parents' adaptation to early diagnosis of sex chromosome anomalies. Am J Med Genet 16:71–79, 1983.
34. Kraft AD, Palombo J, Mitchell D, Dean C, Meyers S, Wright-Schmidt A: The psychological dimensions of infertility. Am J Orthopsychiatry 50:618–627, 1980.

35. Bresnick ER: A holistic approach to the treatment of the crisis of infertility. J Marital Fam Ther 7:181–188, 1981.
36. Platt JJ, Ficher I, Silver MJ: Infertile couples: Personality traits and self-ideal concept discrepancies. Fertil Steril 24:972–976, 1973.
37. Rosenfield DL, Mitchell BS: Treating the emotional aspects of infertility: Counseling services in an infertility clinic. Am J Obstet Gynecol 135:177–180, 1979.
38. Eck Menning B: The emotional needs of infertile couples. Fertil Steril 34:313–319, 1980.

APPENDIX A.* 47,XXY: Anthropometric Data

		Measure and Percentile			Mean		
Case No.	Age (years)	Height (cm)	Weight (kg)	HC (cm)	testicular volume (ml)	Tanner pubic hair	Penis (cm)
13	17–10	184.2 (85)	80.7 (80)	59.4 (>98)	6.5	6	14
18	16–10	185.2 (90)	62.8 (35)	55.4 (45)	2.5	6	15.5
26	15–11	187.3 (>95)	59.2 (35)	56.9 (80)	1.5	6	12.5
31†	13–10	167.5 (75)	50.9 (50)	55.4 (70)	2	4	11
4	12–9	157.2 (60)	38.3 (25)	53.5 (40)	4	2	8
38	12–9	153.4 (45)	38.6 (25)	53.1 (35)	3.5	1	7
36	11–11	163.7 (>95)	42.2 (60)	52.7 (30)	2	1	7
61	11–3	153.7 (90)	38.6 (60)	52.7 (40)	2	1	6.5
58	11–1	150.2 (80)	40.3 (70)	51.2 (10)	2	2	7
57	11–0	150.7 (85)	45.5 (85)	53.2 (45)	4.5	1	7.5
60	11–0	150.5 (85)	38.1 (65)	51.6 (20)	1	1	7.5
55	10–1	144.8 (85)	34.0 (60)	52.2 (35)	1	1	ND
65	10–0	142.0 (75)	32.5 (55)	51.7 (25)	1	1	7
67	9–10	136.3 (50)	27.8 (25)	52.3 (35)	ND	1	6
70	9–0	137.0 (80)	26.8 (35)	52.2 (40)	1	1	7

†Subject has mild gynecomastia.
*Symbols for appendices: ND, no data; Tanner II onset of puberty (breast and pubic hair development): NA, not applicable; *, prepubertal; T, concurrent with testosterone supplement; E, concurrent with estrogen therapy; A, concurrent with oxandrolone therapy; Menarche: NA, not applicable; **, premenarchial; E, concurrent with estrogen therapy; Global assessment rating: 81–90, good functioning in all areas (family, school, peers); 71–80, no more than slight impairment; 61–70, some difficulty in single area; 51–60, sporadic difficulty in several areas; 41–50, moderate interference in most areas; 31–40, major impairment in several areas; comprehensive environment rating (social stratification rating and parents' nurturing skills and stress score divided by 3 [see 33]): 4.0–5.9, good; 6.0–7.9, average; 8.0–11.3, poor.

APPENDIX B. 47,XXY: Elements of Psychosocial Adaptation*

Overall functioning rank order	Case No.	Current age (years)	Birth weight (gm)	Age of walking (months)	Age of combining words (months)	Full-scale WPPSI score	Full-scale WISC-R score	Language impairment	Motor assessment percentile	Special education	Percentile: Height	Percentile: Weight	Percentile: Head circumference	Pubic hair development age at Tanner II	Global assessment rating	Comprehensive environmental rating
Satisfactory	61	11	2,959	17	24	ND	112	Moderate	34	None	90	60	40	*	50	4.3
	58	11	2,539	11	24	109	110	Moderate	27	Partial	80	70	10	11	ND	6.6
	13	18	3,143	13	ND	134	122	None	90	None	85	80	>98	12	47–68	6.0
	26	16	2,863	14	25	94	102	Moderate	95	None	>95	35	80	10 T	68	7.3
	57	12	2,945	14	26	84	92	Moderate	42	Partial	85	85	45	*	ND	5.6
	65	11	3,080	11	24	109	100	Moderate	18	Partial	75	55	25	*	ND	5.0
	70	10	3,459	17	ND	96	102	Severe	27	None	80	35	40	*	ND	10.3
Impaired	31	14	3,250	11	24	109	96	Moderate	38	Partial	75	50	70	11	52	8.0
	60	11	3,171	15	23	81	87	Moderate	2	Partial	85	65	20	*	ND	7.3
	36	13	3,171	10	28	84	99	Moderate	1	Partial	>95	60	30	*	ND	7.3
	4	19	3,171	15	27	91	85	Moderate	ND	Partial	60	25	40	12 T	ND	10.6
	18	17	3,454	15	33	99	102	Moderate	46	Partial	90	35	45	11 T	53	7.0
Markedly impaired	38	13	3,500	14	ND	87	81	Moderate	1	Partial	45	25	35	*	49	9.3
	55	12	2,340	ND	23	76	81	Severe	21	Partial	85	60	35	*	ND	10.3
	67	11	3,180	15	34	71	71	Severe	1	Total	50	25	35	*	ND	9.6

*For abbreviations see Appendix A.

APPENDIX C. 47,XXX: Anthropometric Data

Case No.	Age (years)	Measure and Percentile: Height (cm)	Weight (kg)	HC (cm)	Onset breast Tanner II (years)
2	19–2	175.1 (>95)	53.8 (35)	55.9 (70)	12
15	17–4	170.1 (85)	53.2 (35)	52.2 (2)	9
16	17–1	173.8 (95)	64.1 (75)	53.6 (25)	11
22	16–0	182.3 (>95)	59.5 (65)	56.4 (80)	10
25	15–11	160.8 (40)	46.2 (10)	51.9 (2)	11
28	15–0	172.0 (90)	74.8 (90)	54.5 (50)	*
29	15–0	166.8 (75)	59.2 (70)	53.5 (35)	11
30	14–7	169.5 (90)	59.4 (75)	51.9 (10)	7
10	14–1	173.0 (>95)	47.5 (35)	53.0 (30)	11
45	12–0	168.2 (>95)	66.0 (>95)	52.7 (45)	7
48	11–11	168.8 (>95)	50.7 (80)	54.5 (75)	10

*For abbreviations see Appendix A.

APPENDIX D. 47,XXX: Hormone Data in 2 Year Increments*

Case No.	Age (years, months)	E2 (ng/dl)	FSH (μg/dl)	LH (ng/dl)	T (ng/dl)	Breast and pubic hair stages
45	7,6	3.2	31	4.7	8.0	B2/PH1
30	8,1	3.0	50	2.9	ND	B2/PH2
48	8,1	<2.0	14	<2.5	ND	B1/PH1
45	9,0	2.3	32	11.6	ND	B3/PH1
48	9,0	<1.0	13	3.6	ND	ND
48	9,11	ND	14	<2.5	32	B2/PH1
45	10,5	1.8	26	4.4	ND	B4/PH3
25	11,0	<1.0	27	<2.5	40	B2/PH1
29	11,0	<2.0	14	3.1	ND	B2/PH1
15	11,1	7.7	41	4.2	28	B3/PH3
30	11,1	3.4	18	ND	11	B3/PH3
48	11,1	3.0	27	2.9	17	ND
25	12,0	7.4	51	4.0	28	B2/PH1
16	12,11	12.2	18	2.1	22	B4/PH2
25	13,0	5.5	45	9.2	ND	B3/PH2
29	13,0	ND	18	3.4	18	B4/PH4
15	13,2	5.4	33	5.3	24	B4/PH4
16	13,11	2.0	34	ND	18	B4
22	14,0	4.4	35	17.0	ND	B4/PH5
10	14,1	19.0	41	8.5	47	B4/PH3
15	14,1	2.2	25	6.9	ND	B4/PH4
2	14,11	7.1	67	2.3	34	B4/PH4
2	16,0	5.7	45	ND	28	B4/PH4

*FSH and LH are expressed in μg LER907/dl and can be converted to IU/liter by multiplying the FSH by 0.5 and the LH by 4.5. For abbreviations see Appendix A.

APPENDIX E. 47,XXX: Elements of Psychosocial Adaptation*

Overall functioning rank order	Case No.	Current age (years)	Birth weight (gm)	Age of walking (months)	Age of combining words (months)	Full-scale WPPSI score	Full-scale WISC-R score	Language impairment	Motor assessment percentile
Satisfactory	25	16	3,001	14	31	109	112	None	46
	2	19	3,230	13	ND	103	109	None	10
Impaired	30	14	3,313	17	24	108	105	Severe	<1
	10	18	2,860	12	ND	82	86	Moderate	ND
	15	17	2,350	19	ND	89	80	Moderate	2
	16	17	3,341	11	29	96	70	Severe	<1
	28	15	4,140	14	24	76	90	Severe	3
	48	12	2,780	19	43	73	77	Moderate	1
Markedly impaired	22	17	2,350	16	34	96	73	Severe	<1
	45	12	3,660	22	28	69	76	Severe	<1
	29	15	2,760	21	45	68	50	Severe	<1

*For abbreviations see Appendix A.

(continued)

APPENDIX E. 47,XXX: Elements of Psychosocial Adaptation (Continued)

Special education	Percentile: Height	Percentile: Weight	Percentile: Head circumference	Tanner II onset of breast development (years)	Pubic hair development Tanner II age at	Menarche (years)	Global assessment rating	Comprehensive environmental rating
None	40	10	2	11	13	13	78	7.3
None	>95	35	70	12	11	13	61	6.6
Partial	90	75	10	7	8	12	69	5.6
Partial	>95	35	30	11	12	12	ND	6.3
Partial	85	35	2	9	10	12	68	5.6
Partial	95	75	25	11	12	12	52	8.3
Partial	90	90	50	ND	ND	13	62	8.3
Partial	>95	80	75	10	11	12	46	8.0
Total	>95	65	80	10	10	12	50	7.9
Total	>95	>95	45	7	9	9	39	10.3
Total	75	70	35	11	12	12	35	11.3

APPENDIX F. Female Mosaics: Anthropometric Data

Neonatal chromosome ratio	Case No.	Age (years, months)	Measure and Percentile: Height (cm)	Weight in kg	HC (cm)	Onset breast Tanner II (years)	Menarche (years)
45,X/46,XX							
48:52	17	17,1	155.4 (10)	56.5 (50)	52.6 (10)	11	13
27:73	21	17,0	149.8 (<5)	44.3 (<5)	54.1 (40)	12	13
7.5:92.5	24	16,1	168.1 (75)	62.8 (75)	53.4 (25)	10	13
45,X/47,XXX							
37:63	35	13,0	139.4 (<5)	38.8 (15)	54.0 (55)	11	**
46,XX/47,XXX							
97:3	3	13,1	157.1 (50)	49.0 (60)	54.3 (60)	10	12
90:10	62	11,0	155.0 (90)	39.8 (60)	53.3 (75)	10	**
45,X/46,XX/47,XXX							
17:78:5	12	18,4	171.3 (90)	63.5 (75)	56.5 (75)	13	15

APPENDIX G. Female Mosaics: Hormone Data*

Case No.	Karyotype	Age (years, months)	E2 (ng/dl)	FSH (μg/dl)	LH (μg/dl)	T (ng/dl)	Breast and pubic hair stages
17	45,X/46,XX	12,11	4.9	30	6.8	13	B4/PH4
		14,0	15.5	6.8	8.2	ND	B4/PH4
		14,11	19.1	17	13	ND	B4/PH5
		15,10	25.9	5.6	2.3	ND	B5/PH5
		17,1	4.9	12	5.8	26	B5/PH5
21	45,X/46,XX	12,5	4.4	24	2.4	18	B3/PH2
		13,11	5.5	30	10	ND	ND
		15,11	6.3	14	6.2	14	B5/PH5
		17,0	6.0	23	3.6	29	B5/PH5
24	45,X/46,XX	12,0	6.0	24	4.3	29	B2/PH2
		13,0	7.6	39	ND	23	B4/PH4
		15,0	5.2	24	4.6	ND	ND
35	45,X/47,XXX	8,11	<2.0	12	<2.5	ND	B1/PH1
		10,1	<1.0	28	3.9	ND	B1/PH1
		11,0	ND	23	<2.5	<10	B1/PH1
		13,0	3.8	34	3.8	9	B3/PH2
62	46,XX/47,XXX	11,0	1.9	16	4.0	12	B3/PH1
12	45,X/46,XX/47,XXX	13,1	4.0	9.4	4.9	ND	B1/PH2
		15,0	2.4	39	8.4	ND	B4/PH4

*FSH and LH are expressed in μg LER907/dl and can be converted to IU/liter by multiplying the FSH by 0.5 and LH by 4.5. For abbreviations see Appendix A.

APPENDIX H. 46,XY/47,XXY Male Mosaics: Anthropometric Data

Karyotype	Case No.	Age (years, months)	Measure and Percentile			Mean testicular volume (ml)	Tanner pubic hair	Penis (cm)
			Height (cm)	Weight (kg)	HC (cm)			
46,XY/47,XXY (39:1)	11	17,9	164.4 (<5)	55.5 (5)	57.4 (75)	20	6	Normal
46,XY/47,XXY (35:5)	63	11,0	147.5 (75)	37.0 (60)	54.1 (65)	4	2	7

APPENDIX I. Mosaics—Male and Female: Elements of Psychosocial Adaptation*

Overall functioning rank order	Karyotype	Case No.	Current age (years)	Birth weight (gm)	Age of walking (months)	Age of combining words (months)	Full-scale WPPSI score	Full-scale WISC-R score
Satisfactory	46,XX/47,XXX	62	11	2,577	17	28	127	129
	46,XY/47,XXY	63	11	3,515	13	24	119	118
	45,X/46,XX	24	17	3,539	15	21	90	107
	45,X/47,XXX	35	13	3,005	11	24	116	96
	45,X/46,XX	21	17	3,015	14	26	78	102
	45,X/46,XX	17	17	2,718	12	30	94	81
Impaired	45,X/46,XX/ 47,XXX	12	18	2,888	14	24	117	122
	46,XX/47,XXX	3	19	3,907	14	26	111	104
Markedly impaired	46,XY/47,XXY	11	18	3,630	ND	ND	86	89

*For abbreviations see Appendix A.

Language impairment	Motor assessment percentile	Special education	Percentile: Height	Percentile: Weight	Percentile: Head circumference	Tanner II onset of breast development (years)	Pubic hair development (age at Tanner II)	Menarche (years)	Global assessment rating	Comprehensive environmental rating
Moderate	18	None	90	60	75	10	*	**	ND	5.6
None	ND	None	75	60	65	NA	11	NA	ND	5.1
None	ND	None	75	75	25	10	12	13	72	4.0
None	7	None	<5	15	55	12	13	**	84	6.3
Moderate	ND	None	<5	5	40	12	12	13	75	8.3
Moderate	69	Partial	10	50	10	11	11	13	87	8.3
None	ND	None	90	75	75	13	13	15	66	6.3
Moderate	ND	Partial	50	60	60	10	10	12	ND	9.0
Moderate	ND	Partial	<5	5	75	NA	13	NA	ND	10.3

APPENDIX J. 47,XYY: Anthropometric Data

Case No.	Age (years–months)	Measure and Percentile		
		Height (cm)	Weight (kg)	HC (cm)
37	12–0	149.6 (50)	42.3 (60)	56.3 (90)
41	11–0	148.4 (75)	37.5 (60)	51.8 (25)
201	10–0	144.6 (85)	33.1 (60)	52.9 (50)
69	10–0	148.4 (95)	38.7 (85)	55.0 (85)

APPENDIX K. 47,XYY: Elements of Psychosocial Adaptation*

Overall functioning rank order	Case no.	Current age (years)	Birth weight (gm)	Age of walking (months)	Age of combining words (months)	Full-scale WPPSI score	Full-scale WISC-R score
Satisfactory	201	11	2,600	13	25	97	104
	37	13	3,117	14	29	85	93
Impaired	69	10	4,160	14	35	109	109
	41	12	2,620	15	29	84	101

*For abbreviations see Appendix A.

Language impairment	Motor assessment percentile	Special education	Percentile			Tanner II onset of puberty	Global assessment rating	Comprehensive environmental rating
			Height	Weight	Head circumference			
Moderate	7	Partial	85	60	50	*	ND	6.0
Moderate	1	Partial	50	60	90	12	ND	6.0
Moderate	38	Partial	95	85	85	*	ND	8.3
Moderate	1	Partial	75	60	25	*	ND	8.3

APPENDIX L. 45,X and Partial X Monosomy: Anthropometric Data

Karyotype	Case No.	Age (years, months)	Measure and Percentile: Height (cm)	Weight (kg)	HC (cm)
45,X	8	18,0	134.5 (<5)	37.1 (<5)	51.6 (<2)
45,X	7	14,10	147.0 (<5)	62.8 (80)	55.8 (75)
45,X	39	12,10	136.7 (<5)	42.6 (35)	54.6 (70)
45,X	56	11,4	142.0 (25)	35.4 (30)	52.7 (50)
45,X	66	10,11	121.9 (<5)	30.8 (15)	50.7 (15)
45,X	54	10,10	117.4 (<5)	22.1 (<5)	49.0 (<2)
46,X,del(X)(pter→q21.2:)	23	15,2	152.9 (10)	56.4 (60)	53.4 (30)
46,X,del(X)(pter→q13:)	43	12,0	137.7 (<5)	44.6 (60)	51.0 (15)
45,X/46,X,r(X)(?:p11→q21:) (7:13)	49	12,2	133.3 (<5)	31.9 (5)	51.1 (20)

APPENDIX M. 45,X, Partial X Monosomy and 45,X Mosaics: Physical Anomalies*

Case No.	Karyotype	Short stature	Congenital lymph-edema	Broad chest	Triangular facies	Epican-thal folds	Low posterior hairline	Webbed neck	Cubitus valgus	Deep-set nails	Horseshoe kidney	Coarcta-tion	Dextro-cardia	Primary hyper-tension
	45,X and partial X monosomy													
7	45,X	×	×	×	×	×	×	×	×	×	×			×
8	45,X	×	×	×	×	×	×	×	×	×	×		×	×
39	45,X	×	×	×	×		×	×	×	×		×		
54	45,X	×	×	×	×	×	×	×	×	×		×		
56	45,X		×		×		×			×				x
66	45,X	×	×	×	×	×				×			×	
23	46,X,del(X) (pter→q21.2:)	×				×								
43	46,X,del(X) (pter→q13:)	×					×							
49	45,X/46,X,4(X) (?:p11→q21:)	×				×	×			×				
	45,X mosaics													
17	45,X/46,XX													
21	45,X/46,XX	×		×	×		×							
24	45,X/46,XX													
32	45,X/46,XX													
33	45,X/46,XX													
35	45,X/47,XXX	×								×				
12	45,X/46,XX/47,XXX													

*Adapted from Robinson et al [Pediatrics 73:175–182, 1984] with permission. Copyright American Academy of Pediatrics, 1984.

APPENDIX N. 45,X and Partial X Monosomy: Hormone Data*

Case No.	Age (years, months)	FSH (μg/dl)	LH (μg/dl)	E2 (ng/dl)	T (ng/dl)
45,X					
8	14,3	>150	31	<2.0	14
7	12,1	139	4.3	1.9	29
39	10,0	22	1.9	<1.0	ND
56	9,1	7.4	2.3	4.9	ND
	10,1	154	3.9	2.5	17
	11,4	>150	29	1.9	14
54	7,6	45	<2.5	2.8	<10
	10,10	>150	12	<1.0	<14
117	11,0	38	<2.5	<1.0	ND
	12,2	145	5.7	<1.0	ND
	14,6	>150	25	1.4	<14
	15,0	>150	11	1.4	9
Partial X monosomy					
23	12,4	94	4.2	<2.0	<10
	13,4	31	4.6	1.0	ND
	14,2	100	16	3.7	ND
	15,2	201	26	2.2	ND
43	11,0	526	43	1.1	<14
	12,1	>150	30	1.4	<14
49	7,11	28	<2.5	<2.0	11
	10,0	46	<2.5	ND	<10
	11,4	>150	25	1.1	<15

*FSH and LH are expressed in μg LER907/dl and can be converted to IU/liter by multiplying the FSH by 0.5 and the LH by 4.5. For abbreviations see Appendix A.

APPENDIX O. X and Partial X Monosomy: Elements of Psychosocial Adaptation*

Overall functioning rank order	Karyotype	Case No.	Current age (years)	Birth weight (gm)	Age of walking (months)	Age of combining words (months)	Full-scale WPPSI score	Full-scale WISC-R score
Satisfactory	45,X	56	12	4,531	15	24	105	118
Impaired	45,X/46,X,r(X)	49	12	3,000	17	24	79	84
	45,X	7	19	3,171	12	18	99	109
	45,X	39	12	3,086	12	25	116	99
	45,X	66	11	3,050	10	17	80	73
	46,XXq−	23	17	3,284	13	24	106	92
Markedly impaired	45,X	8	19	2,831	16	ND	63	50
	45,X	54	12	2,640	15	28	71	72
	46,XXq−	43	12	3,240	9	28	65	89

*For abbreviations see Appendix A.

Language impairment	Motor assessment percentile	Special education	Percentile			Tanner II onset of breast develop. (years)	Pubic hair development (age at Tanner II)	Menarche (years)	Global assessment rating	Comprehensive environmental rating
			Height	Weight	Head circumference					
None	34	None	25	30	50	*	*	**	ND	4.3
Moderate	1	Partial	<5	5	20	*	*	**	ND	7.6
Moderate	ND	None	<5	80	75	13E	13E	13E	ND	6.3
Moderate	<1	Partial	<5	25	70	*	12A	**	ND	6.0
Severe	<1	Total	<5	15	15	*	*	**	ND	6.0
None	ND	None	9	60	30	13	13	13	ND	6.6
Severe	<1	Total	<5	5	<2	*	17A	**	51	9.6
Severe	<1	Total	<5	5	<2	*	*	**	ND	7.3
Moderate	ND	Total	<5	60	15	*	*	**	ND	8.3

APPENDIX P. Children's Global Assessment Scale [32] (for children 4–16 years of age)

Rate the subject's lowest and highest level of functioning for the two time periods specified by selecting the lowest and highest range that describes his/her functioning on a hypothetical continuum of health—illness. Use intermediary levels (eg, 35, 58, 62). Rate actual functioning regardless of treatment or prognosis. The examples of behavior provided are only illustrative and are not required for a particular rating.

Lowest level of functioning for any one month within the past 6 months __

Highest level of functioning for any one month within the past 6 months __

Level of functioning within the past month__

91–100: *Superior functioning* in all areas (at home, at school, and with peers), involved in a wide range of activities, and has many interests (eg, has hobbies or participates in extracurricular activities or belongs to an organized group such as scouts, etc). Likeable, confident, "everyday" worries never get out of hand. Doing well in school. No symptoms.

81–90: *Good functioning in all areas.* Secure in family, school, and with peers. There may be transient difficulties and "everyday" worries that occasionally get out of hand (eg, mild anxiety associated with an important exam, occasionally "blow-ups" with sibs, parents or peers).

71–80: *No more than slight impairment in functioning* at home, at school, or with peers. Some disturbance of behavior or emotional distress may be present in response to life stresses (eg, parental separations, deaths, birth of sib), but these are brief and interference with functioning is transient. These children are only minimally disturbing to others and are not considered deviant by those who know them.

61–70: *Some difficulty in a single area, but generally functioning pretty well* (eg, sporadic or isolated antisocial acts, such as occasionally play hookey or petty theft; consistent minor difficulties with school work, mood changes of brief duration; fears and anxieties that do not lead to gross avoidance behavior; self-doubts). Has some meaningful interpersonal relationships. Most people who do not know the child well would not consider him/her deviant but those who do know him/her well might express concern.

51–60: *Variable functioning with sporadic difficulties or symptoms in several but not all social areas.* Disturbance would be apparent to those who encounter the child in a dysfunctional setting or time but not to those who see the child in settings where functioning is appropriate.

41–50: *Moderate degree of interference in functioning in most social areas or severe impairment of functioning in one area*, such as might result from, for example, suicidal preoccupations and ruminations, school refusal and other forms of anxiety, obsessive rituals, major conversion symptoms, frequent anxiety attacks, poor or inappropriate social skills, frequent episodes of aggressive or other antisocial behavior with some preservation of meaningful social relationships.

31–40: *Major impairment in functioning in several areas and unable to function in one of these areas*, ie, at home, at school, with peers, or in the society at large (eg, persistent aggression without clear instigation; markedly withdrawn and isolated behavior due to either mood or thought disturbance, suicidal attempts with clear lethal intent). Such children are likely to require special schooling and/or hospitalization or withdrawal from school (but this is not a sufficient criterion for inclusion in this category).

21–30:	*Unable to function in almost all areas* (eg, stays in bed all day) or severe impairment in reality testing or serious impairment in communication (eg, sometimes incoherent or unresponsive) or judgement (eg, acts grossly inappropriate).
11–20:	*Needs considerable supervision* to prevent hurting others or self (eg, frequently violent, repeated suicide attempts) or to maintain personal hygiene or has gross impairment in all forms of communication (eg, severe abnormalities in verbal and gestural communication, marked social autism, stupor, etc).
1–10:	*Needs constant supervision* (24 hr care) because of severely destructive or self-destructive behavior or gross impairment in reality testing, communication, cognition, affect, or personal hygiene.

Edinburgh Study of Growth and Development of Children With Sex Chromosome Abnormalities III

Shirley G. Ratcliffe, MB, BS, FRCP, Lynne Murray, MA, PhD, and Peter Teague, BSc, MSc

MRC Clinical and Population Cytogenetics Unit, Western General Hospital, Edinburgh EH4 2XU, Scotland, United Kingdom

INTRODUCTION

The longitudinal study of growth and development of children with sex chromosome abnormalities has now been in progress for 17 years, but because of the wide time span of case collection it will be another 8 years before the majority of the children reach maturity. The study compares the physical growth and psychologic development of children with sex chromosome abnormalities with those of a large control group, in both of whom rate of attrition from follow-up has been low, with 93% of the cases and 87% of the controls continuing to attend the clinic.

This latter has been influenced by a number of factors; the facilities of a National Health Service have enabled contact to be maintained when families leave the locality and the payment of travel expenses by the Medical Research Council facilitates the twice yearly visits. In addition, the efforts of the research team to provide the families with information after assessment of the child (both cases and controls) together with continuity of contact with the permanent staff on the team have played a part in maintaining follow-up.

It is important to appreciate that these children have attended a longitudinal study all their lives and in this respect differ from the population at large. This condition is shared by the controls and the cases, but its effect on our observations is difficult to measure when we wish to generalise about the

Birth Defects: Original Article Series, Volume 22, Number 3, pages 73-118

observed alterations produced in children with variations in the sex chromosomes. Any deviation from normal development is likely to be detected earlier than in children not under regular supervision. On the other hand, the knowledge of the chromosome abnormality may well generate anxiety, which is heightened at the time when the child is assessed. The positive reinforcement resulting from confirmation of the good development of the majority of controls may be available to a smaller proportion of the cases. It is impossible to study a child's development without being seen to do so, but we believe that it is important to draw attention to the effects the study itself may have. Over so long a period of study, it is also inevitable that a number of changes of staff and techniques occur, and it seems appropriate to give a full account in this report.

CASE MATERIAL

The newborn cytogenetics survey commenced in 1967 and continued for 12 years [1,2]. Throughout this time, consecutive liveborn infants were screened either by full chromosomal analysis or by examination of buccal mucosal cells for abnormal numbers of X and Y chromosomes followed by full chromosome analysis when an abnormality was found. The survey covered all live births in two of the four maternity hospitals in Edinburgh, representing approximately 35% of births in the city. In addition, a maternity hospital in Fife participated in the study for 2 years from the beginning of 1976, and all twin births were screened at a fourth maternity hospital over a period of 2.5 years.

The total number of consecutive infants screened was 34,380, of whom 17,522 were males and 16,858 females. This resulted in the identification of 70 infants with sex chromosome abnormalities, including 23 with the chromosome constitution 47,XXY, 18 with 47,XYY, 16 with 47,XXX, and one 46,XX male. From the twin survey, 1 XYY was identified in a dizygous twin pair among 137 pairs of twins, one pair of conjoined twins, and one triplet set. There were, in addition, 12 infants with mosaic constitutions or translocations who are listed with the main karyotype groups in Table I. All children with sex chromosome abnormalities were white, and the population from which they were drawn was almost entirely so. The present report refers to the major groups XXY, XYY, and XXX as well as the one XX male and is the third in the series of reports [3,4].

MORTALITY

Two deaths occurred among the cases and one among the controls. One XXY boy was born at 28 weeks of gestation and survived only 7 hr; at

TABLE I. Edinburgh Newborn Cytogenetic Survey 1967–1979.

Edinburgh Newborn Cytogenetic Survey. 1967 – 1979.		
Total number of infants screened		34,380
Chromosome analysis		16,581
Sex chromatin (X and Y) examination		17,799
Abnormalities of the sex chromosomes		70
Males		
(17,522)	47,XXY	23
	47,XYY	18
	46,XY/47,XXY	4
	46,XY/47,XYY	2
	Y – 15 translocation	2
	46,XX (male)	1
Females		
(16,858)	47,XXX	16
	45,X/47,XXX	1
	45,X	1
	45,X/46,XX	1
	45,X/46,XY	1

autopsy, the testes were found to be intraabdominal. An XYY boy had renal agenesis and survived only 30 min. One control boy died suddenly at 4 months, and autopsy revealed Reyes syndrome.

PARENTAL INFORMATION

The parents of the XXX girls were informed of the chromosome constitutions of their daughters shortly after the birth except in one case, when the information was given when the child was aged 10 years, and one child was adopted with information on the known significance of this chromosome constitution given to the prospective parents. All but one of the parents of the XXY boys are aware of the diagnosis and its likely significance. In the case of the XYY boys, it was initially decided not to inform the parents shortly after the birth lest the reports then current of an association with an increased risk of later detention in maximum security institutions should disturb the parent-child relationship. In the intervening years, six sets of parents (Cases 7, 9, 10, 12, 13, 16) have received the information when the child was aged 2 months in one case, 4 years in two cases, and between 7 and 8 years in the remaining three cases. In two cases, disclosure was accidental, one mother reading the case notes when the doctor was briefly out of the consulting room; in the second case, the chromosome constitution was alluded to by an orthopaedic surgeon to whom the child had been referred. In the opinion of the paediatrician in three instances, the parents have been able to assimilate the information without undue anxiety, including the case of one XYY boy who was accepted into a foster home and later adopted. In the other three (in one of whom the information was accidentally given), the mothers have continued to express feelings of rejection towards the child. Follow-up of the XYY boys was agreed to by the parents on the basis of minor perinatal problems, being one of twins, or participation in a growth and development study.

CONTROL GROUP

From 1972 to 1976, children were recruited from the two Edinburgh maternity units as controls after a circular had been made available to mothers in pregnancy advising them of the setting up of the longitudinal growth and development study. At the beginning of 1973, full chromosome analysis on consecutive liveborn infants had ceased, and buccal smear examination for sex chromatin abnormalities (both X and Y) was being undertaken routinely. To enable a control to be recruited with a normal karyotype, full chromosome analysis was undertaken on the first five babies born each week if they met the following criteria: 1) single birth, 2) birth weight 2.5 kg or more, 3) no gross malformation, and 4) residence within the city boundary.

By the end of the week when the initial chromosome results were available, the mother of the first child born after midnight on Sunday was approached and asked if she would consent to the inclusion of the child in the growth and development study. If she refused or had taken early discharge, then the mother of the next infant was approached until a control had been enrolled. Babies for adoption were not included because the prospective parents were not available. Six children initially recruited were later excluded when the results of banding revealed them to have chromosome variants. Two further male infants were recruited in 1978. A final total control group of 112 males and 89 females was thus collected.

FOLLOW-UP

In 1971, a Growth Clinic was set up. Initially cases, and from 1972 controls, attended every three months during the first year of life and thereafter twice yearly or more frequently if clinically indicated. In those instances when the family had moved distantly the visits were annual, either the family coming to the clinic or being visited at their home by a member of the research team.

At each visit, a history of the child's health, development, and behaviour was obtained from the parent by the paediatrician, who was aware of the chromosome constitution, normal or otherwise. Most of the case group and approximately 50% of the controls were seen by the senior author, who was the only permanent member of staff. The child was examined and measured using the methods and equipment designed by Prof. Tanner of the Institute of Child Health, London. Details of measurements obtained are given in Appendix A together with the method of calculating Z scores.

At approximately yearly intervals, the child was also seen by the psychologist and participated in the programme of evaluation shown in Appendix B. Until mid-1982, the psychologist was not informed of the chromosome constitution of the child, control or case. The administration of certain psychological tests over one period did not accord with the protocol in use for the rest of the study, so these data were not used in the final analysis.

STATISTICAL ANALYSIS

Multiple linear regression has been used to assess the effects of social class, mother's education, and handedness on the WISC, McCarthy, and Stanford-Binet tests and the reading age and Bristol Social Adjustment Guide results. These are "explained" by a linear function of the form:

$$Y_j = \sum_{i=1}^{i=p} \beta_{ij} X_{ij},$$

where Y_j is the expected value for an individual of karyotype, j, whose social class, mother's education, etc, are represented by the $X_{1j} \ldots X_{pj}$. The $\beta_{1j} \ldots \beta_{pj}$ can be estimated by maximum likelihood. In the event that a normal error distribution can be assumed, maximum likelihood is equivalent to choosing a set of βs that minimises the residual mean squared error. This is the least-squares method.

In general, we have preferred to err in the direction of "overfitting," that is, to include explanatory variables (including interactions and higher-order terms) even though they may not have statistically significant effects. In practice, the conclusions are little affected by the inclusion or exclusion of any explanatory variable, except social class, mother's education, and tester.

Usually, a regression surface would be fitted to each karyotype and contrasts made between the cases and the appropriate control. With the present data, however, there were insufficient degrees of freedom among the cases to do this. Since there are substantially more controls than cases, the significance of differences may be overestimated, especially when cases are more variable than controls. Instead a "prediction" was calculated for each case using a model based on the appropriate control (XY for XYY, XY and XX for XXY, XX for XXX). Thus:

$$Y_k = \sum_{i=1}^{i=p} \beta_{ij} X_{ij},$$

where Y_k is the predicted value for a case of karyotype K (whose social class, etc, are represented by the values $X_{lk} \ldots X_{pk}$). The regression coefficients $\beta_{lj} \ldots \beta_{pj}$ are derived from the linear regression model for controls of karyotype j (see equation 1). A paired t test was used to compare predicted values with observed values in each of the three case groups. Results are presented as comparisons between the controls and the "main group," referring to singletons of birth weight greater than 2.5 kg, with those for the twins and low-birth-weight children listed separately and not included in the statistical analysis.

RESULTS

The social class distribution of the 57 infants with nonmosaic sex chromosome abnormalities is given in Table II together with that of the controls, the hospital population from which they were drawn, and the population of Edinburgh and Scotland. Social class was categorised on the basis of the father's occupation *at the time of the child's birth* according to the Registrar General's classification, where I is professional and V is unskilled. This is intended as some measure, albeit crude, of the child's environment of rearing. In some instances, this has changed markedly during the childs life; this aspect will be dealt with nearer the completion of the study.

Statistical analysis showed that the social class distribution of the male and female cases did not differ from that of the male and female controls, but the XYYs differed from the population of Scotland, their families having more professional and skilled individuals, ($P < 0.05$). Further examination of the male controls showed that they differed significantly from the hospital population from which they were drawn ($P < 0.025$), but this was not the case for the female controls. Because these controls were recruited by six doctors over a period of 4 years, an examination was made of social class distribution for each recruiter. It was found that one recruiter had an anomalous distribution for the male controls collected, with 58.1% in social classes I and II, 27.9% in social class III, and 13.9% in social classes IV and V. The female controls collected by this recruiter were unremarkable in their social class

TABLE II. Social Class Distribution of the 57 Infants With Nonmosaic Sex Chromosome Abnormalities*

	n	Social class* (%)		
		I and II	III	IV and V
XXY	23	26.1	47.8	26.1
XYY	18	44.4	50.0	5.6
Male controls at recruitment	112	42.0	44.6	13.4
Male controls after attrition	102	44.1	41.2	14.7
XXX	16	25.0	50.0	25.0
Female controls at recruitment	89	33.7	51.7	14.6
Female controls after attrition	74	32.4	52.7	14.9
Hospital population	5,073	30.4	49.6	20.0
Edinburgh population*	8,136	36.2	44.9	18.9
Scottish population	126,101	23.0	51.7	25.3

*See text for details. Registrar General's classification where I is professional and V is unskilled.
†One in ten sample of all persons under age 16 in 1983.

distribution. In retrospect, it was not possible to ascertain how this anomaly arose. A multiple regression analysis of the data collected from the male controls for the Stanford Binet IQ and the school report (the Bristol Social Adjustment Guide) was carried out to determine whether after controlling for social class there was any significant residual effect of the recruiter. This was not found to be present in these two areas with data from different sources, so it was concluded that the data on the male controls from this recruiter could be retained within the whole group of male controls as long as, in any analysis, social class was controlled for.

ATTRITION OF CASES AND CONTROLS

XXY

Of the 23 infants identified at birth, one died at age 7 hr and two emigrated in the first year of life. One boy has not attended the Growth Clinic, but information is available from the general practitioner and from hospital records. Of the remaining 19 XXY boys, 18 have attended the Growth Clinic regularly, and one is visited by the research team at his local hospital.

XYY

Of the 18 infants, one died at birth. The remaining 17, together with the one XYY boy identified in the twin survey attended the clinic throughout childhood. One XYY boy stopped attending at age 10.5 years. Another was not seen from ages 12.5 to 15 years, then resumed attendance at the clinic. A further XYY boy emigrated to Canada at age 10; information is received annually from the mother and school, and he was assessed in Canada at age 12 by the research team.

XXX

All 16 girls have attended for longitudinal study. Hence, of the surviving cases with sex chromosome abnormalities, 93% continue to attend for longitudinal study.

Controls

Ninety-one percent of the male and 83% of the female controls have remained with the study. One control boy developed scoliosis and was withdrawn from the control group.

GROWTH AND SEXUAL DEVELOPMENT

XXY

The age at the last visit, together with height and weight, and their centiles according to the Tanner et al [5] standards are given in Table III. These

TABLE III. XXY Boys. Age, Height, and Weight.

XXY Boys						
Case No.	Age at last visit	Height cms.	Centile	Weight Kg.	Centile	
Singletons						
1.	17.0	190.2	+97th	60.5	+25th	
2.	17.8	185.6	+90th	65.5	+50th	
3.	15.75	180.9	90th	72.5	+90th	
4.	15.0	172.5	75th	60.5	75th	
5.	15.0	181.1	+90th	78.0	+97th	
6.	14.0	168.0	+75th	56.0	75th	
7.	10.8	150.7	+90th	43.2	+90th	
8.	11.5	147.4	+50th	37.5	+50th	
9.	11.0	151.4	90th	46.2	+90th	
10.	8.5	133.8	75th	35.0	+90th	
11.	10.0	140.5	75th	28.5	+25th	
12.	9.0	142.8	97th	32.5	+75th	
13.	9.5	142.3	90th	31.0	+50th	
14.	8.5	149.8	++97th	43.0	++97th	
15.	8.0	131.0	+75th	25.0	50th	
16.	7.5	130.8	90th	26.4	75th	
17.	6.5	129.4	+97th	32.5	++97th	
Birth Weight <2.5Kg and Twins.						
18.	9.5	140.9	+75th	29.7	+50th	B.W. 1.9Kg.
19.	16.5	189.5	+97th	70.4	+75th	Twin B.W. 2.7Kg.

standards have been used for the cases because, at the present time, the control boys range in age from 9 to 13.5 years and hence do not provide comparisons for the older cases. It is appreciated that the use of these standards is not entirely appropriate; there has continued to be a secular increase in height of British children over the 20 years since the data was collected, and the social class distribution of these children is not given. At age 8 the control boys are 1.97 cm taller than the Tanner et al norms.

Twelve of 19 boys, including the twin, are at or above the 90th centile, and none are below the 50th centile. Cases 1, 2, 3, and 19 have reached their final heights at 190.2, 185.6, 180.9, and 189.5 cm, respectively.

Ten boys were at a higher height than weight centile, four were proportional, and five were heavier than expected for their height; however, one of these latter (Case 3) was clearly not obese, having skinfold measures below the 50th centile and very good muscle development.

Because only four XXY boys have reached their final heights, comparisons of height centiles between the XXY boys and their parents are preliminary (Table IV), they show that 15 of 19 XXY boys are at higher centiles whereas three fathers and one mother are at higher centiles than their XXY offspring. All XXY boys are at higher centiles than their siblings at the same age (Table 3).

TABLE IV. XXY Boys. Family Height Comparisons.

XXY Boys
Family Height Comparisons.

Case No.	Height Centile	Mothers Height Centile	Fathers Height Centile	Siblings Height Centile	Age	Sex
Singletons						
1.	+97th	75th	+25th	50th	20.8	M
2.	+90th	90th	+75th	75th	11.6	F
3.	90th	10th	+25th	75th	10.7	M
4.	75th	+10th	50th	90th	5.5	F
5.	+90th	+10th	+25th	50th	16.0	M
6.	+75th	+10th	+10th	—	—	—
7.	+90th	75th	+90th	+50th	7.8	F
8.	+50th	+10th	<3rd	+25th	11.5	M
9.	90th	97th	90th	+75th	7.0	M
10.	75th	+50th	+25th	+50th	3.4	F
11.	75th	50th	+25th	10th	5.0	M
12.	97th	90th	10th	75th	11.0	F
13.	90th	+25th	75th	+50th	11.5	M
14.	++97th	+50th	75th	—	—	—
15.	+75th	50th	+97th	—	—	—
16.	90th	+75th	97th	75th	9.0	M
17.	+97th	97th	90th	—	—	—
Birth Weight <2.5Kg and Twins.						
18.	+75th	+50th	50th	—	—	—
19.	+97th	75th	+75th	50th	16.5	TwinM

In that these boys were significantly smaller at birth [6], there has been a marked increase in height velocity, which we have previously shown to occur between ages 4.5 and 7.5 years using complete longitudinal data [4].

Body proportions were examined using Z scores derived from the study controls, height and weight data having been examined for social class effect and none being present. The absence of a social class effect might, however, be related to the relatively small numbers of controls in social classes IV and V in the control populaton.

The Z scores at ages 3, 5, and 7 years [6] show steadily increasing height and, to a greater extent, increase in leg length, which by age 8 years (Fig. 1) is 1.3 SD above the mean of the control boys. Head circumference was significantly reduced at birth and remains so at age 8 years, being the only measure that is below the mean.

Sexual development at birth was normal in the majority of XXY boys, but bilaterally undescended testes were present in two boys (as well as in the infant who died age 7 hr being at 28 weeks of gestation) and unilateral maldescent in one other. During childhood, deviations from normal growth were noted. Testicular volumes were measured using Prader's orchidometer

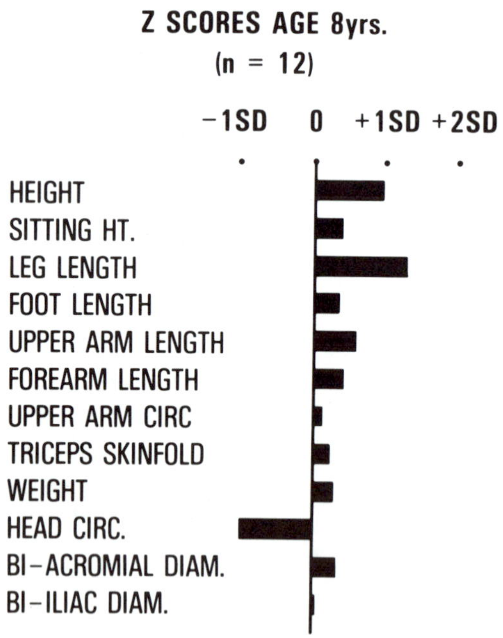

Fig. 1. XXY Boys. Body proportions.

and were less than 1.0 ml up to age 7 years and reached 1.5 ml by age 10 years [7].

In five boys, the testes, which had been in the scrotum at birth and at later examinations, came to lie in the inguinal canal as growth progressed, this being observed by age 3 years in two cases, and in others at ages 5, 6, and 8 years. After several examinations by the paediatrician and surgeon had confirmed the consistent malposition of the testes, orchidoplexy (five unilateral and two bilateral) was carried out in a total of seven (35%) of the XXY boys. Among the control boys, one (0.9%) has required orchidoplexy. The age of onset of puberty, defined by testicular enlargement to 4 ml volume, or the appearance of pubic hair, or penile enlargement, has been examined in the XXY boys and the controls who have reached 13 years of age. Of the 22 control boys, all show signs of pubertal changes at a mean age of 11.8 years (range 10.5–13.0). Similarly, all six XXY boys from the main group of this age show signs of pubertal development at a mean age of 11.9 years (range 10.9–13.0). The one twin XXY showed pubertal changes at age 12.5 years, as did his chromosomally normal cotwin. In the XXY boys, testicular enlargement progressed only to between 3 and 5 ml, except in one boy for whom a volume of 10 ml was recorded, which later regressed to 3 ml.

The size of the penis was normal at birth, but poor growth during childhood was noted in seven cases. Two boys had locally applied 2% testosterone cream, for 4 months in one case, with good response, and for 6 weeks in the second case, with moderate response. Two started on testosterone undecanoate at the spontaneous onset of puberty, and the penis became of normal breadth although of short length. One boy was noted to have a larger than average penis from age 6 years. Circumcision for phimosis was required by two XXY boys, compared to three of the 102 control boys.

Gynaecomastia has been observed in two boys so far. Case 3 had started on testosterone undecanoate 40 mg daily at age 11.5, shortly after the spontaneous onset of puberty. At age 11.7 years, he developed a small nodule under the left nipple which disappeared in two months. Two years later while still on testosterone, he developed a nodule 1 cm in diameter under the right nipple, which persisted for 6 months and then disappeared. At age 14.8 years, a disc of breast tissue 2.5 cm in diameter was palpable on the left side and again spontaneously resolved. At his present age of 16 years, there is a disc of breast tissue of 3 cm diameter on the right side. On account of the variable behaviour of the breast tissue, no action is planned at present. The second boy, Case 5, who has not been on testosterone treatment, developed progressive breast enlargement from age 13.5 years. He was obese, but weight reduction by dietary restriction has not been accompanied by regression of the breast enlargement and he has been referred at age 15 years for plastic surgery.

Hormone studies. Those carried out in the neonatal period have been reported and showed the characteristic burst of testicular activity of the male infant [7]. From age 8.5 years, testosterone, oestradiol, and LH and FSH were ascertained at the twice yearly visits. Testosterone levels in the serum were within the prepubertal range until age 11.5, when rises occurred; these are discussed further under the section on treatment. Results of the LH and FSH estimations are shown in Figures 2 and 3, with those on treatment represented by open circles. Around the age of 12 years, these pass into the abnormal range.

Testosterone treatment in puberty. This was planned for alternate boys at the time of the spontaneous onset of puberty, but because one boy was temporarily in South Africa he was passed and the next boy given testoster-

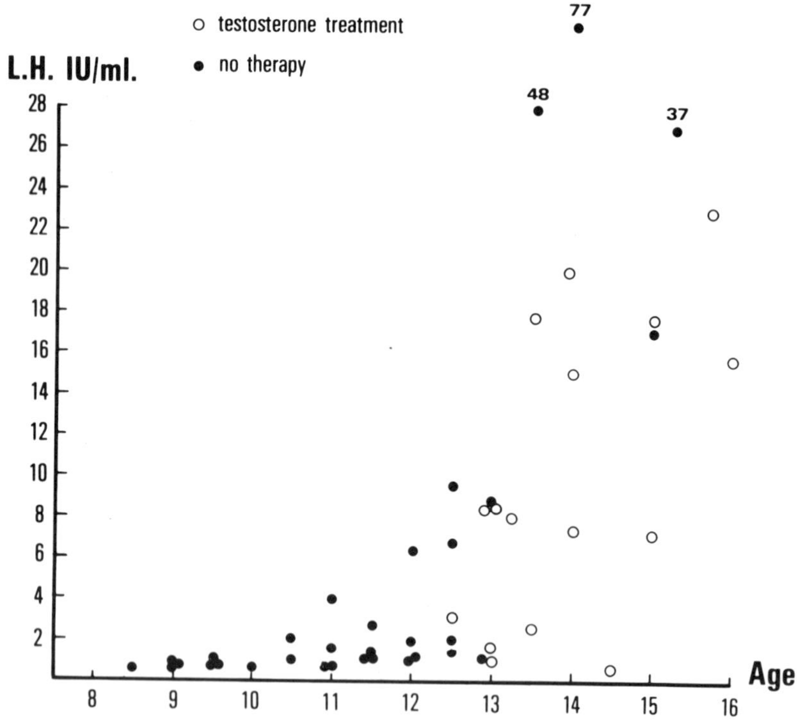

Fig. 2. XXY Boys. Serum LH.

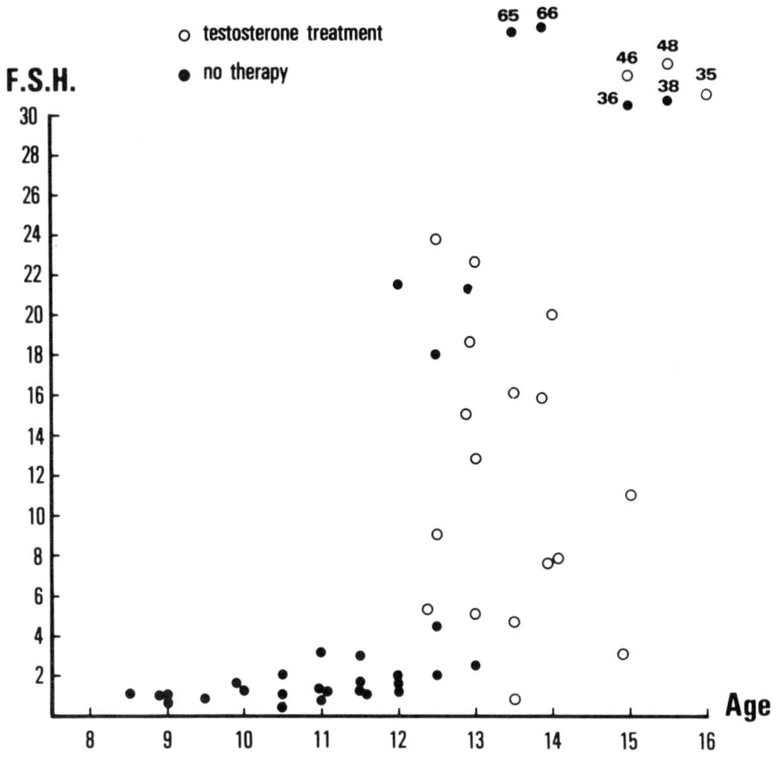

Fig. 3. XXY Boys. Serum FSH.

one. Currently, four boys have received testosterone and three have not, so only preliminary findings can be given.

At the time of the clinical onset of puberty, judged by testicular enlargement to 4 ml volume or penile enlargement or the appearance of pubic hair, treatment was started with 40 mg of testosterone undecanoate daily. Elevation of gonadotrophins was present in the majority 6 months after the clinical onset of puberty. In the four boys who were treated, puberty was recorded as starting at ages 11.5, 12.0, 12.4, and 12.5 years, and in the three boys not treated the onsets were at 10.9, 12.0, and 13.0 years. The treated boys were taller prior to treatment, the mean height at age 10.5 years being 150.1 cm, whereas the untreated boys had a mean height of 146.2 cm at this age. By age 14, after approximately 2 years of treatment, the mean height was 178.2 cm with a velocity of 8.03 cm/year, whereas the untreated boys had a mean height of 170.2 and a velocity of 6.86 cm/year.

The major difference between the two groups appeared in the skinfold measurements which are shown in Figure 4 for the untreated boys and Figure 5 for the boys on treatment. The treated boys had larger skinfold measures during childhood than the untreated, but shortly after starting treatment there was a downward drift, whereas two of the three untreated boys showed marked increases.

The mean serum testosterone at age 14 in the treated boys was 12.6 nmol/liter and 12.3 nmol/liter in the untreated boys. This failure to observe any alteration in the testoserone has been recorded [8] in patients on testosterone undecanoate, the main rise being in the active metabolite 5-dihydrotestosterone, for which we do not have facilities for routine assay.

Gonadotrophins remained elevated in both groups, being 23.8 U/liter in the treated group and 40.1 U/liter in the untreated. Because of small numbers, statistical analysis was not performed.

At the interview with the paediatrician, there were reports from the parents of outbursts of temper in two of the boys not receiving testosterone, but these were set in a background of disturbed family functioning, in one case a single parent family with a physically handicapped mother and a brother showing more markedly disturbed behaviour. In the other case, the outbursts of temper arose initially over going to boarding school and were described by the teachers; they were exacerbated at the time of the parents separation and

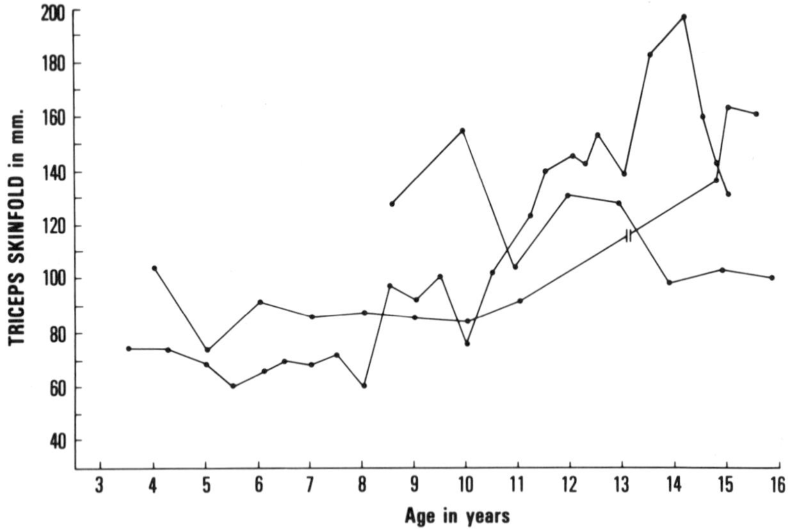

Fig. 4. XXY Boys. Triceps skinfold (no treatment).

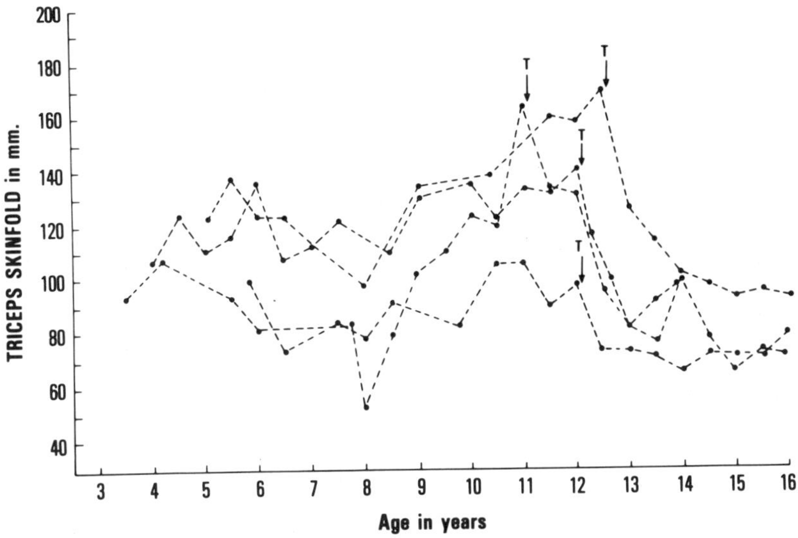

Fig. 5. XXY Boys. Triceps skinfold (testosterone treatment).

divorce when the boy was 15 years old. In the four boys on treatment, no problems were reported, although one boy referred to later in the section on psychologic results had had a brief psychiatric referral at the age of 8 years. Further experience will be required before it is possible to ascertain whether testosterone treatment is beneficial and if modification of the treatment schedule should be made.

XX Male

At birth, this boy weighed 2.8 kg, and his length was 51.5 cm and head circumference 31.5 cm. He was measured from the age of 2.5 years until he reached his final height of 167.7 cm at age 17.5 years, remaining between the 10th and 25th percentiles for males (or 75th percentile for females) and less than expected given his father's height of 185.4 cm (+90th centile) and mother's height of 164.5 cm (+50th centile), (Fig. 6).

There was spontaneous onset of puberty at age 12.5 years, with testicular and penile enlargement and the appearance of pubic hair, but the testes increased only to 5 ml in volume. Gonadotrophins became elevated at age 14 years. The serum testosterone level was within the normal male range at age 17.5, and he was well masculinised with pubic hair extending along the midline of the abdomen to the umbilicus (Stage 6).

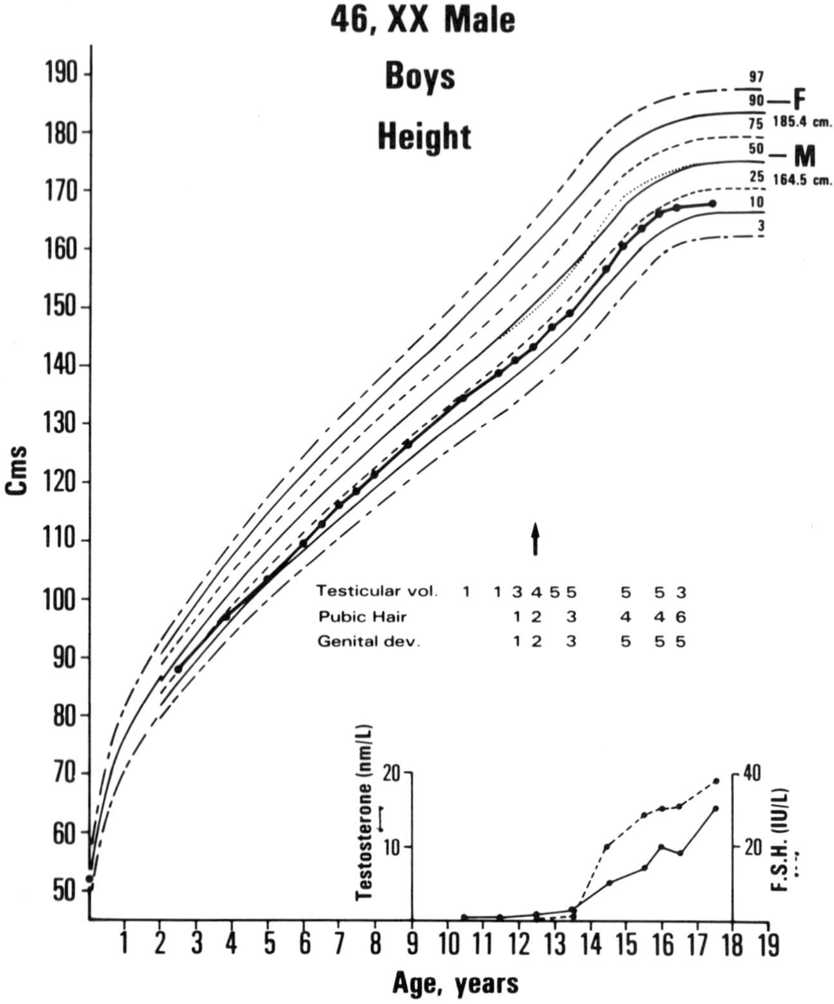

Fig. 6. XX Boy. Growth chart.

Body proportions at age 7 and 9 years are given as Z scores using male study controls (Fig. 7). All measures are below the mean, but sitting height is reduced to a greater extent than leg length, indicating a degree of increased leg length but to a lesser extent than seen in XXY males.

XYY

We have shown previously that the size at birth of the XYY boys who were singletons of birth weight greater than 2.5 kg did not differ from that

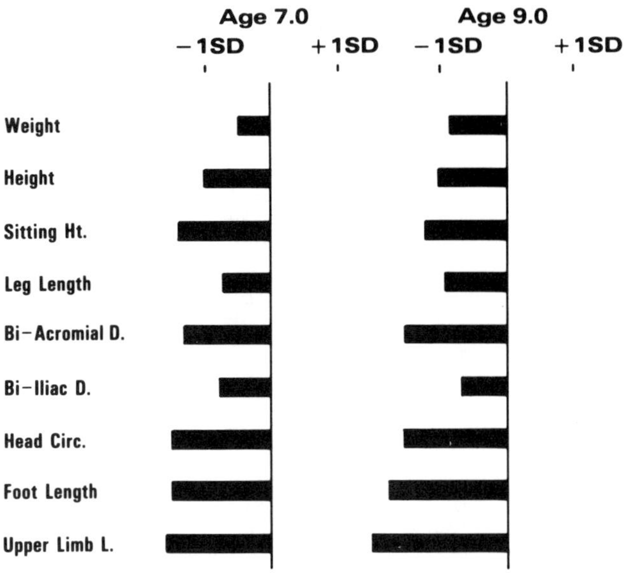

Fig. 7. XX Boy. Body proportions (Z scores).

of the controls [6], and, using only complete longitudinal data, there was a significant increase in height velocity in early childhood [4]. The current height status of the boys is shown in Table V; 10 of 18 (55.5%) are at or above the 90th centile on Tanner's charts, including three of the twins-LBW group. As growth is not yet complete in most of the boys, comparisons with their parents' heights remain preliminary, but here 14 of 18 exceed both of their parents' height centiles (Table VI). This, however, does not allow for the secular change over a generation, so comparisons are also drawn with their siblings when possible. Of 14 comparisons, in eight the XYY boy has a higher centile at the same age (Table VI).

The XYY boys are lighter in weight than would be expected in proportion to their height in 13 of 18 comparisons. The body proportions are further examined in Figure 8 using Z scores in ten XYY boys with complete data at age 9 years (of these boys, 50% were at or above the 90th percentile for height). Stature is significantly increased ($P < 0.05$), with the major contribution from increased leg length ($P < 0.01$), although not to the extent seen in XXY boys. The diameter of the shoulders is only just above zero but the hips are +0.5 SD. The trends demonstrated over the years 3, 5, 7 [6] continue, and at age 9 the body proportions show a similar distribution to that found in seven adult XYY men using the same anthropometric methods [19].

TABLE V. XYY Boys. Age, Height, and Weight.

XYY Boys						
Case No.	Age at last visit	Height cms.	Centile	Weight Kg	Centile	
Singletons						
1.	17.0	188.6	+97th	63.5	+50th	
2.	14.4	158.1	+25th	43.5	+10th	
3.	16.0	185.7	97th	65.0	75th	
4.	16.0	181.0	90th	67.0	+75th	
5.	10.5	152.6	97th	33.8	+50th	
6.	15.1	181.7	+90th	70.0	90th	
7.	14.0	167.3	+75th	43.5	25th	
8.	13.0	166.1	+90th	50.0	75th	
9.	10.8	143.3	75th	32.0	50th	
10.	11.0	148.6	+75th	39.0	+75th	
11.	8.5	130.9	+50	29.5	75th	
12.	8.2	132.9	+75th	30.0	+75th	
13.	7.0	127.5	90th	25.2	75th	
14.	7.0	122.7	+50th	23.5	+50th	
Birth Weight < 2.5Kg and Twins						
15.	7.5	134.5	+97th	29.2	90th	B.W. 2.44Kg
16.	16.4	189.6	+97th	79.5	97th	Twin B.W. 2.10Kg
17.	14.5	160.5	+25th	42.0	+10th	Twin B.W. 2.20Kg
Twin Survey						
18.	16.0	182.0	+90th	71.5	90th	Twin B.W. 2.20Kg

TABLE VI. XYY Boys. Family Height Comparisons.

XYY Boys Family Height Comparisons.						
Case No.	Height Centile	Fathers Height Centile	Mothers Height Centile	Siblings Height Centile		
Singletons					Age	Sex
1.	+97	75	+25	50	8.3	F
2.	+25	25	75	50	7.5	F
3.	97	+25	+10	25	4.5	F
4.	90	+25	25	25	11.8	M*
5.	97	+75	+50	97	5.0	M
6.	+90	50	+75	97	8.5	F
7.	+75	+50	50	90	9.0	M
8.	+90	50	+10	–	–	–
9.	75	+90	+25	25 +75	6.0 10.7	M M
10.	+75	+50	+75	90	8.5	M
11.	+50	<3	+10	50	6.0	M
12.	+75	25	+50	–	–	–
13.	90	+75	+75	–	–	–
14.	+50	25	50	–	–	–
Birth Weight <2.5Kg and Twins.						
15.	+97	97	97	90	3.9	M
16.	+97	75	+75	+25	17.0	Twin F
17.	+25	10	+50th	+25	10.0	TwinM
Twin Survey						
18.	+90	+3	+25	25	15.5	Twin F

*½ sibling.

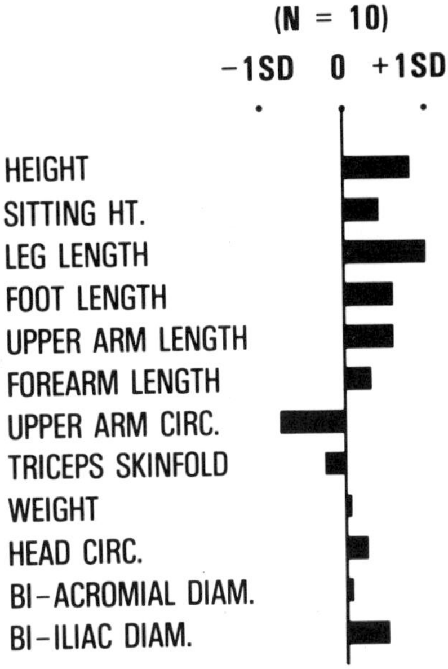

Fig. 8. XYY Boys. Body proportions (Z scores; age 9 years).

The onset of puberty was examined in seven XYY boys from the main group who had reached age 13 years. Six boys had evidence of testicular enlargement to 4 ml volume at a mean age of 11.7 years (range 10.9–13.0). In addition, the three XYY boys from twin pairs had reached puberty at ages 11.0, 12.7, and 13.0 years. This did not differ from the findings in the controls. One XYY boy had an ectopic testis which was replaced in the scrotum at age 8 years. When the boys have completed their pubertal growth spurt, it will be possible to examine whether this contributes a larger than usual component to their final height.

XXX

The age at last visit, the height, the weight, and centile values according to the Tanners et al [5] standards are given in Table VII for the 16 XXX girls. These standards have been used for the cases because at present the control girls range in age from 9 to 13.5 years and hence do not provide comparisons for the older cases. At age 8, the control girls at 127.6 cm are 2.5 cm taller than the 50th centile for girls on Tanner charts; hence, when we are able to use our controls throughout the range of the XXX girls, they will be relatively less tall by comparison. At birth, the XXX girls who were

TABLE VII. XXX Girls. Age, Height, and Weight.

		XXX Girls.			
Case No.	Age at last visit	Height cms.	Centile	Weight Kg.	Centile
Singletons					
1.	17.0	156.2	+10th	49.5	25th
2.	17.0	160.3	+25th	51.0	+10th
3.	15.0	157.5	+10th	44.0	+3rd
4.	14.0	166.6	75th	59.5	+50th
5.	13.0	161.9	+75th	51.0	+50th
6.	11.0	135.2	+10th	30.0	+10th
7.	11.5	153.4	+75th	34.0	25th
8.	9.0	133.4	+50th	26.0	+25th
9.	8.5	134.9	90th	29.0	+50th
10.	8.0	128.3	75th	21.5	+10th
11.	8.0	127.4	+50th	23.5	+25th
12.	6.0	116.4	75th	17.0	+3rd
13.	6.0	113.5	50th	20.0	+25th
Birth weight <2.5Kg and Twins.					
14.	14.5	175.7	+97th	46.0	+10th B.W. 2.39Kg
15.	13.5	155.1	+25th	48.0	50th B.W. 2.10Kg
16.	9.0	139.1	90th	29.5	+50th Twin B.W. 2.7Kg

singletons of birth weight greater than 2.5 kg were significantly smaller than the controls in length, weight, and head circumference ($P < 0.01$ for all measures) [6].

There was a wide range of height centiles at the last visit; three girls (two of whom had reached their final heights at 156.2 and 160.3 cm) were between the 10th and 25th centiles, and three were at or above the 90th centile—these latter included the twin and one low-birth-weight girl. The XXX girls tend to be slim, with 14 of 16 at a lower weight than height centile. Family comparisons are shown in Table VIII but remain preliminary until the cases and their siblings reach their final heights. At present, ten of 16 XXX girls are at higher height centiles than their parents. On the sibling comparisons, six of ten XXX girls have higher height centiles than their siblings.

Body proportions for nine XXX girls from the main group were examined using Z scores based on the controls (Fig. 9). Head circumference was significantly reduced ($P < 0.001$), as was foot length. Sitting height is reduced ($P < 0.05$), whereas leg length is indistinguishable from that of the controls, indicating that proportionally the XXX girls have long legs but to a lesser extent than the XXY boys. They also have significantly longer upper limbs (upper arm and forearm, $P < 0.05$).

In assessing the progress of sexual development, because cases and controls differ in their age ranges, the status as regards breast development at

TABLE VIII. XXX Girls. Family Height Comparisons.

XXX Girls

Family Height Comparisons.

Case No.	Height Centile	Mothers Ht. Centile	Fathers Ht. Centile	Siblings Ht. Centile	Age	Sex
Singletons						
1.	+10	+10	90	–	–	–
2.	+25	+10	+10	+10	F	7.8
3.	+10	+25	<3	25	F	7.4
4.	75	+3	75	+25	F	8.9
5.	+75	adopted		–	–	–
6.	+10	3	50	+25	M	12.8
7.	+75	50	75	90	F	18.0
8.	+50	3	+10	25	F	5.9
9.	90	10	+75	+25	F	10.0
10.	75	+50	25	–	–	–
11.	+50	25	+25	–	–	–
12.	75	50	+25	–	–	–
13.	50	+50	25	+75	F	8.8
Birth Weight < 2.5Kg and Twins						
14.	97	+25	90	75	M	5.1
15.	+25	10	+10	–	–	–
16.	90	+75	25	+25*	F	9.0

*twin

age 12 is compared. Twenty-four of the control girls have reached this age, and 21 (87.5%) have shown signs of breast development (Tanner Stage 2). The mean age at the first clinic visit at which this was observed was 10 years 8 months. Of the XXX girls, nine have reached age 12, and six (66.7%) of them have shown breast development at a mean age of 11 years 4 months. As regards the onset of menstruation, less information is available from the controls; only nine girls have reached age 13 years, and seven of these have menstruated, the mean age of menarche being 12 years 8 months. Among the XXX girls, seven have reached age 13 years, but only two have started to menstruate at a mean age of 13 years 8 months. These findings give some suggestion of a delay in the sexual maturation of the XXX girls, but until the complete cohort of cases and controls has reached maturity, final comparisons cannot be drawn.

PSYCHOLOGIC DATA

XXY

We have previously reported the results of intelligence tests at age 4 years using the Stanford Binet and the Terman Merill norms [4]. In that these norms were published in 1961, it seems more appropriate to give the results using the Thorndike norms of 1972. In the ten XXY boys, including one twin, with results shown in Table IX, the scores ranged from 71 to 121 with

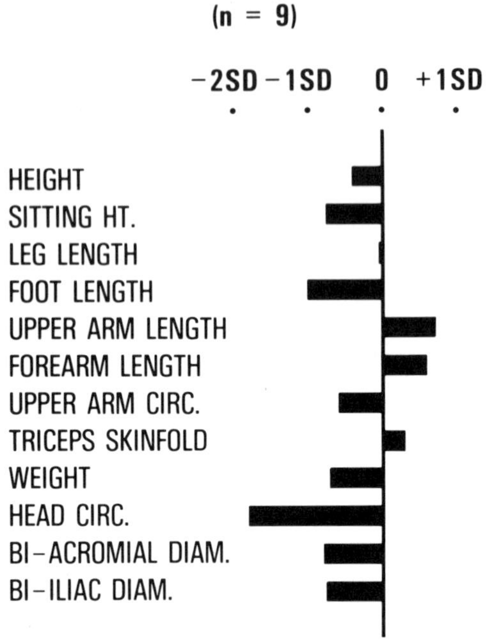

Fig. 9. XXX Girls. Body proportions (Z scores; age 8 years).

a mean of 87.4. Using multiple regression analysis, including variables of social class, mothers education level, and handedness, scores from 81 of the control boys were used to obtain a predicted figure for the nine boys in the main group of 102.5, which differs significantly from the observed score of 87.4 ($t = 4.06$, $P < 0.01$).

Using the McCarthy Scale of Children's Abilities there are only four results available ranging from 83 to 100 (Table IX). The lowest subtest scores were in motor abilities and memory.

Results on the Wechsler Intelligence Scale for Children (WISC) were available for 12 XXY boys including one twin (Table IX). Full-scale scores ranged from 67 to 130, with a mean of 100.8 for singletons and 102.2 with inclusion of the twin's score. Comparison with the scores obtained from 23 control boys on the multiple regression analysis as above gave the results shown in Table X.

The difference between the verbal and performance scores was examined in each child using Sattler's [9] Table p450 for the probability of obtaining the designated differences. Four XXY boys had significantly lower verbal than performance scores, and one boy had a significant difference in the

TABLE IX. XXY Boys. Intelligence Tests.

Intelligence Tests in XXY Boys.						
Case No.	Stanford Binet (age 4)	Mc Carthy (age 5)	WISC (age 6 – 8) V	P	FS	V – P difference
Singletons						
1.	89	—	92	86	88	
2.	—	—	126	128	130*	
3.	85	—	115	100	109	V > P p<0.05
4.	91	—	89	108	98	V < P p<0.01
5.	71	—	84	97	89	
6.	75	—	65	75	67	
7.	114	100	—	—	—	
8.	85	83	—	—	—	
9.	84	85	—	—	—	
10.	93	—	85	97	90	
12.	—	—	91	113	101	V < P p<0.01
13.	—	—	106	127	117	V < P p<0.01
15.	—	—	110	138	125	V < P p<0.01
16.	—	—	99	92	95	
17.	—	95	—	—	—	
Twin.						
19.	121	—	121	108	117	

* Tested at age 9yr. 11m.

TABLE X. Multiple Regression Analysis: WISC Scores for 12 XXY Boys and 23 Controls*

		t	P <
Observed mean verbal score	96.5	4.07	0.01
Predicted	112.9		
Observed mean performance score	105.5	2.3	0.05
Predicted	115.9		
Observed mean full-scale score	100.8	3.83	0.01
Predicted	115.7		

*See text for details.

opposite direction (Table IX). Subtest scores range from 7.45 on information to 12.27 for coding (Table XI).

The results of the XXY boys' scores can also be compared to those of a recent study carried out in Edinburgh by the Scottish Council for Research in Education for standardisation of the WISC-R. To compare our results with these we have adjusted the WISC-R scores to the comparable WISC using the figures recommended by Sattler [9], adding 4 points to the verbal, 6

TABLE XI. XXY Boys. WISC Subtests (ages 6–8 Years).

WISC sub – tests in XXY Boys.
(age 6 – 8).

Case No.	Information	Comprehension	Arithmetic	Similarities	Vocabulary	Digit Span	Picture Completion	Picture Arrangement	Block Design	Object Assembly	Coding
Singletons											
1.	10	9	7	10	10	7	7	6	13	6	8
2.*	14	13	17	15	14	12	15	15	14	15	11
3.	10	16	12	8	17	11	11	7	10	7	15
4.	6	4	10	8	12	9	11	8	10	13	14
5.	7	9	10	4	6	8	12	10	7	10	9
6.	3	4	8	4	3	4	8	5	7	5	7
10.	7	6	7	10	9	7	8	13	7	8	11
12.	6	9	13	11	7	6	10	11	12	10	16
13.	6	10	13	19	10	8	8	14	15	13	19
15.	7	13	10	18	14	8	14	14	13	17	19
16.	6	7	10	18	9	9	10	9	10	9	6
mean	7.5	9.1	10.6	11.4	10.1	8.1	10.4	10.2	10.7	10.3	12.3
S.D.	2.9	3.8	3.0	5.3	4.0	2.2	2.6	3.5	2.9	3.8	4.6
Twin											
19.	12	13	17	11	16	11	9	13	13	12	9

*Tested at age 9yr. 11m.

points to the performance, and 5 points to the full-scale scores; this results in a verbal score of 108.3, performance score of 110.4, and full-scale score of 109.4 for 52 boys from the Edinburgh population. The XXY boys' scores for performance and full-scale IQ are not significantly different from these results, although the verbal score is lower (P = 0.05).

The mean scores obtained by the 23 control boys of verbal 109.8, performance 115.3, full-scale 113.6 also do not differ significantly from the Edinburgh population.

As regards sibling comparisons, results were available for only three at present; thus comparisons for this group will be presented when more data are available.

School achievement. All boys except one attended ordinary schools; the exception, Case 18, with severe speech problems, attended a special school for children with minimal physical handicaps.

Reading ability was assessed in primary school using the Burt Reading Test (1974 Scottish norms), and the results for 15 XXY boys are presented in Table XII, along with information on educational intervention and behaviour in school. In view of the known association of reading difficulties with delayed language development, Table XII includes results on those children

TABLE XII. XXY Boys. Speech and Reading.

Speech and Reading in XXY Boys.								
Case No.	RDLS EL	RDLS VC	Speech Therapy	Reading Age yr.	Reading Age m.	Remedial Reading	School Behaviour UR	School Behaviour OR
Singletons								
1.	—	—	0	5	11	+	–	–
2.	—	—	0	—	—	0	–	–
3.	—	—	0	7	3*	+	A	M
4.	—	—	0	7	5	0	A	M
5.	—	—	0	5	10	+	M	S
6.	−1.5	−3.0**	0	6	3	+	M	A
7.	R	+1.4	0	8	0*	0	M	S
8.	R	R	+	6	8	+	S	A
9.	R	R	+	8	0	0	S	S
10.	−0.8	+0.3	0	—	—	+	M	S
11.	−1.1	−0.3	+	5	3	+	A	M
12.	R	R	+	6	9	+	S	A
13.	−0.5	0.0	0	6	11	0	S	M
14.	—	—	+	6	9	+	S	S
15.	+0.6	+1.4	0	6	8	+	Sev	S
16.	−1.7	−1.7	+	too young			–	–
17.	−1.3	−0.5	+	too young			–	–
				mean 6yr. 8m. (S.D. 0yr. 9.7m).				
LBW and Twin.								
18.	—	−1.5	+	5	5	+	S	S
19.	—	—	0	7	8*	0	A	M

* Estimated reading age calculated from centile score at age tested.
** Tested at age 4.5. R = refused o cooperate in test.
UR = Under Reaction OR = Over reaction S = Stable
M = mild under or over reaction A = appreciable under or over reaction
MA = maladjusted under or over reaction
Sev = severe maladjusted under or over reaction

who were tested at age 3 on the Reynell Language Development Scales, together with the information on whether speech therapy was received by the child, so an individual child's pattern of development can be examined.

Behaviour in school was evaluated using the Bristol Social Adjustment Guide, of which details are given in the section on behaviour. The latter results listed in Table XII include scores between ages 6 and 10 years.

When the reading test was not administered at the target age of 8 years (Cases 3, 7, and 19), an estimated score was obtained from the centile score achieved at the age tested. As a group, the XXY boys performed rather poorly, with reading age falling significantly below their chronologic ages: The mean reading age for 13 boys from the main group was 6 years 8.6 months at age 8 years. The multiple regression analysis controlling for social class, etc, comparing cases to study controls confirmed a significant deficit in the XXY boys' reading ability ($t = 9.08$, $P < 0.001$).

The mean reading age for the male controls was 8 years 5.7 months, and this differs significantly ($P < 0.001$) from the norms for the Scottish population (1974). However, when the social class distribution of the male controls is adjusted to that of the Scottish population, the weighted reading score becomes 8 years 2.3 months, and this does not differ from the 8 year norms.

Eight of the total group of 19 XXY boys had received speech therapy, that is, 42%. Of the 15 boys with reading test results, six had received speech therapy, and of these all but one later received educational intervention in the context of a normal school. The one boy (Case 3) who received remedial help but whose estimated reading age indicated no marked deficit was tested at age 9 years after receiving 3 hr of remedial help per week for 9 months. It seems that the educational system was alert to the needs of those children who had real reading deficits; in the case of the XXY boys, remedial teaching was required for 66.7% of those in primary school vs 17.6% of the control boys.

The level of remedial teaching in the control boys can be compared with that reported in the National Child Development Study (1972; Table A170) [10]. Here, 6.65% of boys at age 7 received special help in the context of an ordinary school, and the teachers recorded (in 1965) that a further 9.55% would benefit from such help, a total of 16.2%. It would appear that education provision in Edinburgh at the present time in the children attending our study is at a level to meet the children's needs. In that the teachers supply an annual report to the research team on both the cases and the controls, it might be that there is a heightened awareness of any educational problems and so remedial help is at the optimal level. In the control group, 13 of 102 boys attended private schools at the time of the reading test, whereas only one of the chromosomally abnormal (XYY) boys did; however, insofar as entry to private school is partly controlled by an ability test, it is not possible to dissociate the child's ability from potentially advantageous classroom conditions.

Behaviour. The behaviour of the XXY boys was examined initially at age 3 years using the Behaviour Screening Questionnaire of Richman and Graham [11], which considers 12 aspects of the young child's behaviour in daily life. The behaviour is scored as 0, 1, or 2 according to the severity of any problems experienced in the previous month and was completed with the parents within 3 months of the third birthday by the medical staff (who were aware of the chromosome constitution). A total score of 10 or more is considered to identify problem children.

The questionnaire was administered to 12 XXY boys from the main group (Fig. 10). Two boys (16.7%) fell into the problem category, scoring equal to or over 10, vs four of 92 (4.3%) control boys; this difference ($p \sim 0.1$) was not statistically significant.

The majority of XXY boys (75%) showed no definite problem in any area; when problems occurred, they were spread across all categories except worries and fears. Furthermore, for each behaviour category in which any

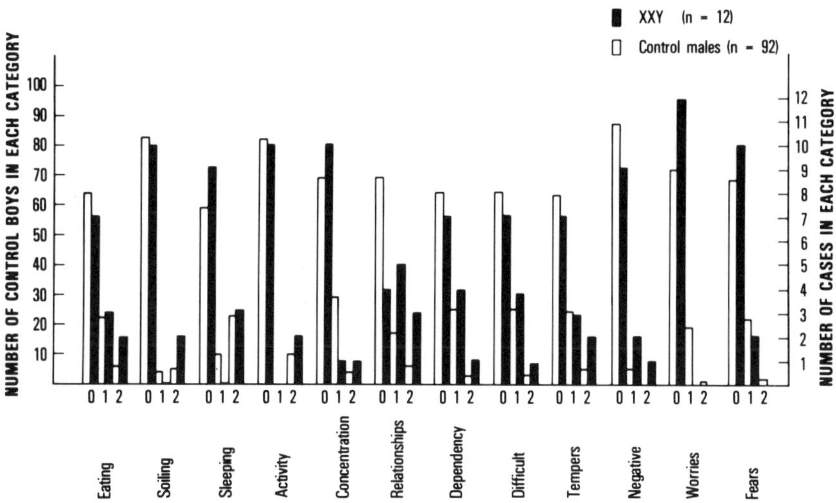

Fig. 10. XXY Boys and controls. Behaviour Screening Questionnaire.

problem occurred, only a minority (21.4%) of the XXY boys had a score of 2.

Behaviour in school was evaluated using the Bristol Social Adjustment Guide (BSAG) [12], which is completed annually by the class teacher. As before, multiple regression analysis was used to produce predicted scores, and because of small numbers the yearly reports at ages 7–10 years were amalgamated. When a child had more than one report, the mean score was calculated. The data was subjected to log transformation during analysis on account of its non-Gaussian distribution. The score increases with the incidence of problems and consists of underreaction and overreaction scores, which are presented separately.

Results are available for 13 of the XXY boys from the main group. The score for underreaction was significantly raised compared to study controls ($t = 3.84$, $P < 0.01$), but there was no significant alteration of the overreaction score. In addition, results are available for one boy of low birth weight and one twin. Comparisons of all cases with the manual's population norms on underreaction show the XXY boys' mean score on this measure to be between the 26th and 32nd centile rank, a score that is classified as "mild underreaction." On overreaction, these boys' mean score places them between the 43rd and 38th centile rank, and this just falls in the "mild overreaction" category.

One boy (Case 1) had psychiatric referral from the Growth Clinic at age 8 following a number of episodes of stealing money from his grandmother and brother, wandering from home, telling lies; there was tearfulness and temper tantrums at home, and a deterioration in his school performance despite remedial help.

He was assessed to have a mild neurotic behaviour disorder with developmental immaturity. The psychiatrist considered that his rather elderly parents had dealt with his problems by overprotection and infantilisation. More involvement with his father was recommended, and after three visits to the psychiatric clinic he was discharged and no further problems occurred despite the trauma of his father's death from heart disease when the boy was 12 years old. Of the 102 control boys so far three boys have been referred to a child psychiatrist and five to the educational psychologist for behaviour problems.

XYY

The first intelligence test was carried out at age 4 using the Stanford Binet (Thorndike norms). Individual results are shown in Table XIII and means with comparisons in Table XV.

Results ranged from 72 to 106, with a mean score for ten XYY boys from the main group of 88.6. In the multiple regression analysis, the scores obtained from the 81 control boys were used to obtain predicted scores for the XYY boys allowing for the variables of social class, mother's educational level, and handeness. The observed mean score of 88.6 for the ten XYY boys differed significantly ($t = 5$, $P < 0.001$) from the predicted score of 106.5.

TABLE XIII. XYY Boys. Intelligence Tests.

Case No.	Stanford – Binet (age 4)	Mc Carthy (age 5)	WISC age (7 — 8.6) V	P	FS	V – P difference
Singletons						
1.	106	—	111	121	117	ns
2.	89	—	95	93	95	ns
3.	72	—	85	93	88	ns
4.	86	—	80	83	80	ns
5.	97	—	121	115	120	ns
6.	85	—	97	113	105	p< 0.05
7.	74	—	106	100	104	ns
8.	87	92	104	128	117	p< 0.01
9.	96	105	—	—	—	
10.	94	113	—	—	—	
11.	—	—	84	96	88	ns
12.	—	—	125	138	134	ns
13.	—	—	102	114	108 R	ns
14.	—	—	108	104	107	ns
LBW and Twins						
15.	—	—	123	115	121	ns
16.	106	—	108	131	121*	p< 0.01
17.	82	—	95	92	93	ns
Twin Survey						
18.	—	—	92	93	92	ns

* Tested at age 11.1 years. R = WISC – R

For the McCarthy Scale of Children's Abilities, results are available on only three children at age 5, the lowest subscale being in the motor category. Using the WISC, results are available for a total of 15 boys from the newborn survey and one from the twin survey. The full-scale scores ranged from 80 to 134, with a mean of 104.8, for the main group (n = 11, omitting the boy tested on the WISC-R) and 106.2 when the twins from the newborn survey and the low-birth-weight boy are included. Comparisons of these scores with the 23 results available from control boys using the regression analysis as above are shown in Table XV. Significant reductions are found in the verbal ($P < 0.01$) and full-scale scores ($P < 0.02$), but the lower performance score ($P < 0.06$) does not reach the accepted level of statistical significance.

Three XYY boys were found to have verbal scores significantly lower than their performance scores (Table XIII). The XYY boys' results on the subtests of the WISC (Table XIV) ranged from the lowest mean score of 8.8 on information to the highest mean score of 12.5 on block design, with no outstanding strengths or weaknesses. In eight cases comparisons with their siblings (Table XV, Fig. 11) showed no significant difference.

The XYY boys do not differ significantly from the Edinburgh population in WISC score, but it has to be borne in mind that the XYY group has a

TABLE XIV. XYY Boys. WISC Subtests.

WISC sub – tests in XYY Boys.

Case	Information	Comprehension	Arithmetic	Similarities	Vocabulary	Digit Span	Picture Completion	Picture Arrangement	Block Design	Object Assembly	Coding
1.	12	12	9	12	13	11	10	14	14	17	10
2.	8	10	13	7	8	9	6	10	11	9	—
3.	5	7	11	9	7	7	8	11	10	8	8
4.	5	7	7	6	9	7	9	8	11	5	5
5.	12	11	13	15	16	13	11	15	17	11	7
6.	6	12	8	7	14	11	11	13	9	13	13
7.	10	12	12	13	11	8	8	8	13	10	11
8.	11	7	10	11	13	11	11	14	16	14	15
11.	7	7	10	8	7	5	11	8	10	10	8
12.	11	14	13	20	18	8	18	13	18	17	11
14.	12	7	13	15	14	6	10	9	11	11	12
15.	7	18	12	19	15	11	15	12	11	10	13
16.	10	10	6	14	11	16	13	14	15	17	12
17.	7	10	12	10	5	11	5	10	9	12	8
18.	6	11	10	9	8	5	7	8	9	7	14
mean cases (1 – 17)	8.8	10.3	10.6	11.8	11.5	9.6	10.4	11.4	12.5	11.7	10.2
S.D.	2.6	3.2	2.4	4.4	3.8	3.0	3.4	2.5	3.0	3.6	2.8

TABLE XV. XYY Boys. Intelligence Tests—Multiple Regression Analysis.

Intelligence Tests in XYY Boys.
Multiple Regression Analysis.

Stanford Binet (1972 norms).

	n	observed mean	predicted mean	t	p
XYY	10	88.6	106.5	5.25	<0.001

Wechsler Intelligence Scale for Children.

	n	observed mean	predicted mean	t	p
XYY	11				
Verbal		101.4	112.9	3.43	<0.01
Performance		107.6	117.7	2.22	<0.06
Full scale		104.8	116.6	3.06	<0.02

	Verbal	Performance	Full scale	
n = 8 Siblings of XYY	111.3 (S.D. 19.8)	115.0 (S.D. 18.1)	114.1 (S.D. 20.3)	
n = 8 XYY with sibs.	108.0 (S.D. 10.8)	114.8 (S.D. 17.8)	112.2 (S.D. 14.4)	N.S.
Edinburgh population	108.3	110.4	109.2	N.S.

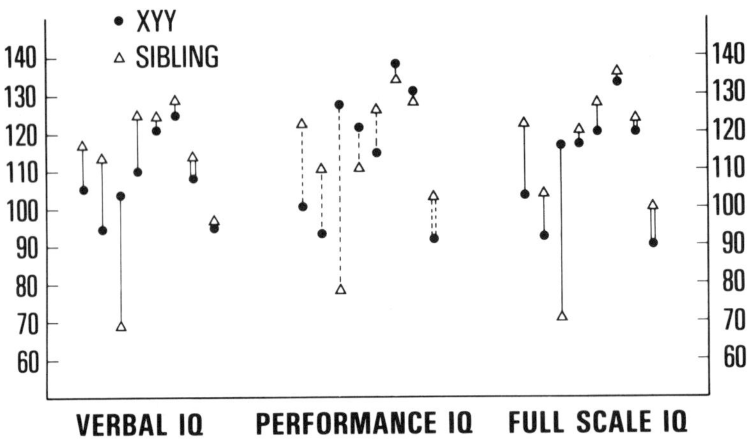

Fig. 11. XYY and Sibling. WISC comparisons.

social class distribution that lacks the expected numbers in the unskilled categories (I and II–3, III–7, and IV and V–1) and therefore might have been expected to score somewhat better.

School achievements. All boys except one attended ordinary schools. Case 13 had a severe delay of speech development and attended a school for speech-disordered children until age 8, when he returned to mainstream schooling.

The XYY boys' scores for reading ability are shown in Table XVI, along with speech ability and therapy, remedial reading, and behaviour in school for boys aged 5–10. Results were available for 15 boys from the newborn survey and the boy from the twin survey, two cases, 14 and 15, being too young as yet to be tested. As with the XXY boys, when the boys' reading age could not be tested at the target age of 8 years, an estimated 8-year-old score was calculated on the basis of the centile score obtained at the later age to permit comparison.

The results of the Burt test show a wide range of reading ability. Two XYY boys had superior ability, a further two were above their chronological ages, whereas three others were markedly depressed. As a result, the mean score of 7 years 8 months for the 13 XYY boys from the main group was only slightly below the chronological age and did not differ significantly from that predicted from the 94 control boys with reading results.

Seven of the 13 boys in the main group, that is, 54%, and one twin received remedial teaching for reading. Except in one case, the teachers were unaware of the chromosome abnormality, and the extra help was initiated by the school. The mean reading age of the group receiving remedial teaching was significantly lower than for the XYY boys who did not need this help ($P < 0.01$), indicating that the teachers' perceptions of the need for help

TABLE XVI. XYY Boys. Speech and Reading.

Speech and Reading in XYY Boys.

Case	RDLS		Speech	Reading Age		Remedial	School[1]	
	EL	VC	Therapy	yr.	m.	Reading	Behaviour	
Singletons							UR	OR
1.	—	—	O	8	5	O	–	–
2.	—	—	O	7	11	O	M	MA
3.	—	—	O	* 7	9	+	M	M
4.	—	—	O	* 5	9	+	S	MA
5.	—	—	O	*10	0	O	MA	S
6.	—	—	O	8	3	O	S	S
7.	—	−0.2	+	* 6	11	+	M	A
8.	−0.2	+0.2	+	7	6	O	M	S
9.	+0.6	−0.5	+	* 7	7	+	–	–
10.	−0.3	+0.9	O	10	6	O	M	S
11.	−1.8	−1.1	+	6	2	+	S	S
12.	−0.9	+1.9	+	6	4	+	M	S
13.	−2.6	+1.8	+	6	10	+	S	A
14.	−0.3	+0.9	O	—	—	—	S	S
			mean	7yr.	8m.			
			S.D.	1yr.	5m.			
LBW and Twins.								
15.	−0.3	+1.2	+	—	—	–	S	A
16.	—	—	O	* 6	1	+	–	–
17.	—	—	O	7	8	O	M	M
Twin survey.								
18.	—	—	+	* 7	5	O	S	A

RDLS Reynell Development Language Score age 3 – 5 years { EL Expressive Language; VC Verbal Comprehension
*Estimated score – see text.
[1] School Behaviour – Bristol Social Adjustment guide.
UR = Under Reaction OR = Over Reaction S = Stable M = mild under or over reaction
A = appreciable under or over reaction MA = maladjusted under or over reaction

coincided for the most part with lower obtained reading scores. There were two boys receiving remedial help who had reading ages only 3 and 5 months behind their chronological ages (Cases 3 and 9); both of these boys had estimated reading ages, one having been tested at 8 years 6 months after repeating 1 year in school, whereas the second boy was tested at age 9 after having remedial help for reading for 1 year. Eighteen of the 102 control boys (17.6%) received remedial teaching for reading in primary scool, these boys having a mean reading deficit of 12.1 months; in all but one there appeared to be a reading deficit as demonstrated by the test.

The relationship of reading difficulty to previous speech problems was also examined. The mean reading score of the six boys who had received speech therapy was 6 years 10.7 months as opposed to 8 years 0.4 months for those eight boys not requiring speech therapy. However, this difference was not significant ($P < 0.1$).

Behaviour. This was first examined at age 3 using the Behaviour Screening Questionnaire. Results from 92 male controls and 9 XYY boys from the main group are shown in (Fig. 12). Because the questionnaire was not available until 1971, it was not possible to use it with the older XYY boys. None of the total scores of the XYY boys were at the level indicating a behaviour problem, whereas such scores were found in 4.2% of control boys. Examination of scores on the separate behaviour items shows that on seven of the 12 items, no XYY boy has a score at the problem level; the areas where two of ten XYY boys scored 2 are limited to the categories

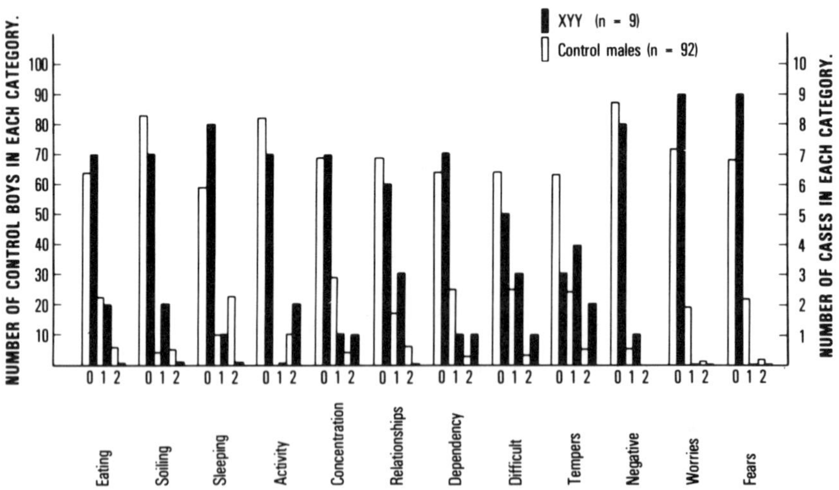

Fig. 12. XYY Boys and controls. Behaviour Screening Questionnaire.

"high activity level" and temper tantrums, whereas only one of ten XYY boys obtained a score of 2 in the areas poor concentration, dependency, and difficult to handle. Overall, in none of these categories do more than 20% of the ten XYY boys score in the problem range.

Behaviour in school was reported annually by the teachers using the Bristol Social Adjustment Guide. As before, multiple regression analysis was used to produce predicted scores, and because of small numbers the yearly reports at ages 7–10 years were amalgamated. When a child had more than one report, the mean was taken.

The score for underreaction for 11 of the 14 XYY boys from the main group with school reports at these ages was significantly elevated over the prediction (t ratio = 3.29, 10-f, $P < 0.01$), but overreaction did not differ significantly from the predicted score. With reference to population norms in the BSAG manual (1970) the mean underreaction score placed the XYY boys in the "mild underreaction" category, between the 32nd and 40th centile ranks. On overreaction, they fall into the "mild overreaction" group, between the 29th and 34th centile ranks.

In the primary school age range, psychiatric referral from the Growth Clinic occurred in two boys from the main group and two boys who were members of twin pairs. These case reports have been published [13]. At the time of the psychiatric referral, none of the parents was aware of the boys chromosome abnormality. The first boy was referred at age 9 on account of disruptive behaviour in school for 2 years and an episode when he played truant from school and broke into and stole from a sweet kiosk resulting in police involvement. There were multiple problems in the home environment, with the parents' separation and the father's ill health. He was considered to be suffering from a moderately severe emotional disorder, reacting to environmental stress by depression and regression, superimposed on a disposition that was vulnerable both temperamentally and as a result of meningitis at age 3 with subsequent unilateral hearing loss. The second boy was referred at age 7 for school-related bed-wetting over a 2 year period. In addition, on waking each day, he would weep without apparent reason and seemed miserable and unhappy at least half of the day. He was diagnosed as having an emotional disorder with a moderate degree of depression.

Of the twins, the first was referred for stealing money at home and at school over a 2 year period. In addition, there were complaints from his single mother of his difficult temperament and daily temper tantrums since age 2. He was considered to have a mild emotional disorder with depression. The second twin was referred at age 9 for persistent temper tantrums and demanding behaviour, which his mother could no longer cope with because of her own physical and marital problems. From the age of 3, he had temper tantrums many times daily. He had severe reading problems and was later transferred to a dyslexia unit at a private school. His electroencephalogram was abnormal with almost continuous spikes in the right temporooccipital region on falling asleep; intermittent spikes were present during wakefulness, and there was an increase in generalised θ components. He was considered to have a moderately severe emotional disorder associated with a right posterior temporal lobe focus. In both these latter cases the chromosomally normal cotwin, although sharing the same environmental problems, showed no evidence of disordered behaviour.

XXX

Ten XXX girls, including two of birth-weight less than 2.5 kg, were tested on the Stanford Binet (Thorndike norms; see Table XVII). Their scores ranged 57 to 94, with a mean score of 75.9 for the singletons of greater than 2.5 kg birth weight. Comparison with the predicted score of 105.8 shows a significant reduction ($P < 0.001$).

Five girls have results available on the McCarthy test; these are in close agreement with the scores that four of the girls obtained on the Stanford Binet, the mean score being 85.2. The lowest score was on the quantitative subscale, with memory the next lowest.

TABLE XVII. XXX Girls. Intelligence Tests.

Intelligence Tests in XXX Girls.						
Case No.	Stanford Binet (age 4)	Mc Carthy (age 5)	WISC (age 7 – 10) V	P	FS	V—P difference
Singletons						
1.	68	—	72	82	75	
2.	57	—	74	82	75	
3.	80	—	79	107	91	V < P (p<0.01)
4.	71	—	99	86	92	
5.	91	87	85	97	90	
6.	94	89	118	111	117[R]	
7.	82	89	—	—	—	
9.	—	—	104	110	107	
10.	64*	72	77	87	80	
11.	—	—	114	113	115	
12.	—	89	—	—	—	
Birth Weight <2.5Kg.						
14.	80	—	80	94	85	
15.	74	—	85	86	94	

* Tested at age 4yr. 6m.
R = WISC – R

On the WISC, there are results available for eight girls from the main group and two who had birth weights of less than 2.5 kg. In addition, one girl (Case 6) was tested on the WISC-R; she achieved the highest score of all the XXX girls with a full-scale score of 117. The remaining eight girls had full-scale scores ranging from 75 to 115, with a mean of 90.6. This was significantly lower than the predicted full-scale score of 110.3 ($P < 0.01$) obtained from eight control girls. However, some caution should be exercised because small differences between the case and control groups in the distribution of social class and mother's education could substantially affect the accuracy when numbers are small.

The mean verbal score of 88.0 was significantly lower than the prediction of 111.2 ($P < 0.01$), but the mean performance score of 95.5 was depressed to a lesser degree in comparison with the predicted score of 107.6 ($P < 0.05$). Only one XXX girl had a significantly lower verbal than performance ability, although in seven other girls the differences were in the same direction (Table XVII). Subtest scores (Table XVIII) ranged from 6.9 on both information and comprehension to 11.5 on coding, a pattern similar to that shown by the XXY boys.

Comparison of the WISC results with those from 67 girls from among Edinburgh school children shows a significant deficit in the XXX girls verbal

TABLE XVIII. XXX Girls. WISC Subtests (ages 7–10 years).

WISC sub – tests in XXX Girls.
(age 7 – 10).

Case No.	Information	Comprehension	Arithmetic	Similarities	Vocabulary	Digit Span	Picture Completion	Picture Arrangement	Block Design	Object Assembly	Coding
Singletons											
1.	6	8	4	6	3	6	8	4	9	7	9
2.	4	5	·6	6	8	—	8	6	9	7	7
3.	7	6	6	5	7	9	9	13	10	7	16
4.	10	6	12	10	9	12	8	5	8	8	11
5.	6	7	8	8	10	7	8	8	7	10	15
9.	10	9	10	16	6	13	11	13	13	11	12
10.	6	4	8	7	5	8	8	8	9	6	10
11.	6	10	17	18	13	9	13	10	11	13	12
mean	6.9	6.8	8.9	9.5	7.6	9.1	9.1	8.4	9.5	8.6	11.5
S.D.	2.1	2.0	4.1	4.9	3.1	2.4	1.9	3.4	1.8	2.4	3.0
Birth Weight <2.5Kg.											
14.	4	7	8	8	6	8	11	6	7	11	11
15.	10	7	10	7	6	6	4	7	10	7	12
6. (WISC R)	12	11	10	14	14	17	12	11	14	12	9

TABLE XIX. XXX Girls. Speech and Reading.

Speech and Reading in XXX Girls.								
Case No.	R.D.L.S. EL	VC	Speech Therapy	Reading Age yr.	m.	Remedial Reading	School Behaviour UR	OR
Singletons								
1.	—	—	0	6	2	+	—	—
2.	—	—	0	6	4	+	S	A
3.	—	—	+	6	10	+	MA	S
4.	—	—	+	6	4	+	A	S
5.	−1.9	−0.0	+	6	8	+	MA	S
6.	—	—	+	6	6	+	S	S
7.	+2.2	+0.4	0	6	10	+	A	S
8.	−1.1	−0.1	0	6	6	+	MA	S
9.	−0.1	+0.8	0	8	4	0	S	S
10.	−1.4	−2.0	+	6	4	+	A	MA
11.	−0.5	−1.5	0	7	3	0	M	S
13.	−0.5	+1.4	+	too young			—	—
				mean 6 yr.	8.8 m.			
				S.D. 0 yr.	7.4 m.			
Birth Weight <2.5Kg.								
14.	—	—	+	6	8	+	M	S
15.	−1.9	−1.0	0	7	1	0	S	S
Twin								
16.	−2.5	−2.3	+	7	0	+	MA	S

R.D.L.S. = Reynell Development Language Scale
EL = Expressive language. VC = Verbal comprehension.
UR = Under Reaction OR = Over Reaction S = Stable
A = appreciable under or over reaction MA = maladjusted under or over reaction.

(t = 3.71, P < 0.001), performance (t = 2.81, P < 0.01), and full-scale IQ (t = 3.46, P < 0.001) scores. The girl controls do not differ in verbal, performance, or full-scale scores from the Edinburgh population results on the WISC, although the numbers available for comparison are small. Because only four IQ results are available on the siblings of the XXX girls these comparisons will be presented when further data has been acquired.

School achievements. All but one of the XXX girls attended ordinary schools, the exception, Case 2, being at a school for the educationally subnormal (ESN).

The Burt reading test was administered mainly at 8 years of age to 14 XXX girls (including one twin and two low-birth-weight girls), and results are shown in Table XIX along with details of previous speech therapy, remedial help, and the categorisation of behavioural adjustment in primary school. As before, when a child was tested at a later age, an estimated score for age 8 was obtained (Cases 1–4).

The mean reading age was 6 years 9 months, the scores ranging from 6 years 2 months to 8 years 4 months. Case 2, at an ESN school, achieved a score of 7 years at age 10. Using the multiple regression analysis as before, there was a significant deficit vs the predicted score of 8 years 0.7 months, calculated from 64 control girls with results on the Burt reading test (t =

4.82, $P < 0.001$). This reading deficit is confirmed by comparison with the Scottish population norms for 1974 ($t = -8.4$ $P < 0.001$).

Seven of the 14 XXX girls with reading scores had received speech therapy previously, and all seven required remedial teaching, as did a further four XXX girls. Conversely, there were three girls who did not require remedial help, and these girls also had not needed speech therapy.

As with the other karyotype groups, the educational help for reading was appropriate given the measured reading scores, all 11 girls having scores 12 months or more below their chronological age, with a mean reading age of 6 years 6.7 months for those receiving remedial help, whereas the three not receiving help had a mean reading age of 7 years 6.6 months. Although the number of girls not receiving remedial help is too small to permit statistical analysis of the relationship between school social adjustment and remedial teaching, it is worth noting that these three girls were all classified as stable on the Bristol Social Adjustment Guide.

Thus the overall picture is one of cognitive ability at the lower end of the normal range, with difficulty in learning to read such that 82% of the XXX girls required extra help in comparison with 12.2% of the 74 control girls. However, in three of the 14 XXX girls (ie, 21%) for whom we have results at this stage of the study, there were neither educational nor behavioural problems in school.

Behaviour. The Behaviour Screening Questionnaire [11] was completed within 3 months of the third birthday at the paediatric interview with the parent(s) of 14 XXX girls, including one twin and two low-birthweight girls. One of the 11 girls in the main group and one of the low-birthweight girls achieved overall scores in the problem category ($\geqslant 10$).

Comparison of the 11 XXX girls from the main group with the results from 59 control girls is shown in Figure 13. As with the XYY and XXY boys, the XXX girls show no difference from the controls in the areas of worries and fears, and no XXX girls had a score of 2 (a definite problem) in these two aspects of behaviour.

There is some incidence of behaviour problems in all the other areas, but when a score of 2 is found it is only in a minority of the XXX girls, and 64% were free from significant disturbances. The extent to which any problems are related to the lower intelligence level of the XXX girls cannot be clarified at present; we do not have sufficient numbers with comparable cognitive ability in the control group, nor are there published data relating the scores on the BSQ to intelligence level.

Behaviour in school was reported annually from 1980 using the Bristol Social Adjustment Guide. Because of the wide time span of case collection,

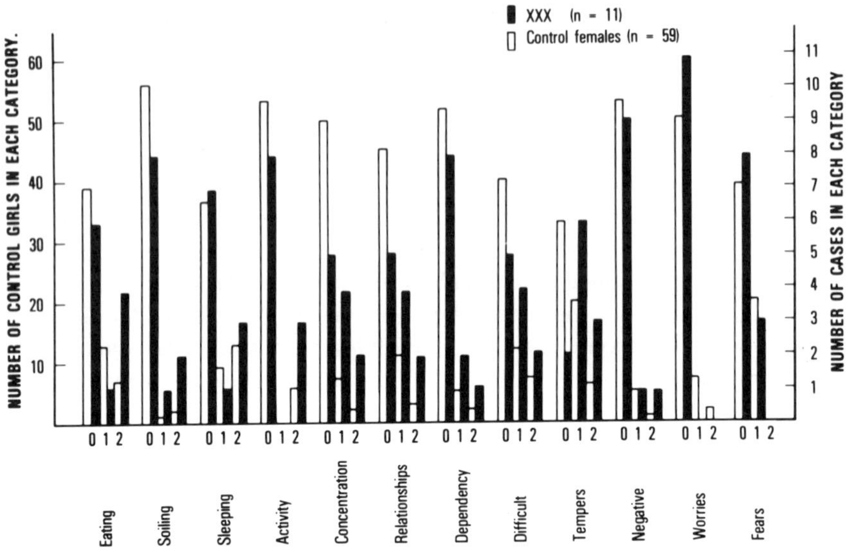

Fig. 13. XXX Girls and controls. Behaviour Screening Questionnaire.

the reports from years 7 to 10 were amalgamated to produce sufficient numbers for comparison with the controls on the multiple regression analysis. Results are available from ten girls from the main group at these ages. The score for underreaction was significantly elevated over that predicted from the controls ($t = 3.86$, $P < 0.01$). The overreaction score did not differ significantly from the prediction. A further three results—one for a twin and two for low-birth-weight girls—are also available. Results of the total XXX group vs BSAG manual norms show their mean underreaction score to fall in the "appreciable underreaction" category between the 13th and 22nd centile ranks. On overreaction, the mean score is classified as stable and falls between the 29th and 36th centile ranks.

One XXX girl had psychiatric referral from the Growth Clinic while in primary school. This girl had shown elective mutism with adults outside the family from starting school at age 4.5 until she was 6 years old. School progress was poor, and she was held back a year. At age 11 she started to get spontaneous hair loss exacerbated by plucking out her hair. Psychiatric counselling at age 12 and the prescription of a wig, as her hair loss was almost total, resulted in slow improvement.

Among the control girls, two had psychiatric referral initiated by their general practitioners. In one case, this was on account of an hysterical reaction to a minor knee injury, and in the second girl for severe tantrums at age 3.5 years. In both girls, the problems resolved after a single counselling session.

Discussion of Control Group

There must necessarily be a degree of self-selection in any population that is prepared to cooperate in a study involving regular visits to a clinic for examination over a period of at least 15 years. In the later course of the study, a number of findings have caused us to reassess the validity of the control group and to alter the way in which comparisons are made between the cases and controls. These have included, first, the differing social class distribution between the male controls and the hospital population. A finding of considerable interest was that this was confined to those enrolled by one recruiter. Whatever the real factors that gave rise to this situation, we have found that controlling the data for social class eliminated the otherwise significant recruiter effect.

Second, the exclusion of twins and infants of low birth weight in the controls obviously did not apply to the finding of the case material. Hence in this paper we have used only statistical comparisons between cases who were singletons of birth weight greater than 2.5 kg and the controls, the remaining cases being presented separately. Chromosomal variants were also excluded from the control group, because, although it is generally assumed that these have no phenotypic effect, a detailed study of reasonable duration has not yet been carried out on such a group.

Furthermore, we have found that attendance at private schools, relatively more common in Edinburgh than elsewhere in the country, is higher for the controls—at 23% for males and 18% for females—than for Edinburgh schoolchildren in general (9.4%). This raised incidence of private school attendance was not restricted to the effect of any one recruiter and therefore might reflect some more general self-selection or recruitment effect. It is impossible to estimate how representative the final control population is of that actually available on the postnatal ward at the time of recruitment, since details of the latter were not recorded. Nevertheless, the significantly higher reading ages obtained by the controls ($P < 0.001$ for males and $P < 0.05$ for females) when corrected for social class were not significantly different from those of the 1974 norms for the Scottish population. The number of children who attended private schools at the age of 8 (13 of 102) when the reading scores were obtained was lower than at present. Although there is generally assumed to be an advantage in private schooling in terms of smaller class size, etc, it is not clear from our limited results that this requires further explanation other than social class.

Being aware of the various difficulties with the controls, we have set up regression-based models that can be used for standardisation of all comparisons between cases and controls. To date, this has been done only for the psychological data and in future reports will be applied to the growth data; however, preliminary analysis shows no effect of social class on this aspect of the children's development. In addition, we have used population norms when available for comparison with the case and the control group. We believe, therefore, that despite the difficulties the data do permit comparisons between the cases and controls, although great caution must be exercised in drawing inferences for the likely development of individual children with sex chromosome abnormalities.

GENERAL DISCUSSION

In the second report of the longitudinal study of growth of children with sex chromosome abnormalities [4] the period from birth through middle childhood was described with few of the children having reached puberty. In the present report, approximately half of the cohort (28 of 54) have reached this landmark and eight children have reached their final height.

In the XXY boy, the early effect of additional X material in slowing brain growth, illustrated by the smaller head circumference at birth, is seen to persist throughout childhood; this was also demonstrated to be present at age 16 in a separate study of 12 XXY boys identified by sex chromatin screening of a newborn population [14]. The Finnish study by Varrela [15] of 28 adult XXY males similarly showed a significantly smaller head circumference in comparison with both controls and male relatives. Smaller head circumference is also found in females with an additional X and in both groups is associated with a cognitive deficit of a small degree.

Similarities are also revealed in the body proportions of these two groups in the longer limbs, the legs being particuarly affected but more so in the males. In XXY males, this has previously been attributed to inadequate testosterone production at puberty, but from our data it is seen to be present from early childhood [6]; and in both sexes, suggesting an effect of additional X chromosomal material on bone growth, although the precise mechanism remains to be clarified. In some XXY boys, the failure of the testes, which were at one time descended, to remain in the scrotum during growth in childhood has been an unexpected finding and implies that closure of the tunica vaginalis has been incomplete. The incidence of orchidoplexy of 35% can be compared with the cumulative lifetime incidence of orchidoplexy in the Oxford Record Linkage Study of 3.38% [16]. Hadziselimovic and Herzog

[17] have concluded from their experimental studies that androgen produced by the Leydig cells is primarily responsible for normal descent of the testes. This raises questions about the adequacy of testosterone production throughout childhood in XXY boys, which we are investigating using serial salivary sampling. The onset of puberty was not delayed in XXY boys, although the increase in testicular size was limited to approximately 5 ml volume.

Growth data from the one XX male found in the newborn survey is described here for the first time and covers the whole period of his maturation. The ultimate height of 167.7 cm was between the 10th and 20th centiles for British males and was less than expected given his father's and mother's heights at the +90th and +50th centiles, respectively. Not only is height less than that for average males but the Z scores for various body measurements show all to be reduced. If, on the other hand, he is compared with females, then his height is at the 75th centile.

The findings in our single case are similar to those recorded by Varrela [18] in four adult XX males using the same anthropometric methods. They imply that at least some of the genes on the Y chromosome concerned with the larger dimensions of males are absent or inactivated in XX males.

In the XYY, the infant was of average size at birth, and acceleration of growth velocity occurred earlier than in the XXY. There was a general increase in most measures of bone growth, with the exception of the breadth of the shoulders, which in our data and those of others [19,20] remain proportionately smaller. At puberty, the increase in the normal male's biacromial diameter occurs in response to rising testosterone levels, and there is little evidence to suggest that this is deficient in the XYY male; indeed, Schiavi et al [21] through a carefully controlled study of Danish military recruits found a small but statistically significant increase in blood testosterone level.

The XYY boys were lighter in weight than expected for height and also had smaller circumferences of upper arm as well as of thigh and calf (not shown in Fig. 8), suggesting less than optimal muscle development. There was a small increase in leg length, which did not reach statistical significance and which was also found in the Finnish study of adult XYY males [19]. The age of onset and progress through puberty in the XYY boys so far have not differed from that in the controls.

For the XXX girls, our findings depart from those of others [22] in that only a minority (three of 16) are at or above the 90th centile for height. The mean maternal height of the mothers of the XXX girls at 158.3 cm was less than that of the control girls' mothers at 159.7 cm, but the difference was not statistically significant.

In an earlier study from this Unit, Milne et al [20] measured 26 XXX women, almost all of whom were identified in surveys of mental subnormality hospitals. Their mean height of 160.4 cm was greater than that of women in the same institutions but did not differ from that of 124 female blood donors from the general population or from the Tanner et al [5] standards for females of 162.2 cm. In the Milne et al study, sitting height was significantly reduced, as it was in our group ($P < 0.05$), indicating proportionately long legs although to a lesser extent than found in XXY males.

The most outstanding difference, mentioned above, was that of head circumference, which was −1.83 SD at age 8, the reduction being of greater magnitude than in the XXY boys, similarly, there is a larger reduction in the IQ scores of the XXX girls.

Nine of the 16 XXX girls have shown signs of sexual development, and at this stage in the study there appears to be some delay in the onset of breast development and of menstruation in the XXX girls compared with, as yet, small numbers of contemporary Scottish girls. However, the XXX girls differ very little from British norms established by Marshall and Tanner [23] from the Harpenden Growth Study of 1948–1972, in which the mean age of reaching breast stage 2 was 11.15 years and of menarche 13.47 years. Clearly, a larger quantity of data is required before definitive statements can be made about the rate of sexual development in the XXX girl.

Turning to the psychologic development of the children with sex chromosome abnormalities, delayed speech development was recorded in the last report [4], the incidence of speech therapy being 42% in the XXY, 44% in the XYY, and 50% of the XXX children. This finding is similar to those in the other newborn survey follow-up studies [22].

The results obtained in testing cognitive ability show a changing pattern over time, the scores obtained on the Stanford Binet test at age 4 being 87.4 for the XXY boys, 88.6 for the XYY boys, and 75.9 for the XXX girls, whereas at age 7 full-scale scores on the WISC were 100.8 for XXY boys, 104.8 for XYY boys, and 90.6 for XXX girls. Although the groups tested were not identical at the two ages, and the tests themselves contain differing items, in addition secular changes mean that these WISC scores expressed in terms of 1949 norms in this case exaggerate the difference between the WISC and Stanford Binet (1972 norms) results. The change might in part be related to improvement in verbal ability over time as the delayed speech development resolved.

The cognitive deficit was greatest for the XXX girls and was associated with a high level of learning difficulties, 79% of all the XXX girls requiring remedial help. In the XXY boys, the cognitive deficit was of smaller extent, but learning difficulties again occured frequently, with 66.7% having educational help. The mild underreaction revealed by the school report appears to be of a nature similar to the increased withdrawal described by Walzer [24] in his study of behavioural style in the XXY boy.

The XYY boys also had a raised incidence of learning difficulties, although the degree of cognitive deficit was less than that found for the other two groups, and indeed their results on the WISC, although differing from social class-matched controls, did not differ from those of the Edinburgh population. In this group, the interaction of any behavioural difficulties with a minimal cognitive deficit might play a more important role than in the other two groups. Whereas the BSQ showed no major difference at age 3 between the behaviours of the XYY children and the controls, differences were apparent in behaviour in school, additionally, the level of psychiatric referral was higher than in controls. When problems were present, they were of a minor nature, and in the main resolved either as a result of developmental progress or in response to standard measures of speech therapy, educational assistance, and psychiatric counselling. Further follow up is required to determine the long-term outcome in these children and to collect accurate information for use in antenatal counselling when sex chromosome abnormalities are discovered prenatally.

ACKNOWLEDGMENTS

The following past and present members of the Unit have contributed to the study over the years, and their work is gratefully acknowledged: David Axworthy, Alison Bailey, Catherine Bissett, Gary Butler, Sheila Callan, Andrew Carothers, John Cox, Jim Davies, Norman Davidson, Neil Fraser, Andrina Ginsberg, Fiona Grace, Valerie Henderson, Neiva Magdelena, Mark McKie, William McLaren, Jonathan Nshaho, Paula Sousa-Lima, Lynn Smith, Anthea Springbett, Ian Tierney, and Jonnie Wilson.

REFERENCES

1. Jacobs PA, Melville M, Ratcliffe SG, Keay AJ, Syme J: A cytogenetic survey of 11,680 newborn infants. Ann Hum Genet 37:359–376, 1974.
2. Buckton KE, O'Riordan ML, Ratcliffe SG, Slight J, Mitchell M, McBeath S, Keay AJ, Barr D, Short M: A G-band study of chromosomes in liveborn infants. Am Hum Genet 43:227–239, 1980.
3. Ratcliffe SG, Axworthy D, Ginsborg A: The Edinburgh study of growth and development in children with sex chromosome abnormalities. In Robinson A, Lubs HA, Bergsma D (eds): "Sex Chromosome Aneuploidy: Prospective Studies in Children." New York; Alan

R. Liss, Inc., for the March of Dimes Birth Defects Foundation. BD:OAS XV (1):243–260, 1979.

4. Ratcliffe SG, Tierney I, Nshaho J, Smith L, Springbett A, Callan S:. The Edinburgh study of growth and development of children with sex chromosome abnormalities. In Stewart DA (ed): "Children With Sex Chromosome Aneuploidy: Follow-Up Studies." New York; Alan R. Liss, Inc., for the March of Dimes Birth Defects Foundation. BD:OAS 18(4):41–60, 1982.
5. Tanner JM, Whitehouse RH, Takaishi M. Standards from birth to maturity for height, weight, height velocity and weight velocity: British children 1965. Arch Dis Child 41:454–471, 613–635, 1966.
6. Ratcliffe SG: Longitudinal growth studies in children with sex chromosome abnormalities. In Papadatos CJ, Bartsocas CS (eds): "Endocrine Genetics and Genetics of Growth." New York; Alan R. Liss, Inc., 1985, pp. 301–309.
7. Ratcliffe SG: The sexual development of boys with the chromosome constitution 47,XXY (Klinefelter's syndrome). Clin Endocrinol Metab 11:703–716, 1982.
8. Skakkebaek NE, Bancroft J, Davidson DW, Warner P: Androgen replacement with oral testosterone undecanoate in hypogonadal men: A double blind controlled study. Clin Endocrinol 14:49–61, 1981.
9. Sattler JM: "Assessment of Childrens Intelligence, Rev. Ed." Philadelphia: Saunders, 1974.
10. Davie R, Butler N, Goldstein H: "From Birth Till Seven: The Second Report of the National Child Development Study (1958 Cohort)." London; Longman, 1972.
11. Richman N, Graham PJ: A behavioural screening questionnaire for use with three year old children: Preliminary findings. J Child Psychol Psychiatr 12:5–33, 1971.
12. Stott DH: "Bristol Social Adjustment Guides, 5th ed." London; Hodder and Stoughton, 1976.
13. Ratcliffe SG, Field MAS. Emotional disorder in XYY children: Four case reports. J Child Psychol Psychiatr 23:401–406, 1982.
14. Ratcliffe SG, Bancroft J, Axworthy D, McLaren D. Klinefelter's syndrome in adolescence. Arch Dis Child 57:6–12, 1982.
15. Varrela J: Effects of X chromosome on size and shape of body: An anthropometric investigation in 47,XXY males. Am J Phys Anthropol 64:233–242, 1984.
16. Swerdlow AJ, Wood KH, Smith PG: A case-control study of the aetiology of cryptorchidism. J Epidemiol Commun Health 37:238–244, 1983.
17. Hadziselimovic F, Herzog B: The meaning of the Leydig cell in relation to the etiology of cryptorchidism: An experimental election-microscopic study. J Pediatr Surg 11:1–8, 1976.
18. Varrela J: Body size and shape in 46,XX males: An anthropometric investigation. Ann Hum Genet 48:261–267, 1984.
19. Varrela J, Alvesalo L: Y chromosome effects on body size and shape in 47,XYY males. Am J Phys Anthropol 68:239–245, 1985.
20. Milne JS, Lauder IJ, Price WH: Anthropometry in sex chromosome abnormality. Clin Genet 5:96–106, 1974.
21. Schiavi RC, Owen D, Fogel M, White D, Szechter R: Pituitary-gonadal function in XYY and XXY men identified in a population survey. Clin Endocrinol 9:233–239, 1978.
22. Stewart DA (ed): "Children With Sex Chromosome Aneuploidy: Follow-Up Studies." New York: Alan R. Liss, Inc., for the March of Dimes Birth Defects Foundation BD:OAS, 18 (4), 1982.

23. Marshall WA, Tanner JM: Variation in the pattern of pubertal changes in girls. Arch Dis Child 44:291-303, 1969.
24. Walzer S: A method for the longitudinal study of behavioural development in infants and children: The early development of XXY children. J Child Psychol Psychiatr 19:213-229, 1978.
25. Reynell JK: "The Reynell Language Scales." Windsor, IL: NFER, 1969.
26. Thorndike RL: "Stanford Binet Intelligence Scale. Third Revision." Boston: Houghton Mifflin Co., 1973.
27. McCarthy D: "McCarthy's Scale of Childrens Abilities." New York: The Psychological Corporation, 1970.
28. Stott DH: "Bristol Social Adjustment Guide. 5th ed." London: Hodder and Stoughton, 1976.
29. Wechsler D: "Wechsler Intelligence Scale for Children." New York: The Psychological Corporation, 1949.
30. Scottish Council for Research in Education: "The Burt Word Reading Test. 1974 Revision." London: Hodder and Stoughton, 1976.
31. Bender L: "A Visual Motor Gestalt Test and Its Clinical Use (Research Monograph No. 3)." Washington, DC: American Orthopsychiatric Association, 1938.
32. Coan RW, Cattell RB: "The Early School Personality Questionnaire (The ESPQ)." Champaign, IL: Institute for Personality and Ability Testing, 1972.
33. Witkin HA, Oltman PK, Raskin E, Karp SA: "Childrens Embedded Figures Test." Palo Alto, CA: Consulting Psychologists Press, Inc., 1971.
34. Porter RB, Cattell RB: The Child Personality Questionnaire. Champaign, IL. Institute for Personality and Ability Testing, 1972.
35. Kagan J: Reflection-impulsivity and reading ability in primary grade children. Child Dev 36:609-628, 1965.
36. Rosenzweig S, Fleming EE, Rosenzweig L: The children's form of the Rosenzweig picture frustration study. J Psychol 26:141-191, 1948.
37. Rutter M, Tizard J, Whitmore K (eds): "Education, Health and Behaviour." London: Longmans Green, 1970.
38. Piers EV, Harris DB: "The Piers-Harris Childrens Self-Concept Scale." Nashville, TN: Counselor Recordings and Tests, 1969.
39. Goldberg DP: "The Detection of Psychiatric Illness by Questionnaire (Maudsley Monographs No. 21)." London: Oxford University Press, 1972.
40. Hess Rd, Shipman VC: Early experience and socialization of cognitive modes in children. Child Dev 36:869-886, 1965.
41. Elliot CD, Murray DJ, Pearson LS: "The British Ability Studies. Revised Edition." Windsor, IL: NFER, 1983.
42. Cattell RB, Cattell MDL: "The High School Personality Test." Windsor, IL: NFER, 1969.
43. Gordon HW, Galatzer A: Cerebral organisation in patients with gonadal dysgenesis. Psychoneuroendocrinology 5:235-244, 1980.
44. Cattell RB, Eber HW, Tatsuoka MM: "Handbook for the 16PF." Champaign, IL: Institute for Personality and Ability Testing, 1970.
45. Wechsler D. "Wechsler Adult Intelligence Scale." New York: Psychological Corporation, 1971.

Appendix I
Anthropometric Study
(all measures except wt. in cms)

Weight (kg)
Supine length and crown-rump up to age 1.5 years.
Stature
Sitting Height
Skin fold thickness, triceps and subscapular.
Bi-acromial and bi-iliac diameters.
Circumference of head, upper arm, thigh and calf.
Length of upper arm, forearm, hand, calf, foot.
(Leg length = stature − sitting height)

$$\text{Z score} = \frac{\text{Measure on case} - \text{measure on control}}{\text{Standard deviation of controls measure}}$$

Appendix II
Psychological Development Programme

Age 2	Gesell Development Scale	
Age 3	Reynell Developmental Language Scale	(24)
	(Verbal comprehension and expressive language).	
Age 4	Stanford Binet Form L-M (1972 norms).	(25)
Age 5	McCarthy Scale of Childrens Abilities.	(26)
	Bristol Social Adjustment Guide—annually	(27)
Age 7	Wechsler Intelligence Scale for Children	(28)
Age 8	Burt Reading Test	(29)
	Bender Gestalt	(30)
	Early School Personality Questionnaire	(31)
Age 9	Childrens Embedded Figures Test	(32)
Age 10	Child Personality Questionnaire	(33)
Age 11	Matching Familiar Figures Test	(34)
	Letter Elimination Test	
	Rosenzveig Picture Frustration Test	(35)
Age 12	Wechsler Intelligence Scale for Children	(28)
	Adverse Life Events Study	
	Rutters Parental Questionnaire	(36)
	Piers-Harris Self-Esteem Questionnaire	(37)
	Maternal General Health Questionnaire	(38)
Age 13	British Ability Scales—word reading and number	(39)
	Mother-child cooperative task	(40)
Age 14	High School Personality Questionnaire	(40)
Age 15	Cognitive Laterality Battery	(41)
Age 16	Adult Personality Questionnaire 16PF	(42)
	Wechsler Adult Intelligence Scale.	(43)

Growth and Development From Early to Midadolescence of Children With X and Y Chromosome Aneuploidy: The Toronto Study

Donald A. Stewart, MD, FRCP(C), John D. Bailey, MD, FRCP(C), Charles T. Netley, PhD, Joanne Rovet, PhD, and Elizabeth Park, PhD

Outpatient Department (D.A.S.), Division of Endocrinology (J.D.B.), Department of Psychology (C.T.N., J.R.), Research Institute of The Hospital for Sick Children (E.P.), Toronto, Ontario M5G 1X8, Canada

INTRODUCTION

Of 83 children who were diagnosed at birth by amnion examination as having abnormal X and/or Y karyotypes, 53 are still being followed at The Hospital for Sick Children in Toronto. Two previous publications by the March of Dimes Birth Defects Foundation [1,2] have reported the follow-up study of these children, first up to about age 8 years and then from age 8 years to about 12 years. This publication describes growth, development, and sexual maturity of these children from 12 to 16 years of age, i.e., through the first part of puberty. The result of a controlled trial on the effects of intramuscular testosterone cypionate given to nine 47,XXY boys is also presented.

The present group of 53 children is made up of 31 47,XXY, nine 47,XXX, three 47,XYY, and five mosaic karyotypes. Also included are two males with 46,XX karyotype and three 45,X mosaics. Over the past 4 years, two children have been lost to the study, and three have been added. The added children had been identified as having an abnormal X or Y karyotype at birth, but their parents had not been informed of the abnormality by the family doctor. When they attended a hospital clinic, the parents were informed of the abnormal karyotype and subsequently they requested that their child be included in the study.

Birth Defects: Original Article Series, Volume 22, Number 3, pages 119-182

In this chapter, we first present a report of the growth of children in each karyotypic group followed by a report on the hormonal and sexual development including a trial of testosterone treatment given to 47,XXY boys. Finally, a report will describe psychological development and behavioural features. A short profile of each child is presented, and comments are made regarding gross psychosocial abnormalities when these have occurred.

PHYSICAL GROWTH

47,XXY Boys

Sample. Of the original 43 boys [1,2], 31 presently aged 12.5–16.5 years have been followed. These boys were seen annually near their birthdays until they were 13 years old and every 6 months thereafter.

Height. Unless otherwise stated, the standards used for physical growth are the British standards of Tanner et al [3]. The data have been treated cross-sectionally. Figure 1A and B shows that the mean height rises from the 30th to 40th percentile, prior to age 2 years, to the 60th at age 8 years, and to the 70th beyond that age. The regression of height percentile on age is highly significant ($r^2 = .585$, $P < .001$). This height increase relative to normal standards is reflected in height velocity (Fig. 1C), which is above the 50th percentile although with some flucutations prior to age 8 years. All but three boys (ages 12, 13, and 13.5 years) had had their pubertal growth spurt. There was a sharp rise in growth velocity at 12.5 years of age on average (range 11–14.5 years). This represents a 6 month delay in the onset of pubertal growth spurt and is consistent with the observation that bone age (Greulich-Pyle) was 6 months behind chronological age at that point in time.

These boys tended to be somewhat tall compared to our normal standards. That they exceeded their genetic expectation for height is shown by the fact that for 26 subjects the mean midparent height percentile was 37.5. The mean height percentile for the boys at age 6 years was 43.6, at 10 it was 65, and at 14 it was 69.

Weight. Weight (Fig. 2) exhibited a pattern similar to height, although the greatest increase occurred after 7 years of age.

Head circumference. After 6 years of age, head circumference (Fig. 3) was below normal standards [4] although not significantly so. There appears to be a tendency to approach the 50th percentile at later ages, but the number of subjects at ages 15 and 16 years is smaller.

Body proportions. Data on body proportions are available from ages 11 to 15 years. Results for age 14 are graphically compared to American [5] and British [6] standards (Fig. 4). Anatomical points used in measuring were the

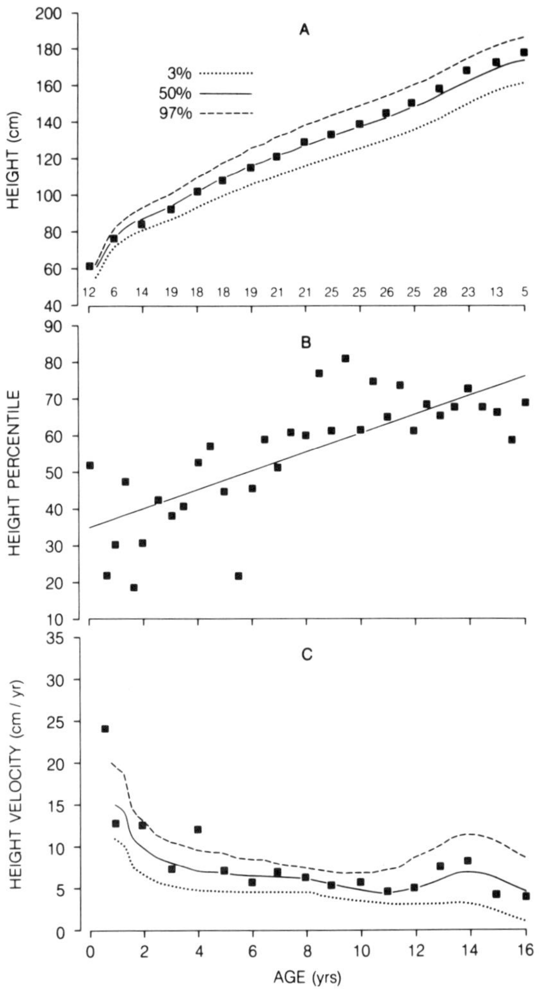

Fig. 1. Height (A), height percentile (B), and height velocity (C) of 47,XXY boys. Numbers above the X axis indicate the number of subjects at each age.

same as in these standards. For these boys, for ages 11 to 15 years, the relationships between measurements stayed fairly constant, although the measurements themselves became relatively larger. The main exceptions to this were weight and span. Weight increased to a greater degree than height. This was reflected in the triceps and subscapular skinfold measurements, which increased from less than the 3rd and 25th percentiles, respectively, at

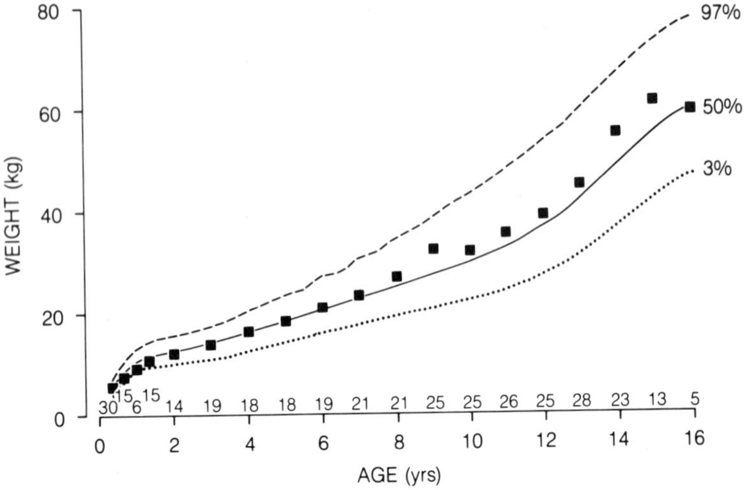

Fig. 2. Weight of 47,XXY boys.

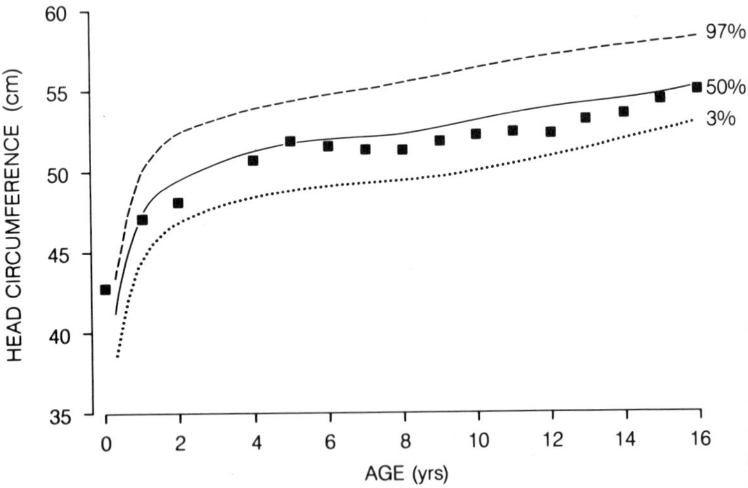

Fig. 3. Head circumference of 47,XXY boys.

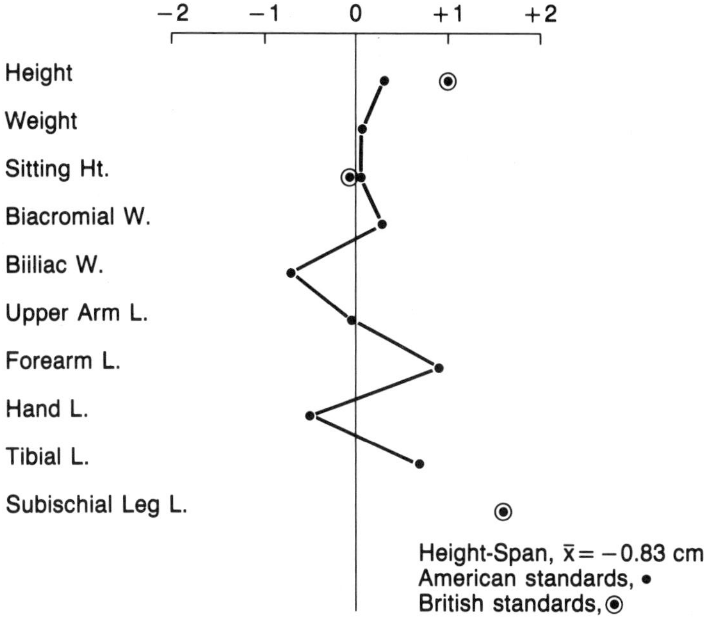

Fig. 4. Z scores for anthropometric measurements of 47,XXY boys aged 14 years.

age 11 to the 85th at age 15. At age 11 (N= 8) height minus span equalled 2.0 cm. This gradually reversed, so that by age 15 (N = 11) height minus span equalled −1.83 cm. In other words, the arms became relatively longer with age.

Sitting height was disproportionately short and tibial and subischial leg lengths disproportionately long. By the age of 5 years [1], well before puberty, and continuing through puberty, these boys had relatively long legs. Z scores for biacromial width and height were approximately the same at all ages with possibly a slight tendency for biacromial Z score to increase more with age. However, biiliac Z score was considerably less than height Z score or biacromial Z score, indicating a tendency toward narrow hips. Arm measurements revealed an average-sized upper arm, long forearm, and short hand.

Skeletal maturation. Subjects had hand-wrist X-rays taken annually. These were all rated by one author (E.P.) using two methods: Greulich-Pyle (GP; Fig. 5) [7] and Tanner-Whitehouse (RUS; Fig. 6) [8]. As expected, the Tanner ratings gave higher values, but comparisons of the two standards showed similar patterns with bone age below average in early childhood and

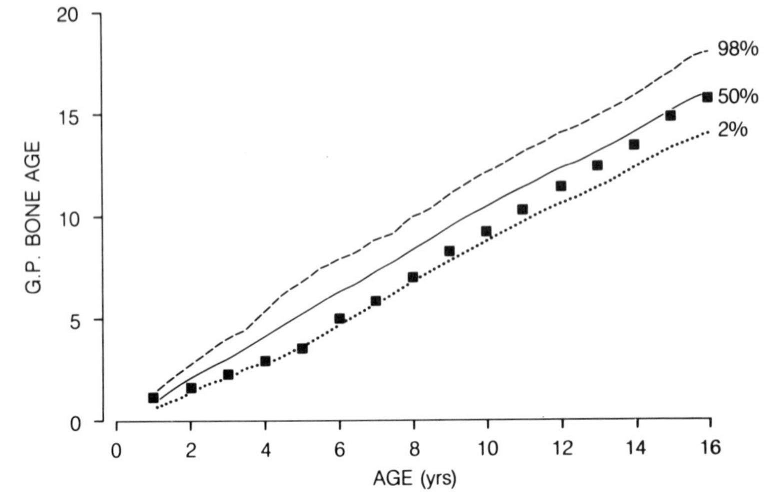

Fig. 5. Greulich-Pyle bone age of 47,XXY boys.

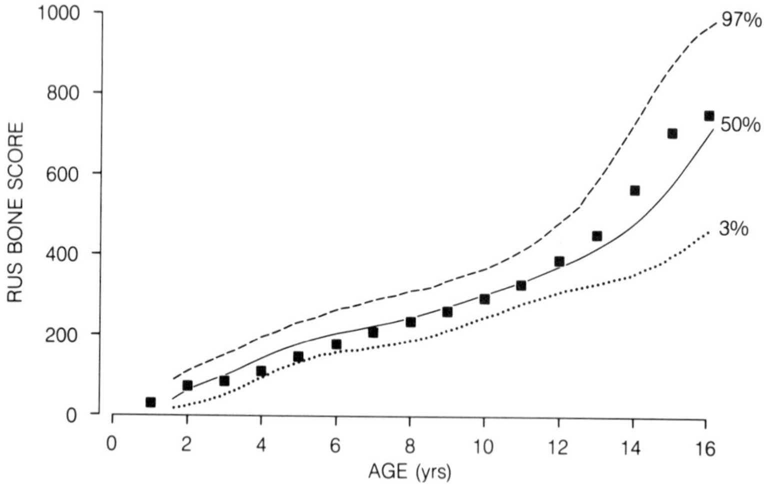

Fig. 6. Tanner-Whitehouse RUS bone score of 47,XXY boys.

then starting to rise relative to normal at ages 9 (GP) and 6 (RUS) years so that by age 14 bone age was near the 50th percentile by GP and well above the 50th by RUS standards.

47,XXX Girls

Sample. Sixteen subjects were identified at birth, of whom six aged 13–16 years have been seen annually.

Height. Figure 7A and B shows that height increased quite dramatically, from around the 40th percentile at age 0–2 years to the 60th at 4 years and the 80th or greater after 6 years. The regression of height percentile on age was highly significant ($r^2 = .663$, $P < .001$). Height velocity (Fig. 7C) was consistently above the 50th percentile until 12 years of age and near the 50th after that. The apparent flattening out of the pubertal growth spurt is an artefact of the cross-sectional handling of the longitudinal data. The average age at the start of the growth spurt was approximately 11 years (range 8.75–12.5). This was only a very rough estimate, because the girls were seen annually. Bone age (GP) was equal to chronological age at onset of the spurt. All six girls have experienced menarche (mean age 12.2 years; range 10.7–13.5).

Mean midparent height was at the 48th percentile. At the most recent visit, the mean height of subjects was at the 79th percentile. Although our girls tended to be tall, given parental height, this was not invariable. One of our subjects was consistently near the 30th, and the midparent height was at the 44th percentile.

Weight. Figure 8 shows weight near the 50th percentile until age 3 years rising to near the 90th after 5 years. It must be remembered that this is a very small sample; divergent values will have a great effect on the average depicted in the graph. This is the explanation for the apparent large drop in weight between ages 13 and 14. Of two girls presently 13 years old, one weighed 72 kg (height 171.3 cm) and the other was very obese at 100 kg (height 167.3 cm).

Head circumference. As can be seen in Figure 9, head circumference was below the 50th percentile although not significantly so, particularly after age 9 years.

Body proportions. Data on anthropometric measurements at ages 11–13 (see Fig. 10; age 13 years) showed the following similarities to 47,XXY boys: height above average, leg length accounting for a disproportionate amount of total height; biacromial width similar in Z score to height and considerably larger than biiliac width; and average upper arm length with long forearm and short hand. In contrast to 47,XXY boys, in whom span

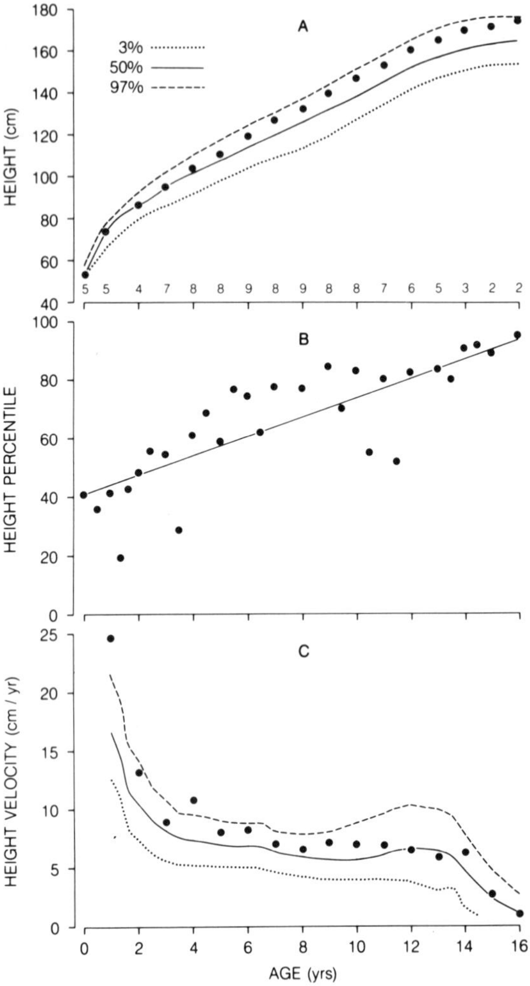

Fig. 7. Height (A), height percentile (B), and height velocity (c) of 47,XXX girls. Numbers above the X axis indicate the number of subjects at each age.

became greater than height, the height-span difference was 2.05 cm (N = 4) at age 11 and 5.22 cm (N = 5) at age 13 years. Also, whereas 47,XXY boys' skinfolds showed a relative increase with age, 47,XXX girls' skinfolds stayed fairly constant, on average in the 60–80% range.

Skeletal maturation. Both GP (Fig. 11) and RUS (Fig. 12) bone ages showed similar trends with bone age rising relative to normal at age 7 years,

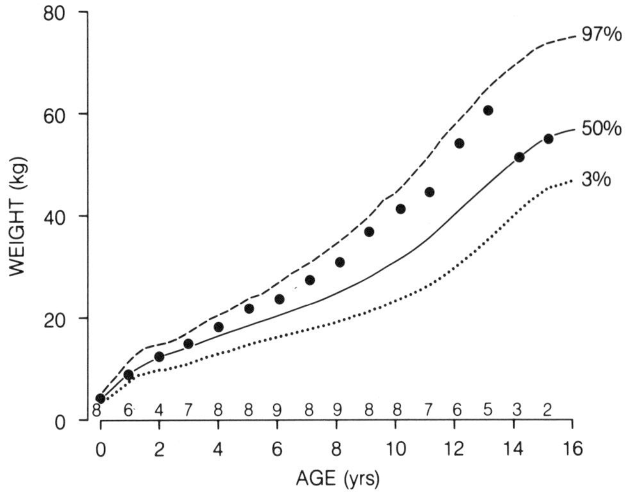

Fig. 8. Weight of 47,XXX girls.

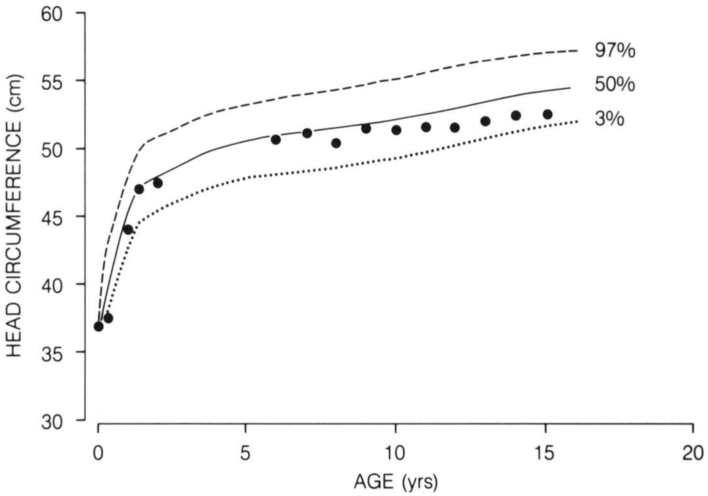

Fig. 9. Head circumference of 47,XXX girls.

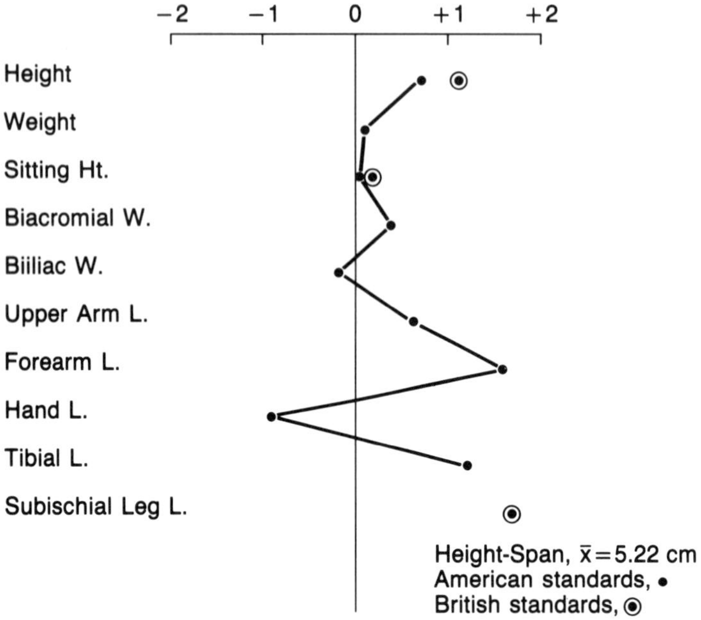

Fig. 10. Z scores for anthropometric measurements of 47,XXX girls aged 13 years.

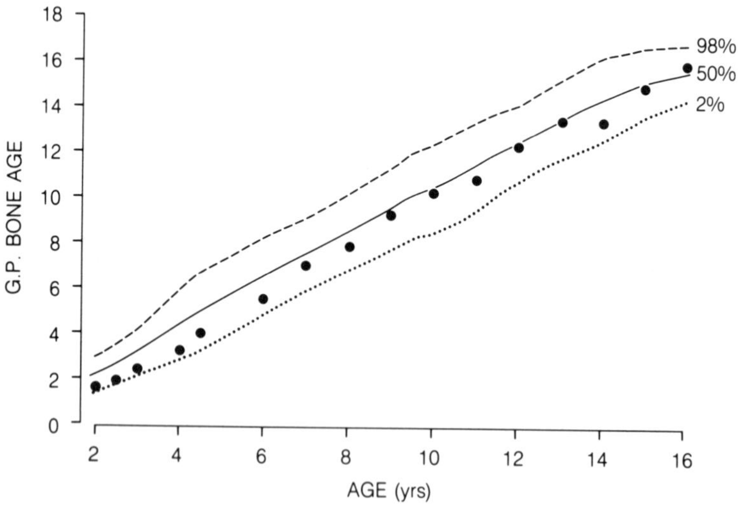

Fig. 11. Greulich-Pyle bone age of 47,XXX girls.

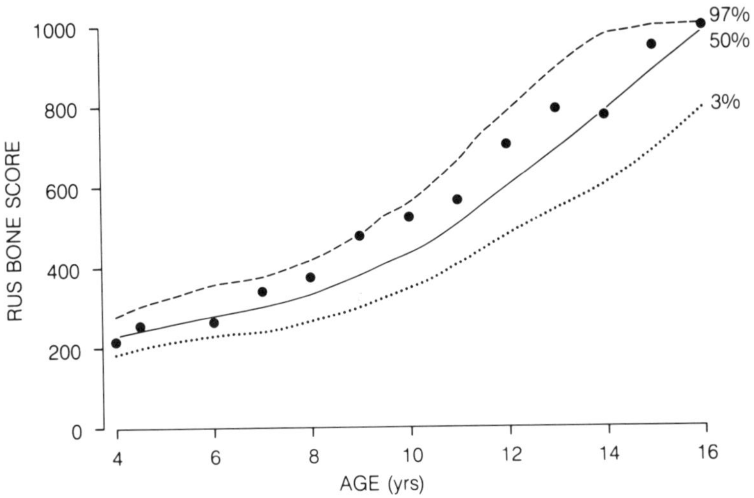

Fig. 12. Tanner-Whitehouse RUS bone scores of 47,XXX girls.

so that after 9 years bone age is near the 50th percentile GP and near the 75th RUS.

Mosaics and Miscellaneous Subjects

46,XY/47,XXY (N = 4). Three of these subjects, followed since birth, have displayed growth patterns with height percentiles increasing with age. The first boy's height was at the 25th until age 7 years and at the 50th by age 11. Parents' heights were at the 50th. The second boy's height rose from less than the 10th percentile in early childhood to the 25th at age 15. Mother's and father's heights were at the 15th and 25th percentiles, respectively. The boy reported as the third subject in our previous publication [2] has been lost to follow-up. We now have two subjects who have been reclassified from the 47,XXY to the 46,XY/47,XXY group. On repeat karyotyping, one of these boys was found to have 77% 47,XXY cells and 23% 46,XY cells. His height was at the 10th percentile in early childhood and followed along the 30th after age 7 years. Parent's heights were at the 80th and 10th percentiles. The fourth subject was diagnosed in the survey at birth as 47,XXY but was then lost to follow-up. He recently rejoined the study and was found on repeat peripheral blood karyotype to be 46,XY (42%)/47,XXY (58%). An early maturer, his height was above the 97th percentile at age 13 and at the 90th at age 15. Parents' heights were 10th and 25th. Weight was either at or below height percentile in these subjects. Bone age (GP) was equal to chronological age in the first two subjects, 1 year delayed in the third subject, and 1.5 years advanced in the fourth subject.

47,XYY (N = 2). Both boys at ages 12 and 14 years were growing at the 60th–70th percentile. Both were considerably overweight, and both had GP bone ages 1–1.5 years behind chronologic age.

46,XX males (N = 2). One subject at age 14 years was at the 75th percentile for height and 50th for weight. Parental heights were 50th and 60th. The other subject at age 15 was at the 40th percentile for height and the 90th for weight. Parents' heights were 75th percentile and above the 90th. Both boys had GP bone age equal to chronological age.

Deleted X Variants (N = 3)

45,X/46,XX_R (N = 1). This girl had been short from birth, being at the 3rd percentile at age 3 months and well below the 3rd at age 12 years. Weight had gone in the opposite direction, from the 10th percentile at age 3 months to the 50th at 12 years. GP bone age was 1 year delayed.

45,X/46,XY_{dic} (N = 1). Early data on this subject were missing, but by age 6 years she was very small with a height at the 2nd percentile and weight at the first. At age 13, height and weight were at the 1st and 6th percentiles, respectively. Bone age was delayed 2 years.

45,X/46,XX→46,XX (N = 1). This interesting girl, previously reported [2], was diagnosed at birth as a 45,X/46,XX mosaic, but subsequent lymphocyte testing at ages 2 and 6 years did not reveal an abnormal cell line. Her height decreased from the 60th percentile at age 1 year to the 2nd at age 13. Weight has decreased from the 25th to the 3rd. She was completely prepubertal at 13 years, and bone age was 2 years delayed.

ENDOCRINE AND PUBERTAL INVESTIGATION: RESPONSE TO TESTOSTERONE TREATMENT IN 47,XXY BOYS

Endocrine investigation of 28 boys aged 9–13 years with 47,XXY karyotype was reported previously [1,2]. The results indicated that these boys showed a normal onset of pubertal changes occurring at the expected chronological age. Signs of pubertal development proceeded in a normal fashion apart from small testicular size. Serum levels of gonadotropin, testosterone, and androstenedione were in the normal range for age and stage of pubertal development. LHRH stimulation testing done at yearly intervals from 9 to 11 years also showed normal prepubertal responses, with three subjects showing a pubertal LH response prior to the onset of definite physical signs of puberty.

Several investigators have advocated the use of exogenous testosterone in treatment of adult males with Klinefelter syndrome [9]. These reports showed that many adult males with 47,XXY karyotype show low normal or subnormal serum levels of testosterone and that exogenous testosterone treatment may improve vocational performance and social adjustment. A few reports of 47,XXY boys showing encouraging results in school performance and social adjustment after treatment with testosterone have been reported [10,11]. There has also been some indication that gynecomastia could be prevented or controlled with this treatment.

To assess possible effects of hormone therapy, alternate 47,XXY boys, on reaching their thirteenth birthdays, were offered testosterone injections. This was to be given in the form of long-acting testosterone cypionate (Depo-Testosterone; Upjohn, Kalamazoo, MI) using a dose of 100 mg intramuscularly every 4 weeks until age 15. The control group was made up of the alternate boys who did not receive the treatment. There were nine boys in the treatment group and 12 in the control group.

In both groups, physical assessment of growth and pubertal development, hormone assays, and personality assessment were recorded every 6 months. Bone age radiographs were repeated yearly. Follow-up information was obtained regarding school performance, and a note was made of behaviour or adjustment problems. Height was measured with the Harpenden stadiometer. Puberty was recorded by Tanner ratings [12] for pubic hair, measurement of stretched penile length (SPL), and measurement of testicular volume with the Prader orchidometer [13]. Presence of facial hair and acne was noted, and any breast development was measured by diameter in centimeters. Bone age and skeletal maturation X-rays were evaluated using the standard Greulich and Pyle (GP) as well as Tanner-Whitehouse (RUS) references. Psychologic testing was carried out using the Piers-Harris and Sentence Completion Self-Report tests [14]. Notation was made of school performance and behavioural problems. The Woodcock-Johnson [15] test of educational achievement was administered. Behavioural problems at home and school were noted. Endocrine studies included blood levels of gonadotropins (follicle-stimulating hormone [FSH] and luteinizing hormone [LH]) as well as testosterone, androstenedione, and estradiol. These were all measured by radioimmunoassay as previously reported. Serum levels of aspartate aminotransferase (SGOT) were measured as a test of liver function.

RESULTS

Growth and Skeletal Maturation

Since very few untreated subjects had reached the age of 15 years at the time of this report, results are presented mainly up until age 14. Results in

Table I show that at age 13 the control boys were by chance taller and heavier and had been growing and gaining weight slightly more rapidly during the previous 12 months than the boys who were about to receive treatment. More of the control boys had begun their pubertal growth spurt by age 13. During the first and the second 6 months of treatment, the treated group grew significantly faster and gained weight more rapidly than the control group. Most of the treated group reached their peak of pubertal growth within the first 6 months of treatment, and all had reached it after 12 months. In the control group, on the other hand, the timing of the growth spurt was more spread out. Some boys had not reached their maximum growth rate by 14 years. Table II shows that at age 13 bone age in the two groups was identical. During the first year of treatment, bone age advanced more rapidly, but not significantly so, in the treated group. Predicted adult height did not change in either group beween ages 13 and 14.

TABLE I. Effect of Testosterone Treatment on Height and Weight in Treated (T) and Untreated (U) 47,XXY Boys

Age	Group (N)	Height (cm)	Height velocity (cm/year)	Weight (kg)	Weight velocity (kg/year)
13	U (10)	160.1 ± 7.99	7.4 ± 2.00	46.9 ± 10.60	4.1 ± 2.97
	T (09)	156.6 ± 5.74	6.9 ± 2.21	44.3 ± 5.76	3.5 ± 2.36
13.5	U (09)	162.3 ± 7.73	5.9 ± 1.52	49.6 ± 11.20	5.7 ± 5.47
	T (08)	161.0 ± 5.08	10.4 ± 2.01†	49.6 ± 6.56	9.9 ± 4.81*
14	U (10)	166.8 ± 8.23	7.2 ± 2.90	52.0 ± 10.50	5.2 ± 2.21
	T (09)	166.8 ± 4.59	9.9 ± 2.06*	54.3 ± 7.60	8.7 ± 5.62

*P < .05.
†P < .01.

TABLE II. Effect of Testosterone Treatment on Bone Age in Treated (T) and Untreated (U) 47,XXY Boys

Age	Group (N)	Greulich-Pyle bone age (years)	Bone age velocity (BA years/year)	Bayley-Pinneau predicted adult height (cm)
13	U (10)	12.6 ± 1.32	1.01 ± 0.346	184.5 ± 3.47
	T (09)	12.6 ± 0.69	1.00 ± 0.256	181.8 ± 7.25
14	U (10)	13.5 ± 1.33	0.98 ± 0.445	184.3 ± 5.70
	T (09)	13.8 ± 0.61	1.31 ± 0.490	181.8 ± 6.51

Pubertal Development

Pubic hair. Table III shows that pubic hair growth advanced more rapidly in the group treated with testosterone. By age 15 years, the untreated boys were catching up but the treated boys were still more advanced.

Stretched penile length. The mean stretched penile lengths (SPL; Table IV, Fig. 13) of the two groups were similar at the age of 13 years. The mean SPL of about 7.2 cm in the group selected for treatment and 7.0 cm in the control group was greater than the mean of 6.4 cm for prepubertal boys aged 10–11 years based on standards of Schonfeld and Beebe [16]; four of the nine treated boys showed an SPL of 8.0 cm or more prior to onset of treatment, indicating that pubertal changes were already occurring. During the first year of testosterone treatment, the mean SPL of the treated group (10.5 cm) showed an increase over controls (8.6 cm), and the difference became greater by age 15.

Testicular volume. Table V shows mean testicular volume compared to bone age of the two groups. There is no significant variation between the groups at any age. Prior to onset of treatment at age 13 years, six of the nine boys showed a testicular volume of 3 ml each or greater. One had a unilateral testis of 4 ml. In the control group, six of the 12 had testis size of 3 ml or more. None of the five in this group showed a testis size less than 2 ml.

Mean testicular volume at 15 years was slightly greater in the treated group than in the control group. This indicates that at least the testicular development was not inhibited by the testosterone treatment. All of the seven boys who had been treated for 2 years had slight but definite testicular

TABLE III. Pubic Hair Ratings in Treated (T) and Untreated (U) 47,XXY Males

Stage	No. 13 year T—9 U—12	No. 14 year T—8 U—12	No. 15 year T—7 U—4
I	T—2 U—4	T—0 U—1	T—0 U—0
II	T—6 U—3	T—0 U—1	T—0 U—0
III	T—1 U—3	T—4 U—7	T—0 U—1
IV	T—0 U—2	T—4 U—3	T—5 U—3
V	T—0 U—0	T—0 U—0	T—2 U—0

TABLE IV. Bone Age (BA) and Stretched Penile Length of Treated (T) and Untreated (U) 47,XXY Males

	BA (years)		Stretched penile length (cm)	
Age	U	T	U (SD)	T (SD)
13	12.5	12.6	7.0 (2.3) N = 12	7.2 (1.2) N = 9
14	13.6	13.8	8.6 (2.1) N = 12	10.5 (1.3) N = 8
15	14.6	15.0	8.6 (1.9) N = 4	11.0† (1.4)* N = 7
		post-T		
16	13.75–18.0	15.4–16.0	8.0–12.0	10.0–11.0

*P < .05.
†P < .01.

enlargement. The unilateral testis had increased to 6 ml, and there were no testes less than 3.5 ml in volume. In the control group there were also slight increases in testicular size in most boys. In both groups, there was a peak testicular volume reached by ages 13.5–14.5 years, but there was no significant change after that age. In a few instances, there appeared to be a slight decrease in testicular volume. It is possible that some had actually undergone contraction of testicular volume, since volumes were slightly greater than those usually reported in men with 47,XXY karyotype [17].

Gonadotropins

Follicle-stimulating hormone (FSH). The mean FSH was elevated (above 14 U/liter) in both groups at age 13 years, and over the next 2 years it tended to rise further (Table VI, Fig. 14). The increase in the treated group mean to 43.1 U/liter was not significantly different than the untreated group mean of 33.2 U/liter at age 15 years, indicating that the testosterone treatment did not suppress FSH secretion. However, there were individual variations, and in one treated boy FSH remained at 6–8 U/liter until age 15, with a sharp rise to 25 U/liter 6 months after treatment was discontinued and a further increase to 68 U/liter at age 16 years. One other boy showed fluctuations between 9 and 12 U/liter during the treatment period. In the control group, six of the 12 showed prepubertal FSH levels of less than 7 U/liter at age 13, and five had serum levels above the upper adult limit of 14 U/liter. At age 15, one of the

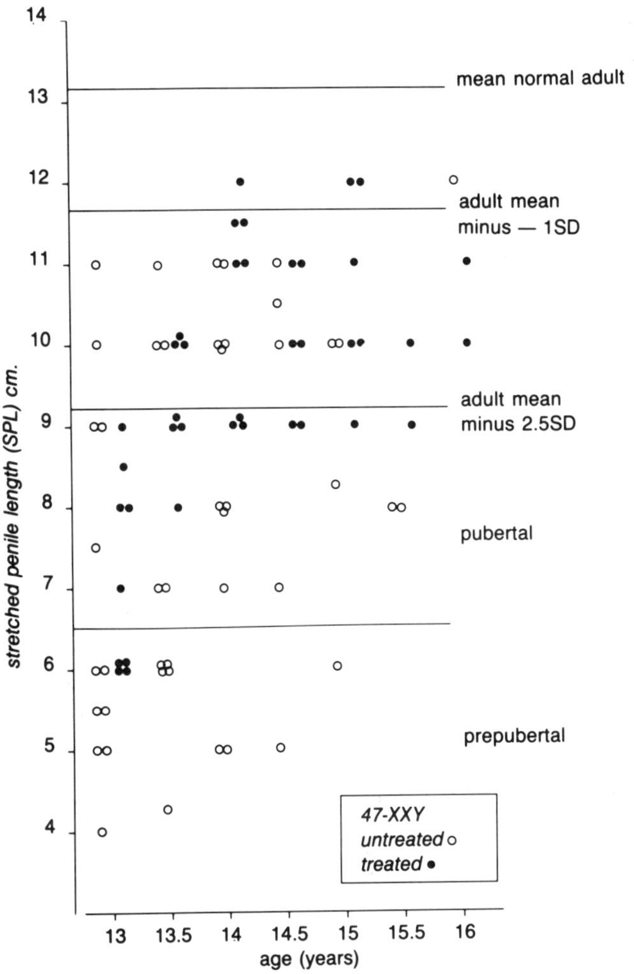

Fig. 13. Stretched penile length in untreated and testosterone-treated 47,XXY boys aged 13–16 years. Normal male standards derived from Schonfeld and Beebe [16].

boys with delayed onset of puberty showed a normal adult FSH of 12 U/liter, and the other three had over 30 U/liter.

Luteinizing hormone (LH). At age 13 years, one boy in the treated group and two in the control group had a serum LH above the normal adult upper limit of 20 U/liter (Table VI, Fig. 15). One in the control group

TABLE V. Bone Age (BA) and Testis Size in Treated (T) and Untreated (U) 47,XXY Males

	BA (years)		Testes (ml)			
			U (SD)		T (SD)	
Age	U (SD)	T (SD)	R	L	R	L
13	12.51 (1.25) N = 11	12.61 (0.7) N = 9	3.29 (1.48) N = 12	3.17 (1.34) N = 12	3.06 (1.05) N = 8	3.11 (0.99) N = 9
14	13.65 (1.27) N = 11	13.84 (0.6) N = 9	4.06 (1.56) N = 12	3.98 (1.85) N = 12	3.93 (1.69) N = 7	3.69 (1.22) N = 8
15	14.6 (2.1) N = 3	15.0 (0.97) N = 7	3.5 (1.0) N = 4	3.5 (1.0) N = 4	4.25 (1.08) N ; 7	4.29 (1.15) N = 7
	post-T					
16	13.75–18.0	15.4–16.0	4.5–6.0	4.5–6.0	5.0–4.0	5.0–4.0

TABLE VI. Bone Age (BA) and FSH and LH Hormone Levels in Treated (T) and Untreated (U) 47,XXY Males (4 weeks after injection)

	BA (years)		FSH (IU/liter)		LH (IU/liter)	
Age	U	T	U (SD)	T (SD)	U (SD)	T (SD)
13	12.5	12.6	22.8 (25.5) N = 12	17.4 (13.2) N = 8	16.1 (27.7) N = 12	12.6 (12.3) N = 9
14	13.6	13.8	33.2 (36.2) N = 9	25.1 (20.1) N = 9	13.2 (12.7) N = 10	10.3 (8.4) N = 9
15	14.6	15	33.2 (17.2) N = 4	43.1 (34.3) N = 7	12.7 (7.09) N = 4	21.4 (18.9) N = 7
	post-T					
16	13.75–18.0	15.4–16.0	55–71	63–68	16–25	17–18

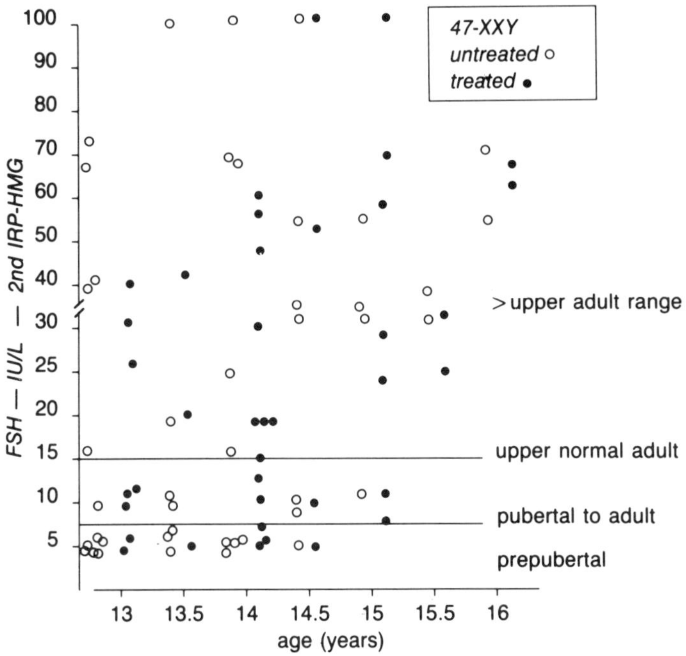

Fig. 14. Serum FSH concentrations in untreated and testosterone-treated 47,XXY boys aged 13–16 years.

showed a very high LH of 101 U/liter, and this was associated with a high FSH level.

During treatment, LH serum levels fluctuated but showed no definite pattern of change. In one of seven treated boys the LH rose at age 15 years to 60 U/liter, but another, whose level was elevated to 43 U/liter at age 13, showed a fall to 26 U/liter. At age 15, only one assay was over 30 U/liter, and only three were above 20. The control group showed similar variability of LH measurements. The boy whose level had been 101 U/liter at age 13 fell to 19 at 15 years, and none of the four showed serum levels above 20 U/liter.

In summary, FSH became elevated between early and midpuberty with significant increases in most boys between 13 and 15 years. In the majority of cases, there was no indication that testosterone injections given in a dose of 100 mg IM every 4 weeks altered this pattern. LH serum levels tended to be closer to the normal range in both groups, with a few subjects showing elevated levels. These did not increase significantly from mid- to late puberty.

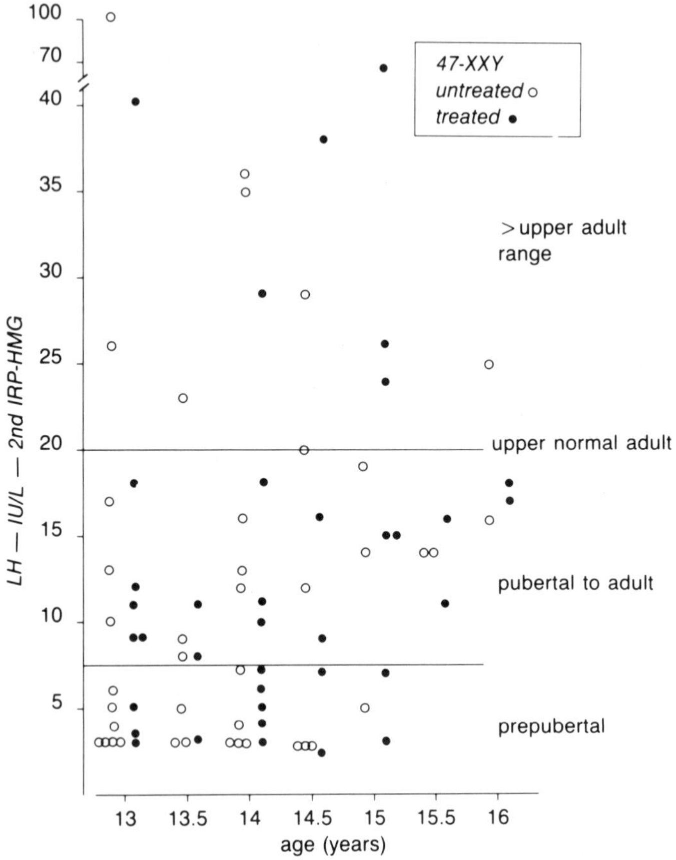

Fig. 15. Serum LH concentrations in untreated and testosterone-treated 47,XXY boys aged 13–16 years.

In fact, in some cases there was a tendency for lower values as adult sexual development was achieved.

Sex Steroids

Testosterone. Serum levels of testosterone were comparable in the two groups at the age of 13 years with a mean of 4.2 nmol/liter in the untreated and 6.35 nmol/liter in the to-be-treated group (Table VII, Fig. 16). At 13 years, three of the boys in the untreated group and one in the treated group had serum testosterone levels in the prepubertal range of less than 1.0 nmol/liter. Only one of the boys in the control group and three in the treated group

TABLE VII. Bone Age (BA) and Hormone Levels in Treated (T) and Untreated (U) 47,XXY Males (4 weeks after injection)

Age	BA (years)		Testosterone (nmol/liter)		Androstenedione (nmol/liter)		Estradiol (pmol/liter)	
	U	T	U (SD)	T (SD)	U (SD)	T (SD)	U (SD)	T (SD)
13	12.6	12.6	4.18	6.35	1.70	2.5	159.1	159.6
	N = 10	N = 9	(4.83)	(6.13)	(1.02)	(1.3)	(57.6)	(54.7)
			N = 12	N = 8	N = 11	N = 7	N = 10	N = 8
14	13.6	13.8	7.17	9.8*	1.18	2.16*	150.75	160.7
	N = 10	N = 9	(3.34)	(4.15)	(0.39)	(0.53)	(57.30)	(69.2)
			N = 9	N = 7	N = 9	N = 8	N = 9	N = 8
15	14.6	15.0	11.23	11.6	1.3	2.46	229.3	191.3
	N = 4	N = 7	(6.06)	(4.87)	(1.00)	(2.60)	(123.8)	(47.2)
			N = 4	N = 7	N = 3	N = 5	N = 3	N = 6
	post Treatment period							
	13.75	15.4	6.6	7.0	1.1	1.6	178	136
16	18.0	16.0	18.6	12.1	1.6	1.5	0	131
	N=2	N=2	N=2	N=2	N=2	N=8	N=1	N=2

*P < .05.

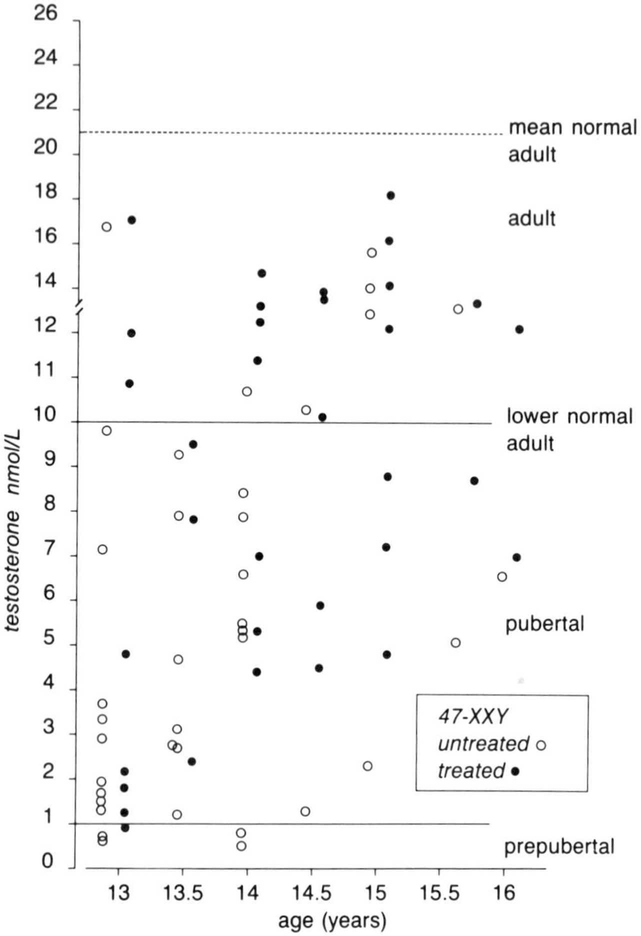

Fig. 16. Serum testosterone concentrations in untreated and testosterone-treated 47,XXY boys aged 13–16 years.

showed serum testosterone levels greater than 10.0 nmol/liter, which is the lower limit of the normal adult range. During the first year of treatment, the boys in the treated group showed a significantly greater increase in mean testosterone level than those in the control group, but at 15 the seven subjects in the treated group showed about the same levels as the four untreated boys. Since serum levels were taken at 4 weeks after an IM testosterone injection, the serum concentrations in the treated group represented endogenous rather than exogenous testosterone.

Androstenedione. Mean serum levels of androstenedione were slightly higher at age 13 years in the treated than the untreated group, and at age 14 this difference was statistically significant (Table VII). The mean serum concentration of androstenedione showed only minor fluctuations throughout the 2 years of observation. Apart from one isolated value all subjects showed serum androstenedione levels consistent with their stage of pubertal development, indicating a normal adrenarche.

Estradiol. Serum concentrations of estradiol showed considerable variation, but there were no significant differences between mean estradiol concentrations of the treated and untreated subjects (Table VII). Some boys in each group showed serum levels that were above the usually accepted upper limit for adult males, but this was not a consistent finding when rechecked on subsequent assays.

Other Observations

Gynecomastia. Four of the untreated boys had significant gynecomastia. One of these had large breasts at 13 years and required surgery. Another boy showed progressive gynecomastia between 13 and 14 years. He was removed from the control group and put on testosterone injections in an attempt to suppress the breast development, but the enlargement continued from ages 14 to 15. One boy in the treated group showed prominent nipple areas from age 13, but there was no progression. Only one boy had developed gynecomastia during treatment. He may require surgery in the future.

There was no indication that testosterone treatment enhanced the development of gynecomastia. Although the observations to date suggest that testosterone treatment suppresses the progression of gynecomastia in some cases, our small numbers do not confirm such a conclusion.

Acne. One of each group developed severe acne between ages 14 and 15 years. Mild acne was more common in the untreated subjects. The incidence and severity of acne did not vary from that seen in normal adolescents.

Facial hair. At 13 years, early beard growth was present in only two boys in the untreated group, and one in the treated group showed moderate facial hair development. Progression of beard growth was slow between 13 and 15 years with no difference in the two groups.

Adverse or Toxic Effects

None of the boys showed recognizable side effects from the testosterone. Most complained about temporary discomfort from the injections. Serum levels of aspartate aminotransferase (SGOT) remained within normal limits in all subjects.

Behaviour

The pretreatment assessment results using the self-report measures at age 13 and those obtained at age 14 for treated and control cases are shown in Table VIII. Only two comparisons indicative of possible treatment effects reached statistical significance. These were 1) increase in freedom from anxiety and 2) more negative attitudes towards parents in the treated group than controls. The remaining comparisons showed no behavioural changes attributable to the treatment.

The Woodcock-Johnson [15] test results for reading and mathematics at mean ages of 14.8 years for controls and 15.6 years for treated cases are shown in Table IX. There was no significant difference in the two groups, indicating that 1.6 years of treatment was not associated with any improvement in educational skills.

Case reviews did not indicate any tendencies towards alterations in the social and emotional functioning of either treated or control cases. Among controls, one boy showed gains in school performance and peer relationships,

TABLE VIII. Scores on Self-Concept Tests of Treated (T) and Untreated (U) 47,XXY Subjects at Ages 13 and 14 Years

	Pre-		Post-		
	T (N = 9)	U (N = 11)	T (N = 9)	U (N = 11)	Factors significance*
I. Piers Harris					
Percentile	43.2	52.1	58.3	57.5	T‡
Behaviour	11.7	12.6	13.4	13.7	T†
Intellectual	7.3	7.9	9.1	7.6	
Physical appearance	6.8	6.5	7.0	7.7	
Lack of anxiety	4.9	6.2	7.1	6.0	T‡, GT†
Popularity	5.4	6.5	7.0	7.3	
Happiness	6.0	7.1	6.9	7.3	
II. Sentence completion					
Self-concept	15.7	18.4	15.2	18.9	G‡
Attitude towards parents	19.3	19.0	17.0	19.4	GT†
Attitude towards peers	20.4	20.6	18.9	21.6	
Achievement	17.8	19.7	17.6	19.9	G†
Learning attitude	19.9	20.2	15.8	19.9	
Body image	18.4	19.2	18.0	18.6	

*G, group; T, time; GT, group × time.
†P < .05.
‡P < .01.
Conclusion: No major differences between treated and control cases in most scales of self-concept tests. Differences restricted to lowered anxiety and more negative attitudes towards parents in series.

TABLE IX. Woodcock-Johnson Psycho-educational Test Results at Follow-up for Treated (T) and Untreated (U) 47,XXY Boys

Grade	U (N = 6)	T (N = 11)
Reading	6.41	4.26
Knowledge	5.41	4.86
Math	5.79	4.76

whereas two others deteriorated in these respects. In the treated group, one boy improved in his adjustment, another became delinquent and involved in criminal activities. These results indicate that treatment with testosterone 100 mg IM every 4 weeks did not produce major changes or improvements in self-perception, school performance, or psychosocial functioning.

Summary of Results

Most males with 47,XXY karyotype showed normal progression of puberty from 13 to 15 years of age. Testicular size increased slowly during this period and remained well below the mean normal for age. FSH rose steadily to abnormally high levels as puberty progressed. Abnormally high levels of LH were less consistent. Testosterone rose to low normal adult levels and remained at this low level, consistent with those reported for 47,XXY adults.

Treatment with testosterone cypionate 100 mg IM every 4 weeks caused significant change in initiating puberty in those boys whose onset of puberty was delayed beyond age 13 years. Testosterone treatment did not produce measurable changes in behaviour or school performance apart from any that might be related to pubertal development. Significant gynecomastia was encountered in only four boys in our control group and one in the treated group.

DISCUSSION

The failure of testosterone to produce significant change in the treated subjects compared to those who were not treated could be interpreted in at least three ways. There might have been no effect at all, or the dosage and interval between injections was inappropriate, or the timing in relation to pubertal development was incorrect.

The immediate conclusion might be that treatment with exogenous testosterone is not beneficial for treatment of teenage boys with 47,XXY karyotype. The lack of significant differences in the treated group in regard to secondary sexual development, hormone concentrations, or behavioural improvements compared to controls would justify such a conclusion, but there are valid reasons for considering the other possibilities.

Failure to show significant change with testosterone administration could be related to inadequacy of the treatment. The replacement of testosterone in physiologic amounts is difficult at any age, and this is especially true of the period of adolescence. An empirical dose of 100 mg every 4 weeks was chosen for pragmatic reasons. Ideally, one would use smaller doses given more frequently at the start of treatment with increases to produce levels similar to what occurs in normal puberty. The time of starting treatment is also open to question. Most 46,XY boys show signs of pubertal development with coincidental increases in gonadotropins and androgens by the age of 13 years. There is considerable variation in the stage of puberty achieved in both treated and untreated groups. By the age of 13 years, a few of the subjects had advanced to midpuberty with abnormal elevation of FSH, but a larger number were still prepubertal. Perhaps the best time to start therapy would be after a boy has shown an initial spontaneous onset of puberty and before he has developed elevated serum FSH levels.

Regardless of the time of starting testosterone treatment or the dosage used, the achievement of a continuous physiologic level of serum testosterone remains elusive. IM injections of a long-acting testosterone preparation given every 2 weeks would certainly provide a more even testosterone level, but there would still be peaks and valleys from the first to the second weeks. There is also a very practical problem in persuading acceptance of IM injections on a 2-week basis. An oral preparation of testosterone undecanoate has been developed to circumvent the disadvantages and complications associated with other forms of oral testosterone therapy [18], but there have been very few long-term studies on the use of this agent in children. There appears to be a variability in absorption and consequently serum levels of testosterone achieved by this form of treatment. An oral preparation would be desirable, although this poses the additional problem of long-term compliance. Further studies will be required before such treatment could be recommended if testosterone replacement or supplementation can indeed be shown to be indicated.

It is quite possible that testosterone treatment can benefit some but not all individuals with 47,XXY karyotype. At least one of the treated boys in our group showed real behaviour and personality improvement, but this was certainly not a frequent observation. There is a wide variation in serum testosterone in adult males with 47,XXY karyotype, with most having levels about 50% those in normal men. In spite of elevated gonadotropins, 47,XXY subjects show a response to stimulation with human chorionic gonadotropin

(HCG). There may be good reason to give exogenous testosterone to those with low basal levels who show inadequate response to an HCG test [19]. Probably, young men with 47,XXY karyotype should have an HCG response test at the age of 17 years [19]. Those who show high serum FSH concentrations with relatively low levels of serum testosterone and with an inadequate response to HCG stimulation probably should be given exogenous testosterone therapy at an interval and dosage that will keep their serum concentration within the midrange for a normal adult male. If such treatment does not result in improvement in physical or behavioural characteristics compared to an untreated group, there may be need for reconsideration of the value of testosterone treatment in males with Klinefelter syndrome.

During their early teen years, boys with 47,XXY karyotype show normal physical signs of pubertal development apart from the failure of normal testicular enlargement. As puberty progresses, serum concentrations of gonadotropins and testosterone gradually reach those noted in adult males with Klinefelter syndrome.

Treatment with IM Depo-Testosterone 100 mg every 4 weeks does not produce beneficial effects of 47,XXY boys from 13 to 15 years of age. Treatment with more physiologic doses of testosterone given at more frequent intervals might produce beneficial effects on psychosocial adaptation, academic and occupational performance, prevention of gynecomastia, and psychosexual adjustment in adolescent 47,XXY boys, but this can be evaluated only by further investigation.

Mosaic Boys

Four boys with 47,XXY/46,XY mosaicism have been followed in the Toronto study (Tables X and XI). One of these boys was started in the testosterone treatment trial before repeat karyotype revealed that 77% of the cells were 47,XXY and 23% were 46,XY. He continued receiving testosterone until age 15 but was not included in the report of the trial.

It is apparent from our studies that the presence of a normal 46,XY cell line has a significant effect in preventing the abnormal rise in gonadotropins and allowing larger testicular growth than in boys with all 47,XXY cells. In the one boy with 90% XY cells, gonadotropins, testosterone, and gonad size were at the normal adult levels by age 15. The boy with 42% XY cells had early pubertal development, and at age 15 his testicular volume was 12 ml, which is in the lower range for normal adult males. The effect of mosaicism on testicular function will not be clear until secondary sexual development is complete and seminal fluid examination can be carried out.

TABLE X. Bone Age (BA), Gonadotropins, and Sex Steroid Levels in Treated (T) and Untreated (U) 47,XXY/46,XY*

	Age 13					
Karyotype	BA (years)	FSH (IU/liter)	LH (IU/liter)	T (nmol/liter)	A (nmol/liter)	E (pmol/liter)
47,XXY untreated	12.5	22.8	16.1	4.2	1.7	159
1 7% 47,XXY 93% 46,XY	12.8	4	3	2.7	0.8	178
2 10% 47,XXY 90% 46,XY	12.0	3	<3	1.3	0.3	242
3 58% 47,XXY 42% 46,XY	14.0	12	4	11.2	1.4	275
4 77% 47,XXY 23% 46,XY	11.0	5	<3	1.2	0.17	117.5

	Age 15					
Karyotype	BA (years)	FSH (IU/liter)	LH (IU/liter)	T (nmol/liter)	A (nmol/liter)	E (pmol/liter)
47,XXY untreated	14.6	33.2	12.7	11.2	1.3	229
1 7% 47,XXY 93% 46,XY						
2 10% 47,XXY 90% 46,XY	14.0	4	<3	13.0	1.9	180
3 58% 47,XXY 42% 46,XY	16.5	13	6	27.0	3.5	293
4 77% 47,XXY 23% 46,XY	13	6	<3	14.7	—	268

*FSH, follicle-stimulating hormone (2nd IRP-HMG); LH, luteinizing hormone (2nd IRP-HMG); T, testosterone; A, androstenedione; E, estradiol.

The difference in pubertal development in mosaic boys emphasizes that every boy with 47,XXY karyotype should have repeat chromosome studies if he shows normal testicular growth while onset of puberty is delayed and FSH remains in the preadolescent range.

46,XX Males

Pubertal development was assessed in two boys with 46,XX karyotype. Both boys showed patterns similar to those with 47,XXY karyotype, having small testes but normal progression of penile and pubic hair growth. Abnor-

TABLE XI. 47,XXY/46,XY Mosaic Males Compared to 47,XXY Males in Relation to Bone Age (BA), Stretched Penile Length (SPL), Pubic Hair Ratings (PH), and Testis Size

	Age 13					Age 15				
				Testes (ml)					Testes (ml)	
Karyotype	BA (years)	SPL (cm)	PH	R	L	BA (years)	SPL (cm)	PH	R	L
47,XXY (mean)	12.5 ±1.25	7.0 +2.3	2+ 2+	3.3 ±1.5	3.2 ±1.3	14.6 (2.1)	8.6 ±1.9	3+	3.5 ±1.0	3.5 ±1.0
7% 47,XXY* 93% 46,XY	12.8	8	2	2.5	2.5					
10% 47,XXY 90% 46,XY	12.0	6	2	3.5	3.5	14.0	11	4	15	15
58% 47,XXY 42% 46,XY	14.0	10	4	8.0	8.0	16.5	12	4	12	12
77% 47,XXY 23% 46,XY	11.0	7.5	1	3.0	3.0	13	13	?	6	6

*This boy was one of the youngest and data are not available at age 15.

mally high levels of FSH were reached at age 13–14 years. One boy was started on Depo-Testosterone at age 14.5 years but has not been followed regularly in our study.

47,XXX Girls

Seven 47,XXX girls have been followed from 10 to 15 years of age. Table XII shows that all developed pubertal changes in breasts and pubic hair at appropriate ages. Menarche occurred at 10.5 to 13.5 years in all these girls. There was no history of metrorrhagia or any significant degree of dysmenorrhea or menstrual irregularity. Serum concentrations of FSH, LH, estradiol, testosterone, and androstenedione were appropriate for age and stage of pubertal development, as is shown in Figures 17–20.

PSYCHOLOGIC DEVELOPMENT

This section presents the results of behavioural investigations and analyses conducted on our X and Y aneuploid children since 1981. They fall into two broad classes. The first section deals with the results of assessments designed to provide descriptively useful information in such areas as intelligence, school performance, and adjustment in each aneuploid group. The second is concerned with identifying the biologic and environmental mechanisms responsible for the specific behavioural deficits and anomalies found in some aneuploid conditions. Each type of study is presented below under an appropriate heading.

Descriptive Studies

The data to be reported in this section were obtained using standardized psychological tests, structured and unstructured interviews, and independent ratings of case behaviours and histories. (These were done by two investigators, both well acquainted with each case.) Much of this information was collected through regular examinations of probands conducted annually in the Hospital. Whenever possible, we identified a same-sexed sib as a control for each proband and attempted to follow the same examination protocol for both case and control. This was not always possible, however, and in some cases probands and controls were discordant for sex; in others, no control sib of either sex was available.

47,XXX

Cognitive ability. The WISC-R was used to assess intellectual functioning of both cases and controls. Data on six 47,XXX girls and four sib controls

TABLE XII. Breast and Pubic Hair Development in 47,XXX Females

Case	Breast stage at age						Pubic hair stage at age						
No.	10	11	12	13	14	15	10	11	12	13	14	15	Menarche
1			4	5					4	5			10.5
2	2	4	5	5			2	3	4	4			10.7
3				4						5			11.2
4	1	1		3	4		1	1		3	5		12.0
5	1	2	4	4	4^+		1	1	4	4	5		13.0
6	1	1		3	5		1	1		3	5		13.5
7	1	1	2	3	4	5	1	1	2	3	3	4	13.5
Mean													12.1

TABLE XIII. WISC-R Test Results for Six 47,XXX Girls and Four Sib Controls

	47,XXX (N = 6)		Controls (N = 4; three females, one male)		
	Mean	SD	Mean	SD	P
Information	7.50	1.37	11.67	1.53	<.003
Similarities	7.83	1.72	13.50	1.73	<.002
Arithmetic	8.67	1.63	12.00	1.00	<.005
Vocabulary	7.67	0.52	12.00	1.63	<.001
Comprehension	8.00	1.67	14.00	1.00	<.001
Digit span	9.00	3.41	10.75	1.89	>.10
Picture completion	9.83	1.33	14.00	3.55	<.06
Picture arrangement	10.33	1.75	12.00	1.73	>.10
Block design	11.17	2.40	15.25	2.98	<.06
Object assembly	10.33	2.65	13.00	0.81	<.05
Coding	13.50	3.39	15.25	4.11	>.10
Verbal IQ–performance IQ	−19.67	15.49	−9.00	5.47	>.20
Verbal IQ	87.00	6.03	116.00	2.31	<.01
Performance IQ	106.67	13.06	125.00	7.70	<.06
Full-scale IQ	95.50	5.54	122.75	4.99	<.003
Age	13.83	1.33	11.50	1.29	.024

(three females, one male) were available for the period from May, 1981, to May, 1984. The results are presented in Table XIII and indicate that the extra X girls had greater deficits in verbal abilities than in performance abilities vs sibling controls.

Educational skills. These were assessed routinely using the Wide Range Achievement Test [20]. More recently, the Woodcock-Johnson Psycho-Educational Battery [15] was added to the assessment protocol. The latest WRAT results for the six 47,XXX girls and three sib controls are presented in Table XIV and the Woodcock-Johnson results in Table XV. The WRAT results indicate that the extra X girls had significantly lower scores in reading and arithmetic than their sibs, findings consistent with their generally below average Woodcock-Johnson results. At present, two of the six probands and none of the four sib controls are receiving special educational programming.

Emotional adjustment. Personality and psychosocial adjustment were evaluated in several ways. Parents of probands completed the Child Behaviour Checklist (CBCL) [21], the Personality Inventory for Children (PIC) [22], and the Middle Child Temperament Scale [23]. Ratings of the behavioural characteristics of each case were also made independently by two investigators involved with some part of the assessment programme (C.T.N.

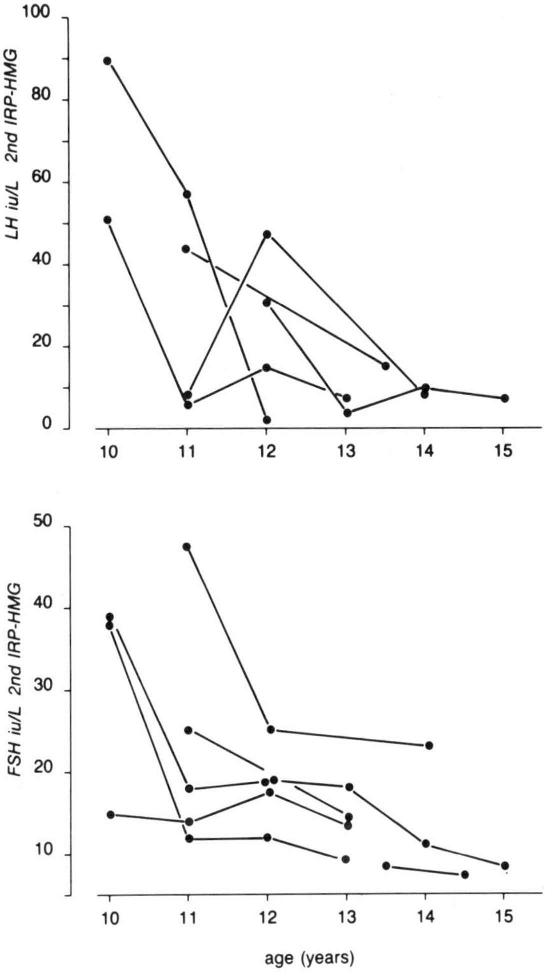

Fig. 17. FSH and LH concentrations in single random serum samples of 47,XXX girls aged 10–15 years.

and J. Pinnonseault, research assistant). These ratings were made on five-point scales for three behavioural characteristics: passivity-activity, social relationships, and activity level. Ratings of the quality of parenting were also made by two investigators well acquainted with the social and emotional functioning of each family (C.T.N. and D.A.S.). Finally, one investigator (C.T.N.) reviewed each case and rated each girl's level of adjustment or

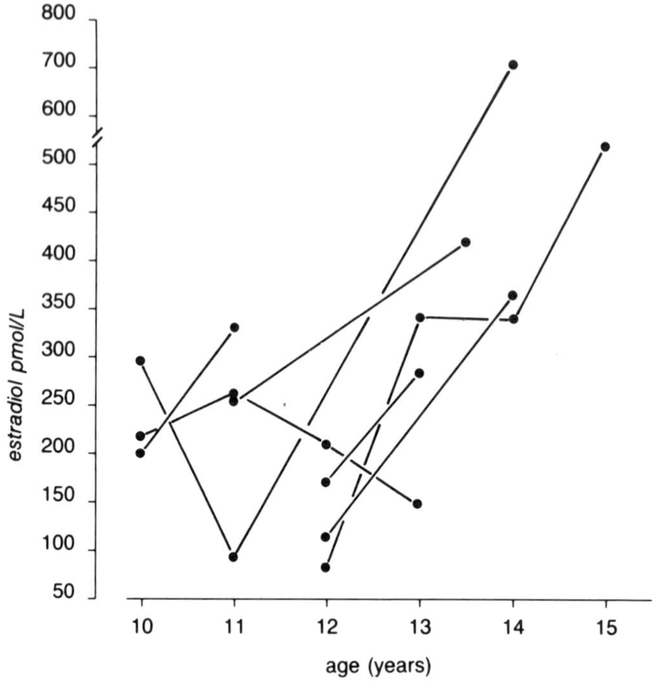

Fig. 18. Serum estradiol concentrations in single random serum samples of 47,XXX girls aged 10–15 years. Variability of results was related to phase of menstrual cycle at the time of sampling.

freedom from psychopathology on a five-point scale. The scales used in all ratings are presented in Table XVI.

The interrater reliabilities and concordances for the three behavioural scales and the quality of parenting scale were examined by means of analyses of the data of the 47,XXY sample and are presented in Table XVII. Since the variances of the behavioural scales were very small, the interrater reliabilities were corrected for range limitation using the family functioning variance as an estimate of unrestricted variation [24]. The reliability of the ratings of adjustment was not examined directly but can be inferred from the observation that ratings on this scale were highly correlated ($r = .775$, $P < .01$) with an independent judgement of quality of parenting (D.A.S.) and 11 of the 12 psychopathology scales of the CBCL that was completed by the parents (only the schizoid dimension failed to meet the .05 criterion of significance). As can be seen in Table XVII, all scales appear to be satisfactory in terms of

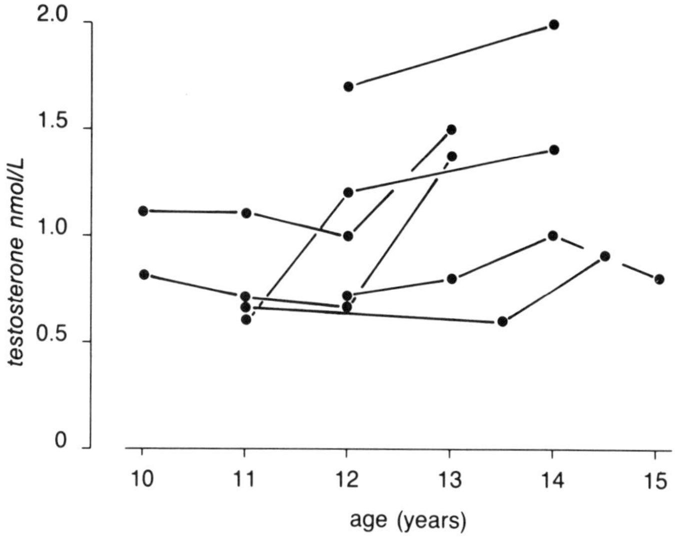

Fig. 19. Serum testosterone concentrations in 47,XXX girls aged 10–15 years.

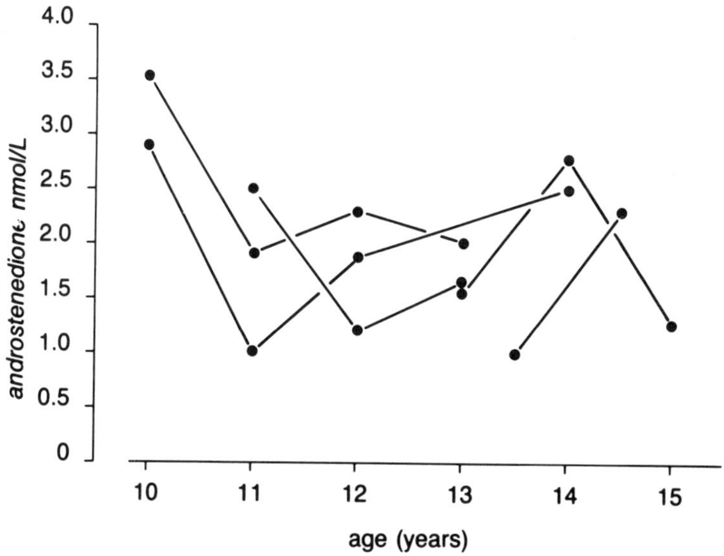

Fig. 20. Serum androstenedione concentrations in 47,XXX girls aged 10–15 years.

TABLE XIV. WRAT Test Results in Reading, Spelling, and Arithmetic for 47,XXX Girls and Four Sib Controls

	47,XXX (N = 6)		Controls (N = 4)		
Percentile	Mean	SD	Mean	SD	P
Reading	30.33	15.81	85.00	20.88	<.01
Spelling	41.50	27.68	82.00	13.86	<.7
Arithmetic	20.50	17.07	67.67	18.15	<.01
Age	13.63	1.57	11.67	1.53	>.10

TABLE XV. Mean XXX Scores on Woodcock-Johnson Psychoeducational Battery (N = 4, age = 14.3, current grade = 7.8)

	Grade	Age equiv.	Percentile
Reading	7.3	12.5	36.3
Mathematics	7.5	13.1	29.8
Written	8.1	13.6	42.0
Knowledge	5.6	10.9	17.0

reliabilities. The results of these ratings will not be presented at this point but in a later section.

The PIC test was administered to only two mothers of 47,XXX cases, a sample too small for meaningful statistical analysis. Similarly, data from only four Middle Child Temperament scales were available for the 47,XXX cases, again too small for analysis. The mothers of all six cases, however, did complete the CBCL. These data were analysed by comparing them to the norms of the test and also to a group of female cystic fibrosis patients in the same age range of 12–15 years. These results are presented in Table XVIII and indicate that both cases and cystic fibrosis patients were rated by their mothers in ways suggesting that they had higher than average levels of maladjustment in a variety of areas. They indicate also that 47,XXX girls differed from the cystic fibrosis patients in being less competent in school and more immature and hyperactive.

Although these data indicate that behavioural disorder might be commoner in 47,XXX girls than in the chromosomally normal, we have been unable to identify a psychopathologic syndrome typically associated with this aneuploid state. Of the six cases examined, two could not be classified by the Edelbrock and Achenbach algorithm [21] for typological assignment because of over- or underresponding on the part of mothers. Of the remaining four, one was classified as normal and one each to the categories of hyperactive-immature, depressed-withdrawn-delinquent, and aggressive-cruel.

TABLE XVI. Rating Scales for Assessment of Quality of Parenting and Personality in Study

1	2	3	4	5
Quality of parenting				
Poor (defined by disorganization, deprivation, or significant psychopathology one or more parents)	Intermediate low	Average	Intermediate high	Excellent (defined by exceptionally strong nurturing and supportive parenting skills)
Psychopathology in probands				
Clear evidence of psychiatric disorder of recognizable type	Significant but mild degree of disorder	Average adjustment	Intermediate between 3 and 4	Exceptionally good adjustment, able to withstand stress, and adaptable
Passivity vs assertiveness among probands				
Exceptionally passive with only rare instances of self-assertion	Intermediate low	Average	Intermediate high	Exceptional degree of assertiveness and dominance
Social relationships among probands				
Very few social contacts, with isolation and withdrawal prominent	Intermediate low	Average	Intermediate high	Many strong social relationships with multitude of friends and group activities
Activity level				
Very low: little involvement or interest in any recreational, social, or academic area	Intermediate low	Average	Intermediate high	Very high: many interests and active participation in each

TABLE XVII. Scale Concordance and Interrater Reliabilities of Ratings in 29 47,XXY Boys

	Concordance		Interrater correlations	
	Within one point	No discrepancy	Correlation	Corrected correlation
Passive-active	92.8	57.1	0.450	0.839
Social relationships	96.2	62.9	0.284	0.805
Activity level	100.0	65.3	0.317	0.803
Quality of parenting	67.4	25.8	0.806	

TABLE XVIII. Child Behaviour Checklist Test Results of Six 47,XXX Cases Compared to Norms and 26 Female Cystic Fibrosis Patients

	47,XXX (N = 6)		Cystic fibrosis (N = 26)			
	Mean	SD	Mean	SD	t	P (diff.)
Behaviour problems						
Internalizing	59.7	12.2	54.4	8.9**	1.00	
Externalizing	58.3	9.0*	51.8	7.4	1.77	*
Total	62.0	9.2†	56.0	9.4**	1.44	
Social competence						
Activities	48.8	8.49	51.3	10.1	0.62	
Social relations	43.3	9.5	46.4	9.6*	0.79	
School	39.0	7.8	49.3	11.2	2.662	‡
Total	43.4	7.9*	46.6	9.0*	0.89	
Anxious-obsessive	59.5	9.6*	55.3	8.3**	0.99	
Somatic complaints	58.0	4.3**	65.0	7.1**	3.13	**
Schizoid	63.7	9.2‡	59.9	6.4**	0.96	
Depressed-withdrawal	62.5	12.4*	55.3	10.8‡	1.31	
Immature-hyperactive	66.8	4.9**	56.0	8.9**	4.07	**
Delinquent	61.0	8.9†	56.7	7.8**	1.09	
Aggressive	60.8	9.8†	53.6	8.0†	1.67	
Cruel	66.0	10.4‡	59.4	3.7**	1.53	

*<.10, †<.05, ‡<.02, **<.01 different from population mean.

To provide a further index of adjustment, each case was reviewed to determine whether treatment had been provided by a mental health professional in the period under review. One of the 47,XXX girls received care of this kind and none of the sib controls. Thus, despite the evidence provided by parental reports and ratings, actual clinical intervention occurred in the case of only one 47,XXX girl.

47,XXY

Cognitive ability. The WISC-R test results of 29 47,XXY boys obtained beween May, 1981, and May, 1983, are presented in Table XIX. The WISC-R scores of 25 sib controls (19 males and 6 females) obtained in assessments conducted between 1977 and 1983 are also presented in this table. Statistical comparisons between these two groups indicate quite clearly that the verbal aptitudes of the extra X males were substantially lower than those of the controls. Their nonverbal abilities were, however, quite comparable.

These results are similar to those reported in the past [2], and, indeed, the correlations between our earlier results, obtained when the aneuploid group was 9.1 years old on average, and those presented in Table XIX, obtained at a mean age of 13.1 years, were quite high (verbal IQ, $r = .922$, $P < .01$; performance IQ, $r = .862$, $P < .01$; verbal IQ-performance IQ, $r = .823$, $P < .01$).

Educational skills. Assessments of the educational skills of the 29 probands and 25 sib controls (19 males and six females) based on the WRAT are presented in Table XX. The results of the Woodcock-Johnson Psychoeducational Battery are given in Table XXI for 26 probands only, since, as yet, no sibs have been assessed with this instrument. Both sets of data indicate

TABLE XIX. WISC-R Test Results for 29 47,XXY Boys and 25 Sib Controls

	47,XXY (29)		Controls (19 males, 6 females)			
	Mean	SD	Mean	SD	t	P
Information	6.85	3.41	9.52	2.62	2.88	<.01
Similarities	7.72	2.49	10.00	3.07	2.55	<.02
Arithmetic	8.19	3.07	9.72	3.19	1.78	NS
Vocabulary	7.15	2.66	10.24	2.52	4.38	<.01
Comprehension	7.12	2.26	9.92	2.15	4.66	<.01
Digit span	6.96	3.46	10.08	3.50	3.28	<.01
Picture completion	11.52	2.15	10.36	2.01	2.01	<.05
Picture arrangement	11.04	2.30	11.38	2.84	0.18	NS
Block design	9.79	2.66	10.76	2.35	1.42	NS
Object assembly	10.54	3.02	11.40	2.74	1.09	NS
Coding	8.11	3.60	9.79	3.11	1.84	NS
Verbal IQ–performance IQ	−18.93	16.63	−5.40	12.90	3.34	<.001
Verbal IQ	82.35	17.41	99.28	11.66	4.13	<.001
Performance IQ	101.31	12.52	104.68	12.17	0.81	NS
Full-scale IQ	90.24	13.42	101.68	10.91	3.33	<.001
Age	13.12	1.78	10.60	4.02	2.88	<.01

TABLE XX. WRAT Test Results in Reading, Spelling, and Arithmetic of 29 47,XXY Boys and 25 Sib Controls

	47,XXY (N = 29)		Controls (N = 25)			
Grade (percentile)	Mean	SD	Mean	SD	t	P
Reading	29.86	28.68	55.87	31.09	3.17	<.01
Spelling	27.59	29.59	48.30	32.74	2.42	<.01
Arithmetic	16.48	17.69	43.39	29.79	3.95	<.01

TABLE XXI. Mean 47,XXY Boys' Scores on Woodcock-Johnson Psychoeducational Battery (N = 26, age = 14.6, current grade = 8.1)

	Grade	Age equiv.	Percentile
Reading	6.1	11.4	27.1
Mathematics	6.0	11.4	21.1
Written	7.0	12.4	34.6
Knowledge	5.7	11.0	21.0

that the cases had levels of educational skills that were below average and, in the case of the WRAT results, below the levels of the sib controls. Not surprisingly, 17 of the 29 cases received special educational assistance or were in full-time special classes in the period under review in contrast to only four of the 25 sib controls (χ^2 8.78, $P < .01$).

Emotional adjustment. Social and emotional functioning of the 29 47,XXY boys was assessed by the methods described previously. The PIC scores of 20 cases, usually based on the responses of their mothers, are presented in Table XXII and compared to normative values obtained in the test-retest reliability study described in the manual of this instrument [22]. These data were obtained when the proband group had a mean age of 11.9 years. Their principal characteristics at this time were their higher than average scores on various scales reflecting problems in academic achievement, intelligence, and development and their lower than average score on the hyperactivity scale of this instrument. The first set of observations is not unexpected given their educational difficulties and psychometrically defined cognitive deficits. The second is less readily explained in terms of intellectual factors and, as other evidence to be presented later will indicate, appears to reflect qualities of temperament characterized by introversion, unassertiveness, and low levels of social comfort and skill.

The CBCL test results of 27 cases at the mean age of 13.5 years are presented in Table XXIII which also gives scores of a male cystic fibrosis patient sample aged 12–15 years. Both groups differed significantly (P <

TABLE XXII. Personality Inventory for Children Test Results of 47,XXY Boys Compared to the Results of Normal Males Examined in a Test-Retest Reliability Study

	47,XXY (N = 20)		Comparison group (N = 46)			
	Mean	SD	Mean	SD	t	P
Achievement	58.60	10.72	50.6	9.7	2.86	<.01
Intellectual screening	62.75	20.47	51.8	8.3	2.31	<.05
Development	56.80	9.28	49.9	10.9	2.63	<.02
Somatic concern	58.30	16.91	52.8	11.1	1.33	NS
Depression	56.10	11.49	54.1	11.7	0.65	NS
Family relations	54.10	10.73	50.3	10.7	1.32	NS
Delinquency	55.40	11.07	54.8	10.8	0.20	NS
Withdrawal	57.05	9.24	54.4	11.4	0.99	NS
Anxiety	54.00	10.05	55.9	13.1	0.64	NS
Psychosis	56.15	14.16	53.7	10.0	0.70	NS
Hyperactivity	43.40	8.41	52.1	12.9	3.25	<.01
Social skills	53.60	10.50	53.5	10.9	0.04	NS

TABLE XXIII. Child Behaviour Checklist Test Results of 27 47,XXY Boys Compared to Norms and 32 Male Cystic Fibrosis Patients

	47,XXY (N = 27)		Cystic fibrosis (N = 32)			
	Mean	SD	Mean	SD	t	P
Behaviour problem						
Internalizing	61.63**	10.37	58.8**	9.5	1.09	NS
Externalizing	60.59**	10.37	56.8**	10.9	1.47	NS
Total	63.25**	11.94	58.6**	10.9	1.56	NS
Social competence						
Activities	43.14**	9.43	51.7	8.7	3.59	<.01
Social relations	38.93**	9.56	40.9**	11.9	0.71	NS
School	34.37**	16.27	49.2	16.3	3.49	<.01
Total	37.07**	8.47	45.1**	8.6	3.60	<.01
Somatic complaints	62.78**	9.74	62.3**	9.0	0.19	NS
Schizoid	61.07**	9.46	57.4**	9.0	1.52	NS
Uncommunicative	61.15**	10.09	58.0**	6.9	1.37	NS
Immature	64.85**	9.86	60.6**	8.9	1.72	<.10
Obsessive-compulsive	57.22**	9.17	57.6**	8.3	0.16	NS
Hostile-withdrawal	63.67**	9.76	58.4**	8.3	2.21	<.05
Delinquent	63.56**	9.86	58.8**	6.2	2.17	<.05
Aggressive	59.14**	10.91	57.5**	11.1	0.57	NS
Hyperactive	62.96**	9.46	59.0**	11.4	1.46	NS

*<.10, †<.05, ‡<.02, **<.01 different from population mean.

.01) from average on each of the behaviour problem scales. The aneuploid group was also rated as lower than average on each of the social competence scales, whereas the cystic fibrosis patient differed from average only in social competence and total competence. Analysis of group differences indicated that the 47,XXY males had lower scores in all areas of competence (except that designated as *social*) and higher scores on the hostile-withdrawal and delinquent scales than the cystic fibrosis patients.

The Middle Child Temperament scale is designed to be completed by parents of children under 12 years of age. Of the 24 47,XXY cases whose parents provided information on this test, a number were older than 11 years 11 months (mean age 13.5 ± 1.03 years). These data must therefore be treated with some caution. However, since they are consistent with other findings, they are thought to be worthy of reporting and are presented in Table XXIV.

A further evaluation of adjustment in the 47,XXY boys was carried out by reviewing each to determine whether he had received treatment from a mental health professional in the period from May, 1981 to May, 1983. Five of the 29 cases had received such help in contrast to two of 24 sibs on whom information was available ($P > 0.10$). Of the five, four were in the care of the Children's Aid Society, because of major dislocations within the family, and one was from a family where the mother was very unstable and the father possibly alcoholic. These findings are similar to those for the 47,XXX females. Both groups were characterized by elevations in various indices of maladjustment, yet only a small number of cases in each received active treatment from a mental health facility. The groups also resembled each other in that no particular psychopathologic syndrome seemed to characterize those who satisfy psychometric criteria for maladjustment. Thus, in the 47,XXY group, the Edelbrock and Achenbach [21] algorithm for classifying deviant cases into categories of disorder resulted in the following assignments of

TABLE XXIV. Middle Child Temperament Scale Test Results for 24 47,XXY Boys

	Mean	SD	P
Persistence	90.62	16.13	<.01
Approach/withdrawal	92.08	11.99	<.01
Adaptability	93.12	15.56	<.05
Predictability	95.08	9.89	<.05
Distractibility	96.71	11.56	NS
Mood	100.67	14.79	NS
Activity	108.17	9.29	<.01
Intensity	109.42	11.33	<.01
Threshold	111.79	11.97	<.01

cases with valid profiles: schizoid, five; uncommunicative, two; somatic complaints, one; immature-aggressive, three; hyperactive, two; uncommunicative-delayed, one; and delinquent, three.

47,XXY Mosaics

Three of our cases with behavioural data for the period under review were classified as 47,XXY/46,XY mosaics. One of these was previously thought to have a pure 47,XXY cell line, but repeat chromosome analyses demonstrated a significant percentage of 46,XY cells. No statistical analyses of data for these cases was undertaken because of small numbers, and we present only basic descriptive findings for them in Tables XXV and XXVI.

47,XYY

Only one 47,XYY case was examined during the interval between May, 1981 and May, 1983. This boy, who was 9 years old at the time of assessment, obtained WISC-R IQs that fell in the average range (verbal IQ 92, performance IQ 95, full-scale IQ 92). His educational skills based on the WRAT and expressed in percentiles were as follows; reading, 82; spelling 93; and arithmetic, 7. He has presented significant behavioural problems and been assessed in an inpatient psychiatric unit because of poor peer relations, marked irritability, and temper outbursts. He has continued to receive outpatient treatment from a child guidance clinic. His one sister, who is 2 years younger, has presented no academic or behavioural problems.

TABLE XXV. WISC-R Test Results for 47,XXY/46 XY Males and Available Controls

	Cases (IQ)		Controls (IQ)	
Case	Verbal	Performance	Verbal	Performance
1	98	129	105	128 (brother)
2	81	96	No control	
3	94	114	103	97 (father)

TABLE XXVI. WRAT Test Results for 47,XXY/46 XY Males and Controls

	Cases (percentile)			
Case	Reading	Spelling	Arithmetic	Controls
1	77	92	9	Not tested
2	37	25	47	No control
3	23	16	12	

Miscellaneous Cases

One female originally classified as 45,X/46,XX mosaic but whose subsequent chromosomal examinations were normal was assessed in the period between May, 1981 and May, 1983. At age 14 years 3 months, her intelligence was found to be average (verbal IQ 98, performance IQ 91, full-scale IQ 94), and her school progress normal. There were no behavioural problems, and she was described by her parents as an active and sociable girl. She has no sibs.

One 46,XX male was examined in the period under review. This boy, when intellectually assessed at age 14 years 6 months, had overall average ability but was found to have a reliably higher performance than verbal IQ (108 vs 91, $P < .01$). His academic skills were somewhat below average on the WRAT (reading percentile, 23; spelling, 7; and arithmetic, 9). He presented no significant behavioural problems.

Theoretical Issues

Cognition. The results just described as well as others to be found in the literature indicate that anomalous numbers of X chromosomes are associated with specific deficits in verbal or spatial intellectual ability. This conclusion seems best established for 47,XXY males, who are verbally impaired but spatially normal, and for 45,X Turner females, who are spatially impaired but verbally normal. The evidence in the case of 47,XXX females is more mixed but does suggest that language development is more likely to be impaired than more nonverbal aspects of cognition.

We have in previous publications [25,26] advanced the argument that these impairments in cognitive development have their origin in prenatally occurring disturbances in growth. These proposals were prompted by the work of Waber [27,28], who postulated that variations in developmental rate around puberty could account for the normally occurring female superiority on verbal tests and the male superiority on spatial tasks [29]. She examined this issue in studies of slow- and fast-maturing males and females and found that, regardless of sex, slow maturers were more able spatially than verbally whereas fast maturers tended to show the opposite pattern of skills. She also presented data suggesting that maturation rates influenced hemispheric organization, with slow development being associated with a greater left vs right hemisphere asymmetry for verbal processing than fast development.

There are a number of observations suggesting parallels, both intellectual and biologic, between X aneuploids and early and late maturers. 45,X females with spatial deficits have bone ages advanced relative to stature and, like Waber's fast maturers, have relatively strong verbal skills [30]. Extra X males and females during their prepubertal years are frequently delayed in bone age maturation and, like Waber's slow maturers, are verbally weak [2]. Moreover, cytological research has indicated that mitotic cell division rates are accelerated in XO females and depressed in extra X states [31, 32].

Although these findings are consistent with the notion that in X aneuploids delayed rates of maturation interfere with verbal development and accelerated rates interfere with spatial development, they constitute only indirect kinds of evidence. More persuasive is the observation that individual differences in bone age maturation in 47,XXY males are related to their degree of verbal deficit [33]. Further evidence for a direct relation between growth and cognitive development is provided by the finding that a dermatoglyphic measure, total finger ridge count (TFRC), frequently interpreted to reflect rates of prenatal growth and to be dependent on rates of mitotic cell division [32,34,35], is related to the severity of the verbal disabilities in 47,XXY males [33].

Like Waber, we have also found that hemispheric specialization for verbal and nonverbal information processing is disturbed in 45,X and extra X states. 45,X females, like Waber's fast maturers, failed to show the typical right ear (left hemisphere) advantage on a dichotic listening test using verbal material [35,36]. This finding has been replicated with both dichotic listening and other lateral asymmetry paradigms [37–40]. Collectively, these results indicate that XO subjects do not use their left hemispheres in processing verbal stimuli to the same degree as XX females. Extra X males have a different anomaly in hemispheric organization. The evidence for this conclusion was obtained in a study of 47,XXY males, all of whom were tested with a battery of tests assumed to reflect left hemisphere specialization for verbal tasks (dichotic listening for syllables and for digits and half-field identification of letters) and right hemisphere specialization for nonverbal processing (half-field dot estimation, dichotic listening for melodies, and a dichhaptic recognition task [41]. The results indicated that 47,XXY males had lower than normal degrees of left hemisphere specialization for language and greater than normal degrees of right hemisphere specialization for nonverbal processing [25].

We have also examined the question of whether growth, in particular prenatal growth rate, has a relationship with asymmetries in hemispheric organization. We did this by computing left hemisphere (L Hem) and right hemisphere (R Hem) specialization indices for each of our 47,XXY boys. This was done by calculating in standard score units the degree to which each case deviated from the means of samples of 79 or 97 normal children for each task of the laterality battery. These values were then averaged to provide

L Hem and R Hem scores for each 47,XXY boy, which were then correlated with their TFRCs. Among 28 47,XXYs this yielded a correlation between R Hem and TFRC of −.508 ($P < .01$), which increased to −.665 ($P < .01$) when four left-handed cases were excluded. No correlations between TFRC and L Hem scores were significant, although among the four left-handed subjects there was a trend, with L Hem and TFRC correlating at −.736.

Although these findings indicate that growth rate is related to the development of specific abilities in X aneuploids in ways that resemble the findings of Waber [27,28], they imply different underlying mechanisms do than her results. Among these differing implications, given the association between TFRCs and verbal ability in 47,XXY males, is that the influence of maturation cannot be limited to the period of pubertal onset. Instead, they suggest that events occurring during prenatal life affect the development of ability during the preadolescent years. They also imply that the mechanism mediating an association between growth and ability operates by influencing the development of hemispheric specialization for cognitive processing.

Our tentative explanation for these associations is that slow rates of prenatal growth retard the development of the left hemisphere, which in turn allows the right to develop free of inhibitory influences from the left. As growth rate normalizes, so do the interhemispheric relationships. However, those 47,XXYs with the slower rates of growth are those whose right hemispheres are more functionally active even to the point where they may assume language functions. It would be expected, however, that these would be inefficient, thereby accounting for their verbal deficits. In fact, this seems to be the case; those 47,XXYs with elevated R Hem scores are those with the greatest deficit in verbal IQ relative to performance IQ (94.3 vs 80.3).

Personality. There is now in the literature a sizeable body of data indicating that 47,XXY males frequently show signs of being atypical temperamentally and are at risk for maladjustment [1,2,44]. Some of these data have been obtained in studies of adults [43] and some in longitudinal examinations of neonatally identified children and adolescents [2,44,45]. Collectively, they indicate that extra X males are less social, assertive, and active than normal XY males. Our latest observations on 47,XXY males, described earlier, are for several reasons generally consistent with these conclusions. First, their parents described them as being less than usually active (PIC, hyperactivity score; CBCL, activities; Middle Child Temperament scale, activity, intensity, and threshold scores). Second, their reports also indicated that they are prone to withdrawal (CBCL, hostile-withdrawal scale; Middle Child Temperament, approach-withdrawal scale).

Although our studies and those of others have produced findings that are reasonably consistent, they have not explained why these anomalies occur in

47,XXY males. They have, however, provided data that tend to make some of the more obvious, simpler forms of explanation unlikely. These can be stated as follows: 1) disturbances in parenting induced by the knowledge of a sex chromosome abnormality, 2) responses of children to the intellectual or physical abnormalities that are sometimes present, 3) hormonal abnormalities emerging in and after puberty, 4) abnormalities in hormonal processes occurring in pre- or perinatal life. The first of these seems unlikely to be correct in that precise parental knowledge of a sex chromosome abnormality is not a necessary condition for the presence of the temperamental anomalies [44]. The second seems improbable because they can be observed during infancy and in the absence of any clearly defined intellectual or physical disorder [45]. Pubertal hormones are also insufficient for much the same reason. The atypical behaviours are evident before puberty and in the absence of any demonstrable hormonal abnormality [1]. The fourth, referring to disturbances in pre- or perinatal hormonal events, cannot be ruled out simply because no data bearing on the issue have yet been collected. However, the absence of any clinically evident physical abnormality, which would arise from significant hormonal dysfunction during the time, makes it unlikely.

Our investigations of 47,XXY boys have provided findings that suggest that a more complex set of influences underlies their anomalies in temperament and adjustment. In these, we have assessed hormonal functioning at age 13 (testosterone, estradiol, follicle-stimulating hormone [FSH] and luteinizing hormone or [LH]), quality of parenting (the previously described family functioning rating), and hemispheric specialization (the variables L Hem and R Hem). Data serving as dependent variables were also obtained on personality and adjustment by means of the PIC and CBCL questionnaires. In addition, the ratings of passivity-activity, social participation, activity level, and adjustment were also used as dependent variables.

We analyzed these data using multiple regression techniques in which family functioning, hormonal status, and the hemispheric measures L Hem R Hem were treated as independent variables. The results are shown in Table XXVII. Not surprisingly, they indicate that the quality of parenting was an important determinant of temperament and maladjustment in 47,XXY boys. However, they indicate also that both hormonal status and hemispheric organization influenced various parameters of temperament in these boys. In the case of left hemisphere functioning, elevations in L Hem were associated with higher parental ratings on the CBCL school activities scale and the PIC hyperactivity scale and with lower scores on the CBCL psychosis scale. Right hemisphere functioning was negatively correlated with a number of the CBCL scales of psychopathology. Higher levels of the gonadotrophic hor-

TABLE XXVII. Significant Predictors of Personality and Adjustment in 47,XXY Boys (N = 16) as Determined by Multiple Regression*

L Hem	R Hem	Testosterone	Estradiol	FSH	LH	Quality of parenting rating
School activities (CBCL) (.497)	Externalizing (CBCL) (−.519)	Hyperactivity (CBCL) (−.598)	Externalizing (CBCL) (−.634)	Passive rating (−.531)	Immaturity (CBCL) (−.400)	
Family relations (PIC) (−.380)	Schizoid (CBCL) (−.614)		Internalizing (CBCL) (−.564)	Hyperactivity (PIC) (−.624)	Hostile/withd. (CBCL) (−.469)	Externalizing (CBCL) (−.508)
Psychosis (PIC) (−.402)	Obsess./compul. (CBCL) (−.428)		Obsess./compul. (CBCL) (−.385)			Social activity (CBCL) (.501)
Hyperactivity (PIC) (.598)	Aggressive (CBCL) (−.584)		Aggressive (CBCL) (−.625)			Immaturity (CBCL) (−.602)
						Obsess./compul. (CBCL) (−.496)
						Hostile/withd. (CBCL) (−.603)
						Delinquent (CBCL) (−.539)
						Family relations (PIC) (−.662)
						Psychosis (PIC) (−.641)

*PIC, Personality Inventory for Children; CBCL, Child Behavior Checklist. All factors significant by F test, $P < .05$; partial correlations in parentheses.

mones FSH and LH and testosterone and estradiol were also associated with variations in various measures of temperament and adjustment.

A recent theoretic paper by Tucker and Williamson [46] has provided a possible basis for explaining the relationships between hemispheric functioning and social-emotional development in 47,XXY males and possibly in normal males. These authors advanced the concept that each hemisphere possesses distinguishable properties that have consequences for a wider variety of behaviours than those that can be described simply as verbal or spatial cognition. In their view, the left hemisphere has a particular role in controlling tonic activation levels, internally focussed serial attentional and motor mechanisms, and motivation. They suggested that most of these effects are mediated by the left hemisphere's tendency to increase redundancy in perceptual analysis, thereby reducing the rate of information change. The right hemisphere's functions, more externally or perceptually driven, are phasic in nature and, in that they facilitate habituation to repetitive stimuli, predispose an organism to respond to novelty. Tucker and Williamson suggested that these hemispheric differences have consequences for a number of functions, including motoric, attentional, and emotional behaviours. In their system, the right hemisphere is regarded as particularly important in regulating emotion, a view generally supported in the literature [47]. The left hemisphere, they propose, is of primary importance for motivation and the serial organization of motor and other behaviours.

This theory has some bearing on observations gathered in 47,XXY males, since, with their lower than normal degrees of left hemisphere specialization, they would be expected not only to present language deficits but also to have poor serial memory, impaired motor functions, and low activity levels, all of which they seem to have [1,2,48]. Their strong right hemisphere development would be expected to have consequences for emotional functioning as well. The present data suggest that these also occur; high R Hem scores are associated with low scores on a variety of emotionally laden scales of maladjustment (see Table XXVII).

The finding that hormonal functioning is related to personality and adjustment in 47,XXY boys is similar to other observations obtained with adults [43]. However, it is probable that these effects have emerged only during adolescence, since prior to this time levels of sex hormones have been very low and within the normal range [1]. It is only at age 13 when significant disturbances in hormonal functioning can be identified [49]. It is likely, therefore, that hormones have exerted effects on qualities of temperament already set in place by the twin influences of family functioning and hemispheric organization. For example, low levels of spontaneous activity appear to be one of the most easily recognized early indicators of atypical temperament in 47,XXY boys [45]. It can be hypothesized that this is the result of lower than normal degrees of left hemispheric functioning. Hormonal influences would not be operative during this time but only later when pubertal changes begin to occur, that is, around 13 years of age. It is interesting in this context that recently obtained evidence suggests a similar sequence of events in normal subjects [50]. This view, if correct, indicates that social-emotional functioning in 47,XXY boys is not the direct result of factors that are invariant in time. Rather, it suggests that during development multiple processes operate, at some points mutually interactive and at others sequential. Such a system is conceptually similar to that proposed by Sroufe and Rutter [51] in a theoretic paper on the origins of psychopathology in which they suggest that development is best understood in terms of interrelated, discontinuous, and emergent processes.

BEHAVIOURAL OUTCOME WITH ITS RELATION TO FAMILY DYNAMICS

The following section is a summary of growth to puberty, pubertal development, school performance, intellect, and behavioural characteristics of each of the 53 aneuploid children in the Toronto Study (Appendix A). Rating of family dynamics and functioning is included and the influence of family dynamics on behavioural outcomes of the children is summarized in Tables XXVIII–XXXI.

The children were seen and data recorded at 6-month intervals starting at age 13 years. Measurements presented here are for the most recent visit, and age is recorded in half-years. At each visit, the accompanying adult, most often the mother, was interviewed regarding the child's behaviour and relationship to sibs and peers, and the parent's impression of personality and development. The proband was also interviewed.

Family functioning was evaluated by two investigators (D.A.S., C.T.N). Rating scales were developed independently by these investigators and the ratings compared. There was agreement in these ratings ($r = .813$), so families were rated according to a family function rating.

In the profile sketches, growth is for height only. Pubic hair rating is according to Tanner [12], and testosterone is recorded in millimoles per liter. Follicle-stimulating and luteinizing hormones are in microunits per liter. FFR indicates family function ratings according to our agreed-upon ratings. VIQ and PIQ indicate verbal and performance intelligence quotients.

A trial to evaluate the effects of injections of testosterone cypionate every 4 weeks was started on alternate 47,XXY boys as they reached their thirteenth

TABLE XXVIII. Family Function Rating

1	2	3	4	5
Broken home	Parent separation not supported by sympathetic parent	Home intact but much argument and disagreement	Family apparently stable but much argument	Stable family
Single, disinterested parent				Support child with special care
Blame anomaly for child's misbehaviour	Child has had temporary foster home placement	Little interest in child's anomaly	Both parents working	Strong nurturing and supportive parental skills
			Little apparent concern for child's development	
Psychiatric problem with one or both parents	Pressure for better school performance	Continuous pressure for better school performance		
Family court intervention	Depressed, nonsupportive parenting	Poor sib discipline	Blames school for poor progress	
Criminality in family		Alcohol or drug abuse in parents or family		
Drug abuse, etc				

TABLE XXIX. Relation of Behavioural Ratings to Family Function Ratings

	Family function rating				
Behaviour rating	1	2	3	4	5
47,XXY (N = 30)					
No problems (22)	1	2	1	4	14
Disturbed behaviour (5)	0	2	3	0	0
Delinquency (3)	3	0	0	0	0
47,XXY/46,XY (N = 5)					
No problems	0	0	0	0	4
Disturbed behaviour	0	0	0	1	0
Delinquency	0	0	0	0	0

TABLE XXX. Relation of Behavioural Ratings to Family Function Ratings

	Family function rating				
Behaviour rating	1	2	3	4	5
47,XXX (N = 9)					
No problems	0	0	0	0	7
Disturbed behaviour	0	1	0	0	0
Delinquency	1	0	0	0	0
45,X/46,X— (N = 3)					
No problems	0	0	0	1	2
Disturbed behaviour	0	0	0	0	0
Delinquency	0	0	0	0	0

TABLE XXXI. Study Conclusions

Poor family functioning contributes to behaviour problems in aneuploid children
Conversely, good family support appears to lessen likelihood of antisocial behaviour
It is questionable whether giving supplemental testosterone during early puberty beneficially affects behaviour in 47,XXY boys

birthdays. Boys so treated are designated T and the untreated boys or controls are designated C. The trial was to evaluate whether or not testosterone given throughout early puberty would in any way influence growth, sexual maturation, or personality development.

Three members of the study team (D.A.S., C.T.N., J.D.B.) independently rated families according to a family function rating (Table XXVIII), which was to identify family pathology. There was very close agreement on ratings of the three evaluators. Correlations of these ratings with severe behaviour problems or antisocial behaviour were made. It is suggested by this study that children with a karyotype having an extra X chromosome (ie, 47,XXY and 47,XXX) are much more vulnerable to antisocial behaviour when they are from a severely disturbed family (Tables XXIX, XXX). Conversely, such children from a stable and supportive family need have no excess of antisocial behaviour. Appendix A presents a short summary profile of subjects in the current Toronto Study.

ACKNOWLEDGMENTS

The investigations of psychologic development were supported by grants from the Ontario Mental Health Foundation to C.T.N. and J.R.

REFERENCES

1. Stewart DA, Netley CT, Bailey JD, Haka-Ikse K, Platt J, Holland W, Cripps M: Growth and development of children with X and Y chromosome aneuploidy: A prospective study. In Robinson A, Lubs HA, Bergsma D (eds): "Sex Chromosome Aneuploidy: Prospective Studies on Children." New York: Alan R Liss, Inc., for the March of Dimes Birth Defects Foundation. BD:OAS XV(1): 75–114, 1979.
2. Stewart DA, Bailey JD, Netley CT, Rovet J, Park E, Cripps M, Curtis JA: Growth and development of children with X and Y chromosome aneuploidy from infancy to pubertal age: The Toronto Study. In Stewart DA (ed): "Children With Sex Chromosome Aneuploidy: Follow-Up Studies." Alan R. Liss, Inc., March of Dimes Birth Defect Foundation. BD:OAS 18(4):99–154, 1982.
3. Tanner JM, Whitehouse RH, Takaishi M: Standards from birth to maturity for height, weight, height velocity and weight velocity. British children 1965. Arch Dis Child 41:454–471, 613–635, 1966.
4. Nelhaus G: Head circumference from birth to eighteen years. Pediatrics 41:106–114, 1968.
5. Simmons C: "The Brush Foundation Study of Child Growth and Development, Vol 2 (Monographs of the Society for Research in Child Development Vol. IX #37)." Washington, D.C.: National Research Council, 1944.
6. Tanner JM, Whitehouse RH: Standards for sitting height and subischial leg length from birth to maturity. British children 1978. London: Castlemead Publications, 1979.
7. Greulich WW, Pyle SI: "Radiographic Atlas of Skeletal Development of the Hand and Wrist. 2nd ed." Stanford, CA: Stanford University Press, 1959.
8. Tanner JM, Whitehouse RH, Marshall WT, Healy MJR, Goldstein H: "Assessment of Skeletal Maturity and Prediction of Adult Height (TW2 Method)." New York: Academic Press, 1975.
9. Becker KL: Clinical and therapeutic experiences with Klinefelter's syndrome: Fertil Steril 23:568–578, 1972.
10. Johnson HR, Myhre SA, Ruvalcaba RHA, Thuline HC, Kelly UC: Effects of testosterone on body image and behaviour in Klinefelter's syndrome: Pilot study. Dev Med Child Neurol 454:460, 1970.
11. Caldwell PD, Smith DW: The XXY (Klinefelter's) syndrome in childhood: Detection and treatment: J Pediatr 80:250–258, 1972.

12. Tanner JM: "Growth at Adolescence, 2nd ed." Oxford: Blackwell Scientific Publications, 1962.
13. Zachmann M, Prader A, Kind HP, Hayfleger H, Budlyer H: Testicular volume during adolescence. Cross sectional and longitudinal studies. Helv Paediatr Acta 29:61–72, 1974.
14. Piers LV, Harris DB: Age and other correlates of self-concept in children. J Educational Psychol 55:91–95, 1964.
15. Woodcock-Johnson Psycho-educational battery. Hingham MA: Teaching Resources, 1977.
16. Schonfeld WA, Beebe GW: Normal growth and variation in the male genitalia from birth to maturity. J Urol 48:759–777, 1942.
17. Boisen E: Testicular size and shape of 47,XYY and 47,XXY men in a double-blind, double-matched population survey. Am J Hum Genet 31:697–703, 1979.
18. Skakkebaek NE, Bancroft J, Davidson DW, Warner P: Androgen replacement with oral testosterone undecanoate in hypogonadal men: A double-blind controlled study. Clin Endocrinol 14:49–61, 1981.
19. Topper E, Dickerman Z, Prager-Lewin R, Kaufman H, Maimon Z, Laron Z: Puberty in 24 patients with Klinefelter syndrome. Eur J Pediatr 139:8–12, 1982.
20. Wide Range Achievement Test. Wilmington, DE: Jastak Associates, 1978.
21. Edelbrock CS, Achenbach TM: A typology of child behaviour profile patterns. Distribution and correlates for disturbed children ages 6–16. J Abnormal Child Psychol 8:441–470, 1980.
22. Wirt RD, Seat PD, Broen WE: Personality Inventory for Children. Los Angeles, Western Psychological Services.
23. Middle Child Temperament Questionnaire. Copyright Hegvik R, McDevitt SC, Carey WB, 1980.
24. Gulliksen H: "Theory of Mental Tests." New York: Wiley, 1950.
25. Netley C, Rovet J: Hemispheric lateralization in 47,XXY Klinefelter's syndrome boys. Brain Cognition, 3:10–18, 1984.
26. Netley C, Rovet J: Relationships among brain organization, maturation rate and the development of verbal and nonverbal ability. In: "Language Functions and Brain Organization." New York: Academic Press, 1983.
27. Waber D: Sex differences in cognition: A function of maturation rate? Science 192a:29–38, 1976.
28. Waber D: Sex differences and the rate of physical growth. Dev Psychol 13:29–38, 1977.
29. Maccoby E, Jacklin C: "The Psychology of Sex Differences." Stanford, CA: Stanford University Press, 1974.
30. Almqvist S, Lindsten J, Lindvall N: Linear growth, sulfation factor activity and chromosomal constitution in 22 subjects with Turners syndrome. Acta Endocrinol 42:168–186, 1963.
31. Barlow P: The influence of inactive chromosomes on human development. Humangenetik, 17:105–136, 1973.
32. Polani PE: Chromosomes and chromosomal mechanisms in the genesis of maldevelopment. In Connolly K, Prechtl H (eds): "Maturation and Development." Philadelphia: Lippincott, 1981, pp 50–72.
33. Netley C, Rovet J: Verbal deficits in children with 47,XXY and 47,XXX karyotypes: A descriptive and experimental study. Brain Language 17:58–72, 1982.
34. Mittwoch U: "Genetics of Sex Differentiation." New York: Academic Press, 1973.
35. Netley C: Dichotic listening of callosal agenesis and Turner's syndrome patients. In Segalowitz S, Gruber F (eds): "Language Development and Neurological Theory." New York: Academic Press, 1977.

36. Netley C, Rovet J: Atypical hemispheric lateralization in Turner syndrome subjects. Cortex 18:377–384, 1982.
37. Gordon HW, Galatzer A: Cerebral organization in patients with gonadal dysgenesis. Psychoneuroendocrinology 5:235–244, 1980.
38. McGlone J, Manion N: Spatial deficits in Turner's syndrome: Can CNS damage and/or speech organization account for the above? Paper presented at Canadian Psychological Association meeting, Winnipeg, June, 1983.
39. Rovet J: Hemispheric specialization in Turner's syndrome. Paper presented at Canadian Psychological Association Meeting, Winnipeg, June, 1983.
40. Waber D: Neuropsychological aspects of Turner syndrome. Dev Med Child Neurol 21:58–70, 1979.
41. Flanery RC, Balling JD: Developmental changes in hemispheric specialization for tactile spatial ability. Dev Psychol 15:364–372, 1979.
42. Rovet J, Netley C: The triple X chromosome syndrome in childhood: Recent empirical findings. Child Dev 54:831–845,1983.
43. Schiavi RC, Theilgaard A, Owen DR, White D: Sex chromosome anomalies, hormones and aggressivity. Arch Gen Psychiatr 41:93–99, 1984.
44. Bancroft J, Axworthy D, Ratcliffe S: The personality and psycho-sexual development of boys with 47,XXY chromosome constitution. J Child Psychol Psychiatr 23:169–180, 1982.
45. Walzer S, Wolff P, Bowmen D, Silbert A, Bashir A, Gerald P, Richmond J: A method of the longitudinal study of behavioural development in infants and children: The early development of XXY children. J Child Psychol Psychiatr 19:213–230, 1978.
46. Tucker DM, Williamson PA: Asymmetric neural control systems in human self-regulation. Psychol Rev 91:185–215, 1984.
47. Heilman KM, Satz P: "Neuropsychology of Human Emotion." New York: Guilford Press, 1983.
48. Robinson A, Salenblatt J, Bender B, Puck M: Follow-up studies of X aneuploids identified in Denver. Paper presented at 4th International Meeting on X Aneuploidy, Edinburgh, June, 1984.
49. Stewart DA, Bailey JD, Park E: Growth and hormonal functioning in X aneuploids. Paper presented at 4th International Meeting on X and Y aneuploidy. Edinburgh, June, 1984.
50. Susman EJ, Nottleman ED, Blue JH, Dorn LD, Inoff GE, Chrousos GP, Cutler GB, Loriaux DL: Hormones and mood variability and behavioural problems during early adolescence. Abstract. 7th International Congress of Endocrinology, Quebec, June, 1984.
51. Sroufe LA, Rutter M: The domain of developmental psychopathology. Child Dev 55:17–29, 1984.

APPENDIX A*

T 1 XXY 16 years T FFR 4

Growth has been constantly along 25th centile. Testes 4 ml. Penis 11 cm. Pubic hair T4. Testosterone 13.4. FSH 63. LH 17. Quite talkative and tends

*Age, at last visit; growth, height only; testosterone, mmol/liter; T, treated (100 mg testosterone every 4 weeks from age 13 to 15); FSH, μU/liter; LH, μU/liter; VIQ, verbal IQ; PIQ, performance IQ; FFR, family function rating; C, control for testosterone treatment.

to be a braggart (much the same as his brother, his only sib). VIQ 67. PIQ 90. Very poor school progress. Trying out in vocational school. No delinquent behaviour but easily led by his peers. Parents have been concerned mostly about his poor school progress.

T 2 XXY 15 years T FFR 1

Growth has been constantly on 50th centile. Testes 3.5 ml. Penis 12 cm. Pubic hair T5. Active sexually since age 14. Testosterone 8.8. FSH 58. LH 24. Has a severe language problem resulting from stuttering. Very passive, indifferent, and very easily led. VIQ 77. PIQ 78. Poor school progress with many failures. Severe delinquent problems, eg, theft, drugs. No family support. In custodial group home since 14.5 years. Mother has had two illegitimate children; they were raised largely by a most incompetent grandmother.

T 3 XXY 15 years T FFR 1

Growth increased from below 50th centile to 75th from age 13.5 to 14.5. Testes 4 ml. Penis 13 cm. Pubic hair T4. Testosterone 7.2. FSH 24. LH 15. Quiet, confused, nonaggressive boy. VIQ 60. PIQ 91. Very poor school progress. No delinquent behavior to date. Tends to be self-isolating and insecure in social situations. Was placed in Children's Aid early in infancy and has been in many foster homes.

T 4 XXY 16.5 years T FFR3

Growth was along 50th centile until age 13, then up to 75th centile. Youngest and tallest in family of three boys. Has been sexually active since age 15. Testes 4.5 ml. Penis 10 cm. Pubic hair T4. Testosterone 8.0. FSH $>$100. LH 29. Has had poor school progress and is in a vocational program. VIQ 62. PIQ 77. Parents separated when he was 7 years old; he was raised in his own home by his paternal grandmother. Timid but inclined to be a braggart, as is his father. No delinquent behaviour but is bad tempered when pushed too hard.

T 5 XXY 15 years T FFR 5

Grew along 75th centile until age 12, then at 13 moved up to 90th centile. Five centimeters taller now than 18-year-old brother. Testes 3 ml. Penis 10 cm. Pubic hair T4. Testosterone 14.1. FSH $>$100. LH 111. School progress has been fair to average with considerable help in communications skills. VIQ 84. PIQ 111. Very pleasant but shy, timid and nonaggressive. No antisocial problems.

T 6 XXY 15 years T FFR5

Grew along 75th centile until age 11.5, then up to 90th at age 12.5. Right testicle agenetic, left now 6 ml. Penis 12 cm. Pubic hair T5. Heavy facial and axillary hair. Testosterone 18.1. FSH 29. LH 15. School progress has been satisfactory. VIQ 106. PIQ 100. Very timid and nonaggressive. No delinquent behaviour.

T 7 XXY 16 years T FFR2

Has a full brother with 47,XYY karyotype (see T 3 XYY). Grew along 50th centile until age 12.5, then at 13.5 years moved up to 75th centile. Testosterone 16.1. FSH 70. LH 26. Testes 5.5 ml. Penis 12 cm. Pubic hair T5. Very poor school progress and has been excused from attendance. VIQ 64. PIQ 90. Sullen and timid but violent if crossed severely. Two of three brothers are labeled habitual criminals and he greatly admires one of them. Parents have been indifferent and they have little control over their boys' behaviours.

T 8 XXY 14 years T FFR 5

Grew along 50th centile until age 13.5 years, then at 14 moved up to 75th. Testes 6 ml. Penis 12 cm. Pubic hair T3. Testosterone 12.4. FSH 30. LH 10. Average to good school progress (same as two sibs). Quiet but well adapted socially. Likes sports, plays hockey. No behaviour problems. VIQ 105. PIQ 106.

T 9 XXY 14.5 years T FFR 2

Full brother of T 25 XXY. Growth curve moved up from 75th to 90th centile at 13–13.5 years. Testes 4.0 ml. Penis 10 cm. Pubic hair T3. Has moderate gynaecomastia. Testosterone 8.6. FSH 26. LH 15. Very poor school progress. His 47,XXY brother also has poor school progress, but his 46,XX sister is an honours student. VIQ 58. PIQ 91. Very shy and gullible boy. No delinquent behaviour. Family in treatment with mental health agency to learn effective parenting skills. Mother is confused and very angry at the boys' education problems as well as father's indifference.

T 10 XXY 14 years T FFR 3

Growth has been along 90th centile. Testes 3.5 ml. Penis 9 cm. Pubic hair T3. FSH 15. LH 5. Needs special education, and school progress has been slow. VIQ 75. PIQ 100. Some minor delinquency problems such as stealing with a gang of friends. Timid, shy, and very easily led. He is the oldest in a family of four in which there have been three different fathers.

T 11 XXY 15 years C FFR 5

Grew along 25th centile to age 15 then moved up to 50th centile at 15.5. Testes 4.0 ml. Penis 9 cm. Pubic hair T4. Testosterone 23.7. FSH 56. LH 15. One year behind at school. VIQ 74. PIQ 92. Quiet, timid, and shy. No violent or delinquent behaviour reported. Recently accused of unsubstantiated sexual assault on a 4-year-old girl.

T 12 XXY 14 years C FFR4

Growth has been along 50th centile. Testes 3.5 ml. Penis 8 cm. No pubic hair. Testosterone 1.2. FSH 5. LH <3. Very quiet and self-isolating. Described as a "follower" by his father. Continues with average performance in age-appropriate special education. VIQ 67. PIQ 118. Both brothers are in special education. No aggressive or delinquent behaviour.

T 13 XXY C FFR 5

Grew along 97th centile until age 13 years. From 13 to 16 years, he has moved to the 75th centile. Testes both 6 ml. Penis 12 cm. Pubic hair T5. Facial hair and axillary hair well developed. Had gynaecomastia requiring surgery at age 16. Testosterone 18.6. FSH 71. LH 25. Very quiet, shy, nonaggressive boy. Many school problems. VIQ 88. PIQ 118. No delinquent behaviour.

T 14 XXY 12 years C FFR 4

Growth has been constantly between 3rd and 10th centile. Testes 2 ml. Penis 5 cm. Pubic hair T2. Testosterone 3.0. FSH 20. LH 7. Quiet, timid, passive, sad boy. No active sports. Having difficulty in school. VIQ 96. PIQ 111. No delinquent behaviour. Parents have had considerable discord.

T 15 XXY 14 years C FFR 5

Congenitally deaf. Has grown along 50th centile. Testes 6 and 5 ml. Penis 10 cm. Testosterone 5.3. FSH 5. LH 3. Had a great deal of difficulty in school in spite of good hearing aid. VIQ 68. PIQ 111. School progress similar to his two older brothers. Very shy and timid. Parents think he is "female" in his behaviour. Self-isolating socially.

T 16 XXY 14.5 years C FFR 4

Grew along 25th centile until age 13, then up to 50th at 14. Testes 2.5 ml. Penis 8 cm. Pubic hair T3. Slight gynaecomastia. Testosterone 8.5. FSH >100. LH 29. Major educational difficulties. VIQ 72. PIQ 86. Quiet, gentle boy with few friends. Mother has felt that he is effeminate, and she constantly remarks on this. No delinquency problems.

T 17 XXY 14.5 years C FFR 5

Grew along 75th centile until age 13, then up to 90th. Testes 6.0 ml. Penis 10 cm. Pubic hair T3. Testosterone 8.0. FSH 53. LH 12. Has had special education with average progress. VIQ 94. PIQ 102. He is an anxious boy with poor self-esteem and has had psychiatric counselling. No delinquent behaviour. Parents are very helpful to him.

T 18 XXY 14 years C FFR 5

Growth along 50th centile to age 13, then up to 75th. Testes 4 ml. Penis 11 cm. Pubic hair T3. Testosterone 7.9. FSH 69. LH 36. Thought of as an excellent boy. Good at school. VIQ 100. PIQ 109. Rather shy with no behaviour problems. Quality of parenting exceptionally good.

T 19 XXY 13 years C FFR 4

Has grown steadily along 50th centile. Testes 2 ml. Penis 5 cm. No pubic hair. Testosterone 0.4. LH <3. FSH 5. Has had severe asthma and missed much school and is 1 year behind. VIQ 96. PIQ 98. Quiet, immature, self-isolating socially. No delinquent behaviour.

T 20 XXY 15 years C FFR 1

Growth steadily along 50th centile. Testes 3.0 ml. Penis 10 cm. Pubic hair T4. Had had very poor school progress with several failures. Extremely poor parenting in a broken home, and his negative behaviour could not be managed at home. Is currently in a residential home for delinquent boys. There were no behaviour problems with his sibs. VIQ 82. PIQ 112.

T 21 XXY 15 years C FFR 5

Growth steadily at 90th centile. Testes 3 ml. Penis 9 cm. Pubic hair T4. Has moderate acne and mild gynaecomastia (3 cm). Testosterone 8.4. FSH 32. LH 14. Very quiet and shy boy with very few friends. No behaviour or academic problems. VIQ 100. PIQ 121.

T 22 XXY 13.5 years C FFR 5

Growth steadily at 50th centile. Testes 4 ml. Penis 10 cm. Pubic hair T3. Testosterone 9.6. LH 9. No significant educational or adjustment problems. Family describes him as easy-going, sweet, and happy. Shy, little social contact with peers. VIQ 96. PIQ 85.

T 23 XXY 14.5 years C FFR 3

Growth steadily along 97th centile. Testes 5 ml. Penis 10 cm. Pubic hair T4. Testosterone 12.6. FSH 33. LH 20. Intellectually slow with VIQ 80, PIQ 74. Very unsettled home with multiple separations and psychiatrically disturbed father. No tendencies to aggressiveness or delinquent behaviour.

T 24 XXY 14 years C FFR 5

Growth steadily on 90th centile. Testicles 5 ml. Penis 10 cm. Pubic hair T3. Testosterone 15.2. FSH 67. LH 16. An unassertive, quiet boy with few friends. Was adopted in early infancy. Easily managed by his undemanding parents. Good school progress. VIQ 97. PIQ 115.

T 25 XXY 16 years C FFR 2

Brother T 9 XXY. Grew along 75th centile until age 15, then up to 90th. Testes 4.5 ml. Penis 8.0 cm. Pubic hair T3. Testosterone 6.6. FSH 5.5. LH 16. Has moderate gynaecomastia. Episodes of aggressiveness and threatened violence. He is the family scapegoat and is being considered for a foster home. Major educational problems. VIQ 53. PIQ 110.

T 26 XXY 13 years C FFR 5

Grew at 75th centile until age 12, then up to 90th. Testes 5 ml. Penis 10 cm. Pubic hair T3. Testosterone 3.8. FSH 19. LH 5. Shy and timid. Tries to compete in sports with peers. No educational difficulties. No behaviour problems. VIQ 101. PIQ 111. Very supportive family.

T 27 XXY 14 years C FFR 3

Grew at 75th centile until age 12.5, then up to 97th. Testes 6 ml. Penis 10 cm. Pubic hair T3. Testosterone 12.8. FSH 28. LH 12. A socially immature boy who does not associate much with his peers. Recently, parents separated. Has a poor relationship with his mother with whom he lives and whom he defies. Recently has had poor academic performance, apparently because of poor motivation and effort. VIQ 102. PIQ 108. Mother very depressed and having a hard time coping.

T 28 XXY 14.5 years C FFR 5

Growth has been along 10th centile. Testes 6 ml. Penis 8.5 cm. Pubic hair T3. Testosterone 2.8. FSH 3. LH <3. Has presented no behaviour problems but is getting special help for his poor school progress.

T 29 XXY 12 years C FFR 5

Growth has been along 50th centile. Testes 2 ml. Penis 6 cm. No pubic hair. Testosterone 0.7. FSH 2. LH 3. Quiet and sensitive boy as described by his mother. Has recurrent asthma. VIQ 106. PIQ 98. No school or behaviour problems.

T 31 XXY 11.5 years C FFR 2

Last seen 2 years ago. Growth was along 50th centile. Testes 2 and 2.5 ml. Penis 5 cm. No pubic hair. Testosterone <10. LH <3. A passive, slow boy, having a good deal of difficulty with school. In special education class. VIQ 65. PIQ 84 (cf brother with VIQ 85 and PIQ 102).

T 1 XXX 13 years FFR 5

Has grown continually on 90th centile. Menarche at 10.5 yr. Estradiol 146. FSH 9.0. LH 7. Doing reasonably well at school. VIQ 90. PIQ 102. Quite shy and self-isolating, with few friends.

T2 XXX 11 years FFR 5

Growth has been along 50th centile. No menarche. No pubic hair or breast development. LHRH stimulation test indicated normal gonadotropin release. School progress slow. VIQ 79. PIQ 93. No delinquency problems. Very shy and overprotected at home.

T 3 XXX 15 years FFR 5

Growth along 90th centile until age 13, then up to 97th. Menarche at 13 years. An attractive, upper middle class, stable family. No behaviour or emotional problems. Mother says she is "unpredictable" and impulsive. Has mild educational problems and is currently in a private school for learning disabled. VIQ 97. PIQ 96.

T 4 XXX 15 years FFR 2

Growth has been along 50th centile. Menarche at 13.5 yr. Estradiol 564. FSH 8. LH 7. She is a disturbed girl with a long history of psychiatric contacts. She shows some delinquent, aggressive, and, occassionally, violent behaviour. Mother is very neurotic and incapable of managing this difficult girl. Has had much special education and is 2 years behind in school. VIQ 86. PIQ 92.

T 5 XXX 13 years FFR 4

Has grown continually along 97th centile. Very obese. Menarche at 10.5 yr. Pubic hair T5. Breast B5. No gynaecologic problems. Estradiol 283.

FSH 14. Very poor school progress. VIQ 79. PIQ 106. Pleasant, affable girl with many friends.

T 6 XXX 14.5 years FFR 5

Steady growth along 75th centile. Menarche at 12 yr. Pubic hair T5. Breasts B4. FSH 7. Has done very well at school. VIQ 83. PIQ 126. Very supportive family that feels chromosome anomaly is meaningless.

T 7 XXX 13 years FFR 5

Has grown along 50th centile. No menarche by age 13. Breast development B2. Pubic hair T2. Shy and stubborn. Does not accept authority well. Parents are tolerant and understanding and have managed her delinquency tendencies very well. No major educational problems. VIQ 85. PIQ 118.

T 8 XXX 12 years FFR 1

Stopped her follow-up 3 years ago after being placed in a detention home. Growth has been along 75th centile. No menarche. Very hostile, stubborn, and violent girl. Mother has very limited parental skills. Three sibs have had several court appearances for drugs, sexual offenses, violence. Patient now incarcerated after robbery with violence. Formal education ceased at age 12 years. VIQ 62. PIQ 73.

T 9 XXX 14 years FFR 5

Growth has been along 50th centile. No menarche. Poor school progress has caused a great deal of family anxiety. Now is in a private school for children with learning disabilities. She is a shy and anxious girl with a poor self-image.

T 1 XYY 12 years FFR 3

Has not been seen for 2 years. Growth has been along 50th centile. Testes 2.0 ml. Penis 6 cm. No pubic hair. FSH 4. LH <3. School progress is fair. VIQ 92. PIQ 95. Has no discipline or guidance at home. Called "lazy" at home. No delinquency problems.

T 2 XYY 13 years FFR 1

Growth has been along 50th centile. Had R cryptorchidism, but this has been corrected. R testis 2.5 ml., L 6.0. Penis 6 cm. Pubic hair T2. He is a disturbed boy who has had psychiatric care and is currently in a home for disturbed boys. Has severe temper outbursts if frustrated. VIQ 92. PIQ 95. Education has been significantly poor because of his poor attitude. Family has been broken all his life. Twelve-year-old sister has no delinquency problems.

T 3 XYY 25.5 years FFR 1

Was not identified in our neonatal survey but was identified as 47,XYY when sibs of neonatal aneuploid children were examined. He is a brother of T 7 XXY. At age 25.5 years, height was 186.6 cm. Testes 20 ml. Penis 15 cm. He was a very shy, self-isolating man. He works successfully at a sheltered workshop. VIQ 74. PIQ 90. If frustrated, he has severe, violent temper outbursts. He has no friends and is largely noncommunicative.

T 1 XXY/XY 15 years T FFR 4

At birth was thought to be XXY but on rechecking at age 11 he was found to have a mosaic karyotype. Of 30 cells examined, 23 were XXY and seven were XY. He grew along 25th centile until age 13 and at 14 and 15 was on 50th. Testes 6 ml. Penis 13 cm. Pubic hair T4. Active sexually. Testosterone 14.7. FSH 6. LH <3. Very outgoing, attractive boy. School progress has been good. VIQ 94. PIQ 114. No delinquent behaviour.

T 2 XXY/XY 13 years NT FFR 5

Growth has been along 25th centile. Testes 2.5 ml. Penis is 6 cm. No pubic hair. Testosterone 2.5. FSH 0. LH 3. Mosaicism noted at birth and at age 9 years; 2 cells of 30 counted were XXY, 28 were XY. No school or behaviour problems. VIQ 95. PIQ 120.

T 3 XXY/XY 14.5 years FFR 5

Growth was along 10th centile until age 14, and at 14.5 was between 25th and 50th. Most recent chromosome test shows 21% of cells are XXY. Testes 15 ml. Penis 8 cm. Pubic hair T3. Testosterone 12.1. FSH 6. LH 3. Quiet, shy boy. School progress was normal. VIQ 81. PIQ 96. No delinquent behaviour.

T 4 XXY/XY 9 years FFR 5

Has not been seen for 4 years. At birth, 26 of 66 lymphocytes were XXY and 40 were XY. At age 9, 23 of 30 cells were XY and seven XXY. Until that age, his school progress was satisfactory. VIQ 105. PIQ 123.

T 5 XXY/XY 15 years NT FFR 5

Growth steadily along 90th centile. Testes 12 ml. Penis 12 cm. Pubic hair T4. Testosterone 27. FSH 13. LH 6. Parents describe him as carefree and easy-going. No behaviour or academic problems. VIQ 96. PIQ 93.

T 1 XX male 15 years FFR 4

Has grown along or just above 50th centile. Testes 3.0 ml. Penis 10 cm. Pubic hair T4. Testosterone 10.3. FSH 78. LH 21. Has repeated one grade at school and has had low average progress since. VIQ 91. PIQ 108. A quiet, passive boy with few friends. Parents feel he presents no behaviour or antisocial problems.

T 2 XX male 15 years FFR 5

Growth has been along 25th centile. Testes 10 ml. Penis 9 cm. Pubic hair T4. Slight gynaecomastia. Testosterone at age 14 was 7.8. No school or behaviour problems.

T 1 XXYY 15 years

Has not been seen for 5 years. Known to have much difficulty with school. VIQ 74 7 years ago. PIQ 65. Was a negative, destructive boy.

T 1 XO/XXR 13.5 years FFR 5

Growth well below 3rd centile. Actual height 128 cm. Breasts stage 1–2. Pubic hair T2. Has horseshoe kidney. Of 30 cells counted, 19 were 45X and 11 were 46XXR. VIQ 66. PIQ 78. Very slow at school. Passive. No behaviour problems.

T 2 XO/X dic(Yq) 12.5 years FFR 4

Of 30 cells, 28 are 45X, and two have 46 chromosomes, one an X and one a dicentric Y. Has always been well below 3rd centile in height. One year ago had laparotomy and removal of streaked gonads. Now FSH > 100, LH 57 and estradiol 212. Has had very poor school progress. VIQ 73. PIQ 71. She is a happy and very active girl with no behaviour problems.

T 3 XO/XX 14 years FFR 5

Six of 28 lymphocytes examined at birth were XO. Repeat tests at ages 22 months and 6 years and cultured fibroblasts at age 13 years showed only 46XX karyotype. Growth fell gradually from 10th centile at age 6 years to well below 3rd at age 13, and at age 14 it has come to just below the 3rd. No menarche. Very early breast bud development only. FSH 13. LH 6. Estradiol 341. No stigmata of Turner syndrome. She is a very quiet, sober girl with no school or behaviour problems. VIQ 98. PIQ 91.

Physical and Psychologic Parameters in Children With Sex Chromosome Anomalies: Further Follow-Up From the Winnipeg Cytogenetic Study of 14,069 Newborn Infants

J.A. Evans, PhD, FCCMG, R. de von Flindt, PhD, C. Greenberg, MD, FCCP, FCCMG, S. Ramsay, RN, and J.L. Hamerton, DSc, FCCMG

Department of Human Genetics (J.A.E., C.G., S.R., J.L.H.) and Departments of Pediatrics and Psychiatry (R.d.v.F.), University of Manitoba and Children's Hospital, Winnipeg, Manitoba, Canada R3E OW3

INTRODUCTION

This chapter reports the ongoing follow-up of children with sex chromosome anomalies (SCA) detected in the Winnipeg Cytogenetic Study of 14,069 consecutive live births [1–5]. The SCA group is currently comprised of 13 children, four with 47,XXY, four with 47,XYY (including one mosaic 45,X/46,XY/47,XYY), four with 47,XXX, and one with a mosaic karyotype 45,X/46,XX/47,XXX with the monosomic line predominating. With the exception of one family with a child with 47,XXY (NB5420), all parents had been previously informed of their child's chromosomal anomaly. These children have been studied at relatively regular intervals since birth and are now entering puberty. Previous reports [4,5] have documented earlier findings in these children.

This report presents our findings from two subsequent follow-up studies investigating 1) physical and psychologic parameters in some of the chromosomally normal sibs of the children with SCA, in age- and sex-matched controls selected from the newborn cohort, and in some of the controls' sibs, and 2) a 3 year follow-up of the SCA children themselves.

Birth Defects: Original Article Series, Volume 22, Number 3, pages 183–207

METHODS

The 13 children with sex chromosome anomalies were evaluated in detail with respect to both physical and psychologic variables. For the control/sib study, the methods and measures were the same as in the previous report [5].

The chromosomally normal controls were matched by sex and date of birth with the cases. They were not matched by ethnic origin or socioeconomic status. The distribution of these variables was similar in both groups. The cases included more children with Canadian Indian or Metis background (31% vs 9%). However, this group also had more professional or managerial fathers (38% vs 27%). None of these differences was significant.

The children with SCA were aged between 7 years, 5 months, and 10 years, 11 months, whereas the controls were between 8 years, 1 month, and 11 years, 3 months. We also carried out the same evaluations on five sibs of the SCA children who were available for study (aged between 6 years, 8 months, and 11 years, 7 months) and the available sibs of the control group (aged between 6 years, 3 months, and 12 years, 3 months). In all, 12 controls (8 male and 4 female) and seven control sibs were studied. The control of the child with 45,X/46,XX/47,XXX was not studied. Although all 13 children with SCA were studied initially, two were not available for the follow-up evaluations made approximately 3 years later. Therefore, the sample size was reduced to 11 in these last assessments.

In each of the follow-up studies, data analysis was conducted from two perspectives. The first involved investigation of children with SCA as a unique group, and the second approach investigated the three primary subgroups (46,XXX, 47,XXY and 47,XYY). Analysis of the data concentrated on three main aspects of physical and psychologic evaluation. First, we studied the patterns of minor dysmorphic findings in these children. A previous report [5] documented specific minor anomalies in these children. In these follow-up studies, we concentrated on determining the average number of dysmorphic findings per child. In all, 31 minor dysmorphic signs were chosen for analysis because they occurred in at least two of the total study population (37 children). These were double parietal hair whorl, absent or poorly formed parietal hair whorl, upswept anterior hairline, medial flare to eyebrows, slight upslant to palpebral fissures, slight epicanthus, telecanthus (innercanthal distance >97% for age), deep-set eyes, heterochromia irides, broad nasal bridge, malar flattening, aberrant dentition, aberrant ear shape, attached ear lobes, rotated ears, hyperextensible elbows, six or more digital whorls, aberrant *c* or *d* triradii, *t′* axial triradius unilaterally or bilaterally, interdigital patterns, capillary hemangioma(ta), pigmented nevi, and sacral dimple. The mean number of dysmorphic findings per child in each subgroup was then calculated and compared using a two-tailed weighted Student's t test (Table I).

To determine the relative similarity within and between SCA children, their controls, and sibs with regard to dysmorphic findings, the coefficients of similarity between individuals in the subgroups were calculated. The coefficient of similarity used was Czekanowski's [6] given by 2a/(2a + b + c), where a represents the number of findings present in both children of a pair, b represents those present only in the first child, and c those found only in the second child. This coefficient thus emphasizes the concordance of dysmorphic findings. A gross index of similarities between the children was derived by calculating and comparing the ranges and means of the coefficients of similarity between different subgroups (Table II). These data are not amenable to statistical testing.

The second aspect to be considered was growth. Twelve growth parameters were studied in all children. These were height, weight, head circumference, inner and outer canthal distances, interpupillary distance, ear length, total hand length, midfinger to total hand length ratio, chest circumference, internipple distance, and nipple to chest ratio.

We expressed all growth parameters as Z scores using published growth curves (National Centre for Health Statistics, U.S., for height and weight;

TABLE I. Frequency of Minor Dysmorphic Findings

		No. of dysmorphic findings/child	
Subgroup	No.	Mean	SD
Sex chromosome anomaly cases	13	4.38	2.99
Sibs of cases	5	5.00	2.55
Controls	12	5.33	2.15
Sibs of controls	7	5.29	1.89

TABLE II. Coefficients of Similarity for Minor Dysmorphic Findings

Individuals compared	No. of comparisons	Coefficient*	
		Mean	Range
47,XXY *vs* 47,XXY	6	0.184	0–0.286
47,XYY *vs* 47,XYY	6	0.133	0–0.400
47,XXX *vs* 47,XXX	6	0.262	0–0.615
Cases *vs* sibs	5	0.226	0–0.500
Controls *vs* sibs	7	0.370	0.222–0.667
Controls *vs* controls	66	0.255	0–0.615

*Czekanowski's coefficient of similarity [6] = 2a/(2a + b + c).

Montreal Children's Hospital for head circumference [Preus, personal communication]; and Feingold and Bossert [7], for all other values. Data for whites were used in all cases because standards for Canadian Indian children are not yet available. When all the data had been transformed, the means and SDs for all subgroups were calculated and compared using a two-tailed, weighted Student's t test (Table III).

The last area of investigation involved psychologic functioning in the children and consisted of two major focal points: intellectual and academic functioning as measured by the Wechsler Intelligence Scale for Children—Revised (WISC-R) [8], the Peabody Picture Vocabulary Test (PPVT) [9], and the Wide-Range Achievement Test (WRAT) [10]; and personality functioning as reported by mother on the Personality Inventory for Children (PIC) [11,12]. For details of these measures, see our previous study [5]. For the follow-up study the Vineland Social Maturity Scale [13] and the Drawing Subtest of the McCarthy Scale of Children's Abilities (MSCA) [14] were not administered. The Vineland was considered to be too general a measure of social maturity, and all children at the first follow-up scored well above chronological age expectations. At the time of the most recent examination, all children exceeded the upper norm limits of the MSCA, and all performed at or above the age norms in the previous study.

In the intellectual/academic achievement area Student's *t* test comparisons were made between the SCA group, the control group, the SCA sib group, and the control sib group, for the WISC-R Full-Scale IQ, Verbal IQ, Performance IQ, the PPVT IQ, and the WRAT Standard Scores on reading, spelling, and arithmetic. The randomization test for two independent samples [15] was used for all subgroup comparisons on these same measures and the PIC scale comparisons. In addition to the statistical analysis, the subtest scatter on the WISC-R was studied using Kaufman's [16] approach to the analysis of subtest profiles. All statistical tests were two-tailed unless otherwise noted.

The results of the intergroup comparisons are given in the following section. Details of findings on individual children with sex chromosome anomalies are given in Appendix A.

RESULTS

Minor Dysmorphic Findings

The minor dysmorphic findings noted above indicate that these children, like most others studied in detail, have a range of subtle dysmorphic signs. The most common findings in both the cases and controls were medial flare

TABLE III. Growth Parameters in Children With Sex Chromosome Anomalies, Sibs, and Controls Expressed as Z Scores

Growth parameters	Cases (N) XXY (3) Mean	SD	XYY (3) Mean	SD	XXX (4) Mean	SD	XO/XX/XXX	Sibs of cases (N = 5) Mean	SD	Controls (N = 12) Mean	SD	Sibs of controls (N = 7) Mean	SD
Height	0.55	0.87	0.20	1.34	3.10	1.08*	−2.66	−0.62	0.53	0.04	0.96	−0.33	0.81
Weight	0.67	0.45	0.44	0.47	1.31	0.44	−0.66	−0.46	0.46	0.41	0.50	0.20	0.86
Head circumference	0.33	1.27	0.86	1.34	0.42	1.86	−0.20	−0.57	1.23	1.10	0.93	1.01	0.97
Innercanthal distance	1.10	0.57	0.00	1.50	1.05	1.43	1.36	−0.18	0.55	−0.13	0.82	−0.14	0.94
Outercanthal distance	0.61	0.46	0.24	1.37	0.23	1.04	0.55	−0.25	0.59	0.11	0.77	0.16	0.89
Interpupillary distance	0.71	0.17	−1.35	1.44	−0.09	1.40	−0.91	−0.09	0.64	0.24	0.91	0.32	1.39
Ear length	−0.18	0.26	−0.39	0.30	0.22	0.70	0.91	0.41	0.67	0.23	0.77	0.14	0.53
Hand length	−0.04	0.93	−0.66	1.12	1.03	0.63	−1.67	0.01	1.30	0.62	1.00	0.86	1.23
Midfinger/hand ratio	−0.55	1.03	0.39	0.44	−0.39	0.63	0.08	−0.64	0.97	0.02	3.19	0.07	3.52
Chest circumference	0.49	0.13	0.11	0.91	1.71	1.22	0.51	−0.23	0.77	−0.37	1.65	0.13	1.08
Internipple distance	0.00	0.88	−0.30	0.77	0.67	1.50	0.94	−0.19	0.81	0.13	0.92	−0.09	1.12
Nipple/chest ratio	−0.80	0.89	−0.71	0.98	−1.17	0.87	0.59	0.11	0.85	0.04	1.03	−0.22	1.06

*Significant difference between cases and controls at P = 0.05–0.02

to the eyebrows, aberrant ear shape, hyperextensible elbows, and interdigital patterns. As anticipated, the children with SCA showed no consistent pattern of minor anomalies that could indicate a specific phenotype associated with any of the 47,XXY, 47,XYY, or 47,XXX karyotypes [17]. When the mean number of such dysmorphic signs per child is compared between subgroups (Table I), it can be seen that children with SCA do not differ significantly from their sibs or controls. This suggests that aneuploidy in these cases is not associated with generalized dysmorphology.

However, when individual dysmorphic signs were compared in sib pairs it was found that eight of 34 (23.5%) minor findings occurring in cases and their sibs represented concordances in sibs compared to 28 of 72 findings (38.9%) in controls and their sibs. This suggests that controls tend to resemble their sibs more closely with regard to minor dysmorphic traits than do SCA. To study this further, we calculated and analyzed the coefficients of similarity between children as previously described in Methods (Table II). There was a tendency for control sib pairs to resemble each other more closely than cases and their sib or unrelated controls, though these differences were small. Among the SCA children themselves, a greater dissimilarity, ie, greater variability in minor dysmorphic signs, was seen among 47,XXY and 47,XYY than 47,XXX cases. Although these data are based on small numbers, it is possible that sex chromosome aneuploidy, although not having a major effect on morphogenesis, does allow more variable development than that seen in individuals and families in which all karyotypes are normal.

Growth Parameters

In Table III, which expresses growth parameters as Z scores, it can be seen that almost all parameters in all groups of children fall close to zero. In only one parameter did an SCA group differ significantly from controls. This was height in 47,XXX; these girls were significantly taller. All 47,XXX girls had heights above the 90th percentile compared to only one child in each of the 47,XYY and 47,XXY groups. Although increased height is well documented in 47,XXY and 47,XYY individuals, these boys were prepubertal at the time of examination and had not had their anticipated growth spurts. At the time of their most recent follow-up examinations at a mean age of 11.7 years, four of the six boys with either 47,XXY or 47,XYY had moved to a higher percentile for height (Appendix A).

Other nonsignificant trends in growth parameters were for relative telecanthus in 47,XXY; close-set eyes in 47,XYY; and increased weight, innercanthal distance, hand length, and chest circumference in 47, XXX girls. The 47,XXX cases also tended to have a narrow chest resulting from a reduced internipple to chest circumference ratio. The one child with a predominant 45,X line was relatively short and telecanthic and had short hands. The control children and their sibs had relatively larger heads.

Psychologic Evaluation

Intellectual functioning. No significant differences were found between the SCA sibs or the age- and sex-matched controls on any of the intellectual measures (Table IV). However, the 47,XXX cases scored significantly lower than their controls on both WISC-R Verbal IQ and Full Scale IQ. The 47,XXY cases also scored significantly lower on Verbal IQ but not on Full Scale IQ.

The verbal-performance IQ discrepancy (Table V) was most pronounced in the 47,XXY cases, with three of the four scoring significantly lower on the verbal scale. In the 47,XXX group, only one child scored significantly lower on the verbal scale. In the 47,XYY group, three of the four cases demonstrated a significant discrepancy; however, two of these reflected higher performance on the verbal scale.

In the initial study, 54% of the SCA cases demonstrated a significant discrepancy in verbal-performance IQ, primarily in the direction of lower verbal functioning (five of seven) in contrast to 25% of the control cases and 20% of the sibs. All the significant discrepancies in the control and sib groups were in the direction of higher verbal functioning. In the recent follow-up study (Table VI), the number of SCA children showing significant discrepancies increased to 73%, with the majority of discrepancies remaining in the direction of lower verbal functioning (six of eight).

The Kaufman analysis of subtest scatter did not indicate any particular patterns of strengths or weaknesses on the WISC-R subtests (Table VII) in the SCA subgroups in either the initial or the follow-up study. In addition, there was little consistency within cases on follow-up regarding previously identified strengths or weaknesses. The fact that digit span and mazes were not administered in the initial study obviously precludes comparisons on these subtests. However, the general distribution within the other subtests does not suggest that this omission was of major significance. As might be expected, this lack of consistency in strengths and/or weaknesses in the respective SCA subgroups was also reflected in the absence of any consistent indication of "shared and unique abilities and/or influences" relating to cognitive abilities.

Academic achievement. No significant differences were found between the SCA cases and their controls on score in reading, spelling, or arithmetic (Table VIII). The SCA cases did, however, differ significantly from their

TABLE IV. Intelligence Test Results in Children With Sex Chromosome Anomalies, Sibs, Controls, and Control Sibs

		WISC-R						PPVT IQ	
		VIQ		PIQ		FSIQ			
Group	N	Mean	SD	Mean	SD	Mean	SD	Mean	SD
47,XXY	4	94.50*	7.94	107.50	9.85	100.00	2.16	97.00	13.98
47,XXY sibs	1	108.00	—	117.00	—	113.00	—	107.00	—
Controls	4	113.00	15.77	104.25	6.40	109.50	12.58	116.25	13.05
Control sibs	2	125.50	27.58	115.50	10.61	123.00	14.14	135.00	14.14
47,XYY	4	108.50	22.52	106.50	23.10	108.25	22.17	112.75	14.73
47,XYY sibs	3	112.00	5.20	102.00	8.72	110.00	7.94	118.00	4.00
Controls	4	109.50	13.87	105.75	10.34	108.50	11.47	109.00	7.26
Control sibs	3	107.67	5.03	114.33	9.50	112.00	8.54	111.00	17.35
47,XXX	4	89.00*	15.25	92.75	14.82	89.75*	14.61	93.00	19.53
47,XXX sibs	1	112.00	—	85.00	—	99.00	—	105.00	—
Controls	4	105.00	8.76	103.75	10.01	104.50	8.70	109.75	12.87
Control sibs	2	101.00	5.66	107.50	4.95	104.00	5.66	114.50	5.66
45,X/46,XX/47,XXX	1	102.00	—	100.00	—	101.00	—	112.00	—
Totals									
SCA	13	98.15	16.64	102.46	15.97	99.85	15.42	100.54	16.54
SCA sibs	5	111.20	4.09	103.60	12.97	108.40	7.80	113.20	7.19
Controls	12	109.17	12.36	104.58	8.27	107.50	10.54	106.58	12.14
Control sibs	7	110.86	15.79	112.71	8.12	112.86	13.26	109.14	22.55

*Cases significantly lower than controls $P = 0.043$ (one-tailed test)

TABLE V. Differences Between Verbal and Performance Scale Scores on the WISC-R for Children With Sex Chromosome Anomalies, Their Sibs, and Controls

Group	Age (years)	FSIQ	VIQ	PIQ	Diff.	Significance level*
47,XXY						
NB9333	8.7	101	92	111	+19	<.01
NB7755	9.2	97	86	111	+25	<.01
NB6196	9.7	102	95	115	+20	<.01
NB5420	9.8	100	105	93	−8	
47,XXY controls						
C9333	9.2	122	130	108	−22	<.01
C7755	9.8	112	113	109	−4	
C6196	10.3	112	117	105	−12	<.05
C5420	10.4	92	92	95	+3	
47,XXY sib						
S7755	7.9	113	108	117	+9	
47,XYY						
NB13549	7.4	125	112	133	+21	<.01
NB8416	8.9	129	139	114	−25	<.01
NB10763	8.4	84	92	78	−14	<.01
NB13681	7.6	95	101	91	−10	
47,XYY controls						
C13549	8.1	107	107	106	−1	
C8416	9.8	109	118	96	−22	<.01
C10763	8.9	123	122	120	−2	
C13681	8.0	95	91	101	+10	
47,XYY sib						
S13549	9.3	101	106	96	−10	
S13549	11.6	116	115	112	−3	
S8416	6.7	113	115	108	−7	
47,XXX						
NB13003	7.8	102	107	96	−11	
NB10565	8.5	73	72	78	+6	
NB11407	8.2	82	82	85	+3	
NB1115	10.9	102	95	112	+17	<.01
47,XXX controls						
C13003	8.3	109	111	106	−5	
C10565	9.1	113	108	117	+9	
C11407	8.9	93	92	96	+4	
C1115	11.3	103	109	96	−13	<.05
47,XXX sib						
S1115	9.6	99	112	85	−27	<.01
45,X/46,XX/47,XXX						
NB9342	8.8	105	108	101	−7	

*For determination of level of significance see Kaufman [16, p 24].

TABLE VI. Differences Between Verbal and Performance Scale Scores on the WISC-R for Children With Sex Chromosome Anomalies (Follow-Up Study)

Group	Age (years)	FSIQ	VIQ	PIQ	Diff.	Significance level*
47,XXY						
NB7755	12.0	87	78	101	+23	<.01
NB6196	12.6	100	91	109	+18	<.01
NB5420	12.5	87	90	87	−3	
47,XYY						
NB13549	10.3	122	114	126	+12	<.05
NB8416	12.0	128	135	115	−20	<.01
NB13681	10.7	81	75	91	+16	<.01
47,XXX						
NB13003	10.8	105	115	92	−23	<.01
NB10565	11.2	71	62	85	+23	<.01
NB11407	11.0	85	82	91	+9	
NB1115	13.6	96	88	106	+18	<.01
45,X/46,XX/47,XXX						
NB9342	11.0	101	102	100	−2	

*Kaufman [16, p 24].

sibs on arithmetic. This may well be an artifact attributable to the small number of sibs and/or the age differences between sib pairs that required using different levels of norms on the WRAT. This latter speculation gains some support from the fact that the WRAT scores generally declined (nonsignificantly) in the follow-up testing, and the majority of the children were above the age limit for the level 1 norms.

Personality. Problematic behaviour as determined by the personality inventory for children was reported 2.5 times more frequently (15:6) by mothers of the SCA cases than by mothers of the control group (Table IX). In terms of general areas of problem behaviour, 47% of the reported problems are in cognitive development and academic performance, as opposed to only 17% of problems in the same area reported by mothers of the control group. In addition, only four of the mothers of the SCA group reported no significant problem behaviour, as opposed to seven of the mothers of the control group. Regarding the breakdown within the SCA group, 60% of the reported problems occur in the 47,XXY cases and 30% occur in the 47,XXX cases. The majority of problems in both groups are almost equally divided in the areas of cognitive development and academic performance.

The 47,XXY cases differed significantly from the controls on four of the clinical scales: adjustment, achievement, development, and delinquency (Table X). In the first study, three of the four 47,XXY cases exceeded clinical

TABLE VII. Significant Strengths (S) and Weaknesses (W) on WISC-R Subtests for Children With Sex Chromosome Anomalies in the Control (C) Study and the Follow-Up (F) Study*

	I	Sim	A	V	C	DS	PC	PA	BD	OA	Cd	M
47,XXY												
7755 (C)		W′		St		x						x
7755 (F)	S								S′		W	
6196 (C)						x						x
6196 (F)		S		S′		W						
5420 (C)		W				x	S^1					x
5420 (F)	W					S			W			
9333 (C)						x	S′					x
46,XYY												
13549 (C)			W			x	W′				S′	x
13549 (F)						W′			S		W′	
8146 (C)	S′				W	x	S					x
8146 (F)									S′			
13681 (C)		W				x	S				W′	x
13681 (F)											W′	S
10763 (C)						x						x
47,XXX												
13003 (C)				S′		x						x
13003 (F)	W		S′									S
10565 (C)	W′					x						x
10565 (F)		W′				S′	W					S′
11407 (C)						x		S				x
11407 (F)			S′									
1115 (C)						x		S				x
1115 (F)			S									S
47,X/47,XX/47,XXX												
9342 (C)						x	W′				S	x
9342 (F)												

*Abbreviations: I, information; Sim, similarities; A, arithmetic; V, vocabulary; C, comprehension; DS, digit span; PC, picture completion; PA, picture arrangement; BD, block design; OA, object assembly; CD, coding; M, Mazes; x, not administered. All differences noted are significant at $P < .05$; a prime indicates signifance at $P < .01$.

significance on the adjustment, achievement, and development scales, as opposed to none of the four in the control group. Although a significant difference was also evident between the 47,XXY cases and their controls on the delinquency scale, it should be emphasized that none of the delinquency scores reached clinical significance.

In the follow-up study, the scores on the adjustment scale declined; however, they continued to exceed clinical significance. A similar decline was evident on other scales, and only two cases were marginally above the

TABLE VIII. Academic Achievement Test Results in Children With Sex Chromosome Anomalies, Their Sibs, Controls, and Control Sibs

		WRAT (standard scores)					
		Reading		Spelling		Arithmetic	
Group	N	Mean	SD	Mean	SD	Mean	SD
47,XXY	4	95.75	10.14	96.00	7.48	94.25	3.20
47,XXY sib	1	111.00	—	105.00	—	115.00	—
Controls	4	105.25	6.60	101.25	3.86	96.00	7.83
Control sibs	2	124.50	14.85	109.50	26.16	108.00	29.70
47,XYY	4	121.00	36.21	111.50	27.81	105.00	10.98
47,XYY sibs	3	117.33	15.70	114.33	16.29	113.00	7.00
Controls	4	104.50	14.16	98.50	5.97	96.50	6.76
Control sibs	3	115.67	11.68	116.67	8.50	105.67	5.69
47,XXX	4	101.25	15.54	100.00	15.47	89.75	9.00
47,XXX sib	1	105.00	—	102.00	—	97.00	—
Controls	4	110.25	7.32	109.25	13.00	89.00	4.55
Control sibs	2	114.50	6.36	117.50	17.68	109.50	6.36
45,X/46,XX/47,XXX	1	111.00	—	99.00	—	90.00	—
Totals							
SCA	13	106.77	23.22	102.46	17.62	96.54	9.72
SCA sibs	5	113.60	12.40	110.40	12.84	110.20	8.93
Controls	12	106.67	9.39	103.00	9.09	93.83	6.90
Control sibs	7	118.71	12.27	115.00	14.46	107.43	12.95
Follow-up evaluations							
SCA	11	97.55	13.99	96.18	18.24	86.36	13.05

TABLE IX. Numbers of General Areas of Reported Problem Behavior (PIC) in Children With Sex Chromosome Anomalies, Their Sibs, Controls, and Control Sibs

	47,XXY	47,XYY	47,XXX	Sibs of cases	Controls	Sibs of controls
N	4	3	4	5	12	7
General adjustment	3 (2)*	0	2	1	1	2
Cognitive development	4 (2)	0	3 (3)	1	1	1
Personality/family	2 (1)	1		2	4	3
Total No. of areas	9 (5)	1 (1)	5 (3)	4	6	6
No clinically significant problems reported	0 (0)	2 (2)	1 (1)	2	7	3

*Follow-up evaluations shown in parentheses.

TABLE X. Personality Inventory Clinical Scale Test Results in 47,XXY Children, Sibs, Controls, and Control Sibs

	PIC (T scores)							
	47,XXY (N = 4)		Controls (N = 4)		47,XXY sib (N = 1)		Control sibs (N = 2)	
	Mean	SD	Mean	SD	Mean	SD	Mean	SD
Adjustment	66.50†	8.23	44.50	6.19	71.00	—	51.50	10.61
Achievement	63.00†	4.00	47.50	6.81	69.00	—	45.00	4.24
Intellectual screening	53.25	6.85	48.00	5.72	36.00	—	51.50	19.09
Development	63.50†	1.73	46.50	4.12	59.00	—	44.00	7.07
Somatic concerns	55.75	9.74	48.75	5.62	64.00	—	44.00	4.24
Depression	58.50	5.26	52.75	5.68	57.00	—	60.00	9.90
Family relations	57.50	14.36	48.50	9.98	45.00	—	43.00	2.83
Delinquency	67.75*	9.91	49.75	9.29	64.00	—	40.00	0.00
Withdrawal	49.50	3.32	47.50	3.32	61.00	—	59.50	13.44
Anxiety	57.00	10.95	56.25	5.12	75.00	—	72.00	16.97
Psychosis	51.75	3.50	45.50	5.45	61.00	—	53.50	10.61
Hyperactivity	49.75	3.40	44.00	3.83	54.00	—	33.50	7.78
Social skills	51.25	5.44	46.00	6.48	70.00	—	55.00	5.66

*Significantly different from controls at P = .057.
†Significantly different from controls at P = .029.

clinical significance levels on the development scale, whereas all of the scores on the delinquency scale declined.

The 47,XXX cases differed significantly from the controls only on the achievement scale, and the scores of two of four 47,XXX cases reached clinical significance as opposed to none of four in the control group (Table XI).

In the follow-up study, three of the four achievement scale scores declined, and all scores were below the level of clinical significance. The 47,XYY cases differed from the control group on only one clinical scale—hyperactivity, with the control group scoring significantly higher on this scale (Table XII). In addition, three of the four controls reached clinical significance on this scale, as opposed to none of the 47,XYY cases. No difference on the hyperactivity scale was evident in comparison of the 47,XYY cases with their three sibs. The 47,XYY cases all scored higher than their sibs on the depression and withdrawal scales, but none of the scores reached clinical significance.

Little change in this group was evident on follow-up, and scores mostly declined or remained the same. None of the scores for any of the 47,XYY cases reached clinical significance on the clinical scales in either assessment.

TABLE XI. Personality Inventory for Children. Clinical Scale Test Results in 47,XXX Children, Sibs, Controls, and Control Sibs

	PIC (T scores)							
	47,XXX (N = 4)		Controls (N = 4)		47,XXX sibs (N = 1)		Controls sibs (N = 2)	
	Mean	SD	Mean	SD	Mean	SD	Mean	SD
Adjustment	62.00	10.39	54.00	3.46	52.00	—	48.50	3.54
Achievement	66.50*	15.61	44.25	5.85	50.00	—	46.50	17.68
Intellectual screening	68.75	19.59	48.75	8.66	45.00	—	61.50	7.78
Development	63.50	9.15	48.50	8.81	49.00	—	51.00	16.97
Somatic concerns	57.50	5.20	54.00	12.25	58.00	—	51.50	14.85
Depression	53.75	13.89	52.00	8.04	40.00	—	46.00	11.31
Family relations	49.00	5.89	46.50	2.52	43.00	—	46.00	4.24
Delinquency	60.75	11.18	49.75	7.93	40.00	—	48.00	1.41
Withdrawal	54.75	7.72	47.50	6.76	39.00	—	56.00	18.38
Anxiety	55.25	8.96	56.76	10.59	45.00	—	54.00	4.24
Psychosis	53.50	9.95	53.00	2.31	43.00	—	41.00	2.83
Hyperactivity	50.25	8.38	57.00	3.27	58.00	—	50.00	1.41
Social skills	58.25	8.66	50.50	7.72	36.00	—	43.00	1.41

*Significantly different from controls at P = .057.

DISCUSSION

A unique aspect of the Winnipeg study has been the evaluation of dysmorphic features in children with sex chromosome abnormalities. Our follow-up evaluations confirm that these fortuitously ascertained children with sex chromosome anomalies do not have a high frequency of abnormal dysmorphic findings. However, our relatively crude measures of dysmorphology do indicate a trend towards greater variability between sibs when one has a sex chromosome anomaly compared to control sib pairs. Although no obvious phenotype has been described for children with 47,XXY, 47,XYY, or 47,XXX, it is possible that detailed investigations on larger numbers of similarly ascertained children will reveal that they resemble each other more than has been appreciated previously.

In growth parameters, we can confirm the increase of stature in 47,XXX girls [17], which is present from early childhood. In addition, there are signs suggesting that the 47,XYY boys become increasingly tall with age, as has been reported in other studies [18–20]. Although the 47,XXY boys also show a tendency for increased height, this is not as marked as that seen in 47,XYY and 47,XXX.

The finding of deficits in verbal ability in the 47,XXY and 47,XXX cases, both in terms of significantly lower verbal IQs and the greater preponderance

TABLE XII. Personality Inventory for Children Clinical Scale Test Results in 47,XYY Children, Sibs, Controls, and Control Sibs

	PIC (T scores)							
	47,XYY (N = 4)		Controls (N = 4)		47,XYY sibs (N = 3)		Controls sibs (N = 3)	
	Mean	SD	Mean	SD	Mean	SD	Mean	SD
Adjustment	50.00	3.46	61.75	12.20	51.67	5.77	47.33	2.89
Achievement	47.67	9.29	55.75	10.78	41.00	3.46	43.33	5.69
Intellectual screening	42.00	14.73	52.00	12.94	43.00	12.29	52.67	5.03
Development	48.00	10.15	53.75	10.15	46.00	11.79	43.67	5.03
Somatic concerns	56.00	14.42	61.25	18.08	60.33	19.40	45.33	9.61
Depression	47.00	5.57	64.25	15.09	40.67	0.58	43.00	5.57
Family relations	42.67	4.51	56.50	12.37	44.33	3.06	50.33	16.17
Delinquency	54.00	6.24	49.75	7.50	60.67	14.43	46.67	0.58
Withdrawal	49.33	7.77	46.50	6.76	41.33	2.31	44.67	5.69
Anxiety	54.00	12.00	67.50	14.39	45.67	5.69	47.00	3.61
Psychosis	51.33	7.37	60.25	8.46	43.00	4.00	44.33	2.31
Hyperactivity	43.00	4.00	61.75*	10.87	44.33	7.77	50.00	7.55
Social skills	43.67	6.66	56.25	11.09	42.33	3.51	47.00	5.20

*, Significantly different from cases at P = .057.

of significant verbal-performance discrepancies in favor of higher performance IQs, is consistent with previous findings [17,18,21,22]. However, as Kaufman [16] noted, "... virtually the entire V[erbal]-P[erformance] literature... is beset by contradictions and a lack of success in identifying characteristic patterns for various groups." Although this certainly should not minimize the fact that a markedly higher proportion of children with significantly higher performance IQs have been identified in the 47,XXY and 47,XXX groups, this finding in itself may not provide sufficient information to hypothesize on more specific commonalities in these groups, for either diagnostic or remediation purposes. For example, although our sample was quite small, no consistent shared abilities or influences that might account for this discrepancy were evident in the SCA group. It is hoped that further analysis of the existing data in other studies using the approach proposed by Kaufman can explore this area of potential information.

Difficulties in academic achievement were also most prominent for the 47,XXY and 47,XXX groups, although these difficulties were reflected more in mothers' reports than in actual comparisons with the control group. The mean scores in reading and spelling were lower than those of the controls but not significantly so.

It should also be noted that the 47,XYY group performed directly opposite to the other two groups in all areas of intellectual functioning and academic achievement. Not only was there no significant difference between the 47,XYY group and the controls in the intellectual measures, the 47,XYY group demonstrated a verbal-performance discrepancy significantly higher on the verbal scales, achieved higher mean scores than the controls on the academic achievement scales, and were not reported by mothers as displaying problems in the cognitive and achievement areas.

The previously discussed problems in intellectual and academic functioning in the 47,XXY and 47,XXX groups were also reflected in mothers' reports of psychologic functioning, and, although some other problem behaviours were reported, the majority were clearly in the areas of achievement and cognitive development. Here also, the 47,XYY group results were in striking contrast, with the absence of reported problem behaviours in these areas. It would appear, therefore, that from a psychologic perspective the 47,XXY and 47,XXX groups would be at the greatest risk, particularly with respect to learning and/or school related problems.

It must be emphasized that the individual follow-up studies on children with sex chromosome anomalies being carried out in various centers suffer from problems of small sample sizes and that comparison between them may be difficult because of the different instruments and methodologies being used. It is important to note, however, the similarities in growth, behaviour, and academic functioning reported for these children. Through such studies, additional useful information for the management of such children and the counselling of their families is still forthcoming.

ACKNOWLEDGMENTS

We would like to thank the children involved in these studies and their parents for their willingness to participate in the follow-up examinations. J.A.E. wishes also to thank the National Health Research and Development Programme for continued support through a National Health Research Scholarship.

J.L.H. acknowledges funding from the Department of Health and Welfare, which allowed the initial chromosome study.

REFERENCES

1. Hamerton JL, Ray M, Abbott J, Williamson C, Ducasse GC: Chromosome studies in a neonatal population. Can Med Assoc J 106:776–779, 1972.

2. Hamerton JL, Canning N, Ray M, Smith S: A cytogenetic survey of 14,069 newborn infants. I. incidence of chromosome abnormalities. Clin Genet 8:223–243, 1975.
3. Evans JA, Canning N, Hunter AGW, Martsolf JT, Ray M, Thompson DR, Hamerton JL: A cytogenetic survey of 14,069 newborn infants. III. An analysis of the significance and cytologic behavior of the Robertsonian and reciprocal translocations. Cytogenet Cell Genet, 20:96–123, 1978.
4. Grant WW, Hamerton JL: A cytogenetic survey of 14,069 newborn infants. II. Preliminary clinical findings on children with sex chromosome anomalies. Clin Genet 10:285–302, 1976.
5. Evans JA, de von Flindt R, Greenberg C, Ramsay S, Hamerton JL: A cytogenetic survey of 14,069 newborn infants. IV. Further follow-up on the children with sex chromosome anomalies. Birth Defects 18(4):169–184, 1982.
6. Clifford, HT, Stephenson W: "An Introduction to Numerical Taxonomy." New York: Academic Press, 1975.
7. Feingold M, Bossert WH: Normal values for selected physical parameters. An aid to syndrome delineation. Birth Defects X(13), 1974.
8. Wechsler D: "Wechsler Intelligence Scale for Children—Revised." New York: The Psychological Corporation, 1974.
9. Dunn L: Peabody Picture Vocabulary Test. Minneapolis: American Guidance Service, 1959.
10. Jastak J, Jastak S: "Wide Range Achievement Test." Wilmington, DE: Jastak Associates, Inc., 1978.
11. Wirt R, Lachar D, Klinedienst J, Seat P: "Multidimensional Description of Child Personality: A Manual for the Personality Inventory for Children." Los Angeles: Western Psychological Services, 1977.
12. Lachar D, Gdowski C: "Actuarial Assessment of Child and Adolescent Personality: An Interpretation Guide for the Personality Inventory of Children Profile." Los Angeles: Western Psychological Services, 1979.
13. Doll E: "Vineland Social Maturity Scale." Minneapolis: American Guidance Service, 1965.
14. McCarthy D: "McCarthy Scales of Children's Abilities." New York: The Psychological Corporation, 1972.
15. Siegel S: "Nonparametric Statistics for the Behavioral Sciences." New York: McGraw-Hill Book Company, Inc., 1956.
16. Kaufman A: "Intelligent Testing With the WISC-R." New York: John Wiley & Sons, 1979.
17. Stewart DA, Netley CT, Park E: Summary of clinical findings of children with 47,XXY,47,XYY, and 47,XXX karyotypes. Birth Defects 18(4):1–5, 1982.
18. Stewart DA, Bailey JD, Netley CT, Rovet J, Park E, Cripps M, Curtis JA: Growth and development of children with X and Y chromosome aneuploidy from infancy to pubertal age: The Toronto study. Birth Defects 18(4):99–154, 1982.
19. Ratcliffe SG, Tierney I, Nshaho J, Smith L, Springbett A, Callan S: The Edinburgh study of growth and development of children with sex chromosome abnormalities. Birth Defects 18(4):41–60, 1982.
20. Nielsen J, Sørensen AM, Sørensen K: Follow up until age 7 to 11 of 25 unselected children with sex chromosome abnormalities. Birth Defects 18(4):61–97, 1982.
21. Bender B, Pennington M, Salbenblatt J, Robinson A: Speech and language development in 41 children with sex chromosome anomalies. Pediatrics 71:262–276, 1983.

22. Walzer S, Graham J, Bashir A, Silbert A: Preliminary observations on language and learning in XXY boys. Birth Defects 18(4):185–192, 1982.

APPENDIX A. INDIVIDUAL DATA ON CHILDREN WITH SEX CHROMOSOME ANOMALIES

Scores on psychologic factors obtained in the follow-up study are immediately followed in parentheses by the scores obtained in the initial study [for further details of the initial studies see 3 and 5]. Interpretative statements (given in quotes) are determined by the T-scores on the PIC scales and follow the proposed interpretative hypotheses set forth by Lachar and Gdowski [12].

47,XXY

Case 1 (NB7755)

Physical findings. This boy was studied at 9 years, 3 months, and 12 years of age. In that time interval, his growth parameters were essentially unchanged. He remained at the 25th–50th percentile for height and head circumference (OFC) and at the 50th–75th percentile for weight. His nipples were close-set and he had tapering fingers, genu recurvatum, and very lax elbows. He has a male sib aged 22 months younger who also has lax elbows. Case 1 is relatively smaller in most growth parameters than his brother but is relatively more hyperteloric and telecanthic.

Personality. "The description of this child's behaviour suggests that psychologic psychiatric evaluation could assist in the remediation of current problems. Inventory responses suggest the possibility of poor school performance and associated academic retardation. Teacher observation of limited concentration or difficulty in completion of classroom assignments may be associated with below age expectation performance in reading, mathematics, or spelling."

Both of the hypotheses were evident at the first follow-up; however, a related hypothesis reflecting possible distractability in the classroom and the need for specific remedial action in the school was not reported this time.

Intelligence. WISC-R: verbal IQ 78 (86); performance IQ 101 (111); full-scale IQ 87 (97); PPVT IQ 106 (107).

Academic achievement. WRAT: reading 87 (89); spelling 71 (86); arithmetic 79 (93).

Case 3 (NB 6196)

Physical findings. This boy was studied at 9 years, 8 months, and 12 years, 7 months. He has shown a slight fall off in growth in height, weight,

OFC, outercanthal distance, and hand and ear length. Height and weight are now at the 75th–90th percentile and OFC at the 25th–50th percentile. He has slight clinodactyly of the right fifth finger, mild syndactyly of the second and third toes, mild diastatis recti, and slightly lax elbows. He has five whorls and five ulnar loops on the fingertips and a right I4 pattern.

Personality. "The description of this child's behaviour suggests that a psychological/psychiatric evaluation could assist in the remediation of current problems."

This hypothesis was evident at the first follow up, however two other hypotheses relating to school achievement and one hypothesis relating to family problems were not reported at the most recent assessment.

Intelligence. WISC-R: verbal IQ 91 (91), performance IQ 109 (115), full-scale IQ 100 (102); PPVT IQ 111 (108).

Academic achievement. WRAT: reading 90 (96), spelling 86 (96), arithmetic 76 (92).

Case 4 (NB 5420)

Physical findings. This boy was examined at 9 years, 10 months, and 12 years, 6 months. He has shown a relative increase in both height and weight. His height is now over the 90th percentile and his weight at the 75th–90th percentile. His head circumference remains above the 90th percentile. Ocular measurements showed a relative decrease. He was the smallest 47,XXY boy at birth with weight length and OFC around the 25th percentile. He had several minor dysmorphic findings including inverted nipples with a left accessory nipple, stellate patterned irides, a slightly droopy left eyelid, a pigmented nevus on the left eyelid, a nevus on the right shoulder, one café-au-lait spot, and slightly lax elbows. The dermatoglyphics were unusual with seven ulnar loops, one whorl, and two tented arches; a t' triradius on the left with a hypothenar loop and the *d* triradii were absent.

Personality. "This inventory was completed by an informant who may have been hesitant to describe problematic child behaviours and might be resistant to viewing existing behaviour problems as being caused by psychologic or interpersonal factors. The resulting scale elevations may minimize existing problems. Inventory responses suggest the possibility of poor school performance and associated academic retardation. Teacher observation of limited concentration or difficulty in completion of classroom assignments may be associated with below age expectation performance in reading, mathematics, or spelling. Teachers are likely to observe that this child is easily distracted away from the tasks at hand by ordinary classroom stimuli. Classroom performance may reflect poor study skills. A psychologic psychiatric evaluation may result in specific recommendations to school personnel.

The home is likely to reflect the impact of divorce or separation and may be characterized by instability and conflict. Parental inconsistency in setting limits is suggested. The resulting parent-child interaction may contribute to the development of child behaviour problems."

The only change in hypotheses obtained at the most recent evaluation is the addition of hypotheses suggesting increased defensiveness on the mother's part in reporting. (This family was in the midst of some serious psychologic problems at the time these data were obtained.)

Intelligence. WISC-R: verbal IQ 90 (105), performance IQ 87 (93), full-scale IQ 87 (100); PPVT IQ 108 (95).

Academic achievement. WRAT: reading 94 (110), spelling 90 (104), arithmetic 80 (93).

47,XYY

Case 5 (NB 13681)

Physical findings. This boy was slightly premature and had a birth weight below the 5th percentile for term, OFC at the 10th percentile, and length at the 50th percentile. Early growth was good. His follow-up visits were at 7 years, 7 months, and 10 years, 8 months. He has shown relative increases in height, weight, OFC, and outercanthal measurements. Height and weight are now at the 50th–70th percentile and OFC above the 90th percentile. He is of Canadian-Indian descent and has slight epicanthus, low lateral anterior hairline, broad nasal bridge, slight anteversion, and slightly flat maxilla in keeping with his ethnic origin. His ears had prominent antihelices. He had medial flare bilaterally and slight ptosis. Myopia was present, and he wore corrective lenses. The fifth fingers were slightly short and there were I4 patterns bilaterally.

Personality. Personality inventory for children was not administered.

Intelligence. WISC-R: verbal IQ 75 (91), performance IQ 91 (101), full-scale IQ 81 (95); PPVT IQ 107 (100).

Academic achievement. WRAT: reading 79, (74) spelling 72 (78), arithmetic 78 (91).

Case 7 (NB 8416)

Physical findings. This boy was examined at 8 years, 11 months, and 12 years of age. Height, head circumference, and chest circumference have all shown relative increases. Height is now at the 90th percentile, weight at the 50th–75th percentile, and head circumference at the 75th–90th percentile. He had a narrow forehead with a widow's peak, a grade 2/6 systolic ejection murmur, and very lax joints and was also myopic. He had a sister 27 months younger. Both had hyperextensible elbows. Case 7 was relatively taller than his sister, but she had longer hands and greater head and chest circumferences for her age.

Personality. There was no indication based on the PIC responses that further psychologic/psychiatric evaluation was necessary at this time. This is the same hypothesis generated at the last visit.

Intelligence. WISC-R: verbal IQ 135 (137), performance IQ 115 (114), full-scale IQ 128 (128); PPVT IQ 131 (133).

Academic achievement. WRAT: reading 123 (149), spelling 132 (146), arithmetic 114 (111).

45,X/46,XY/47,XYY

Case 8 (NB 13549)

Physical findings. This boy's 47,XYY cell line was 84% at birth. OFC and weight at birth were at the 10th percentile and length at the 25th percentile. He was last studied at 7 years, 5 months, and 10 years, 4 months, of age. In this time, he has shown substantial acceleration of growth in height and weight, and his chest measurements are also increased. Height and weight have increased from below the 10th percentile to the 25th percentile. Head circumference has remained at the 25th–50th percentile.

His innercanthal and interpupillary measurements were at the third percentile giving a close-eyed appearance. He had flat helices and slight clinodactyly of the fifth fingers. Dermatoglyphics were normal with a small hypothenar loop on the right.

Case 8's two sibs were also studied. They were both girls, one 42 months older and one 17 months older. All had slightly unusually shaped ears with overfolded helices similar to those of their father. One sister had no parietal hair whorl; in the other it was poorly formed. One sister was missing the *c* triradius bilaterally and the mother had an absent *c* triradius on the right hand only. Case 8 was relatively smaller than either of his sisters in all growth parameters except head circumference and chest measurements, although all these children had growth parameters below the 50th percentile.

Personality. There was no indication on the PIC responses given that further psychologic/psychiatric evaluation was necessary at this time.

This hypothesis was not evident at the first assessment; significant concerns about somatic problems were reported, but this degree of concern was not evident at this time.

Intelligence. WISC-R: verbal IQ 114 (111), performance IQ 126 (133), full-scale IQ 122 (124); PPVT IQ 104 (114).

Academic achievement. WRAT: reading 108 (105), spelling 99 (113), arithmetic 106 (102).

47,XXX

Case 9 (NB 1115)

Physical findings. This girl was studied at 10 years, 11 months, and 13 years, 7 months. Her growth parameters were basically unchanged except for a slight reduction in weight and a relative increase in chest circumference leading to a reduction in internipple to chest circumference ratio. Her height remains above the 90th percentile. Her weight is at the 75th to 90th percentile, and her head circumference is relatively small at the 10th–25th percentile. The only minor dysmorphic findings were slight clinodactyly of the fifth fingers and a tendency for keloid formation after mild trauma. The left lateral incisor has still not erupted. She has had regular menses since 11 years of age.

The girl's brother, 19 months younger than herself, was also examined. They had no dysmorphic signs in common. Case 9 was considerably taller and heavier for her age than her brother, with largest chest circumference and hand length measurements, but her OFC and ocular measurements were relatively smaller.

Personality. "Inventory responses suggest the possibility of poor school performance and associated academic retardation. Teacher observation of limited concentration or difficulty in completion of classroom assignments may be associated with below age expectation performance in reading, mathematics, or spelling."

Although similar concerns about school achievement were evident at the first evaluation, those reported a follow-up were markedly less problematic. Also, concerns relating to general adjustment, developmental problems that might be manifest in school difficulties, and concerns about depression were not evident.

Intelligence. WISC-R: verbal IQ 88 (95), performance IQ 106 (112), full-scale IQ 96 (102); PPVT IQ 93 (85).

Academic achievement. WRAT: reading 88 (101), spelling 89 (95), arithmetic 89 (83).

Case 10 (NB 11407)

Physical findings. Since birth, almost all growth parameters in this child have been above the 50th percentile. When examined at 8 years, 3 months, her innercanthal and interpupillary distances were over the 97th percentile, giving a wide-eyed and telecanthic look. At 11 years, 1 month, she was slightly less hyperteloric. Height, weight, and OFC were all above the 90th percentile. She was of Canadian-Indian ancestry and had slightly upslanting palpebral fissures, slight epicanthus, and broad nasal bridge attributable to her ethnic origin. Her ears had a thin inferior helix and the antitragi were hypoplastic. This latter finding was apparently a familial trait. There was a slight medial flare and a wandering right eye. Dermatoglyphics showed ten fingertip whorls, again in keeping with her native ancestry, and she had an I3 pattern on the right hand. Neuromuscular findings in this child have been abnormal since birth, and she continues to be slightly hypotonic and hyporeflexic. She started menstruating at 11 years of age.

Personality. "Inventory responses suggest that the informant ascribed the most virtuous of behaviours and denied minor, commonly occurring problems in the description of this child. This style of responding may have affected other scale elevations such that the resulting profile interpretation minimizes current problems. Children seen for evaluation who obtain similar scores are less likely to be characterized by disobedient, aggressive, or antisocial behaviour.

Concern about limited intellectual endowment may be raised by a psychological evaluation. Poor school performance may be associated with an inability to retain and apply basic skills taught in previous grades."

Both of these hypotheses were evident at the first evaluation; however, the concerns about intellectual ability were noticeably less on follow-up, and previous concerns about school achievement were not reported.

Intelligence. WISC-R: verbal IQ 82 (82), performance IQ 91 (85), full-scale IQ 85 (82); PPVT IQ 85 (95).

Academic achievement. WRAT: reading 95 (100), spelling 103 (102), arithmetic 79 (91).

Case 11 (NB 13003)

Physical findings. This child's growth parameters have been well above the 50th percentile since birth. She was seen at 7 years, 9 months, and 10 years, 9 months. Her growth in all parameters has continued to accelerate, especially chest circumference and internipple distance. Weight and OFC are above the 75th percentile and height above the 90th percentile. Bilateral trigger thumbs were repaired surgically, and she has no other major anomalies. Her parietal hair whorl was poorly formed and the posterior hairline somewhat low. She had squared helices, mild medial flare, mild malar flattering, a broad palate, and slight genu valgum. She has been receiving specialized orthodontic care as well as therapy for nasal speech. She has slight axillary and pubic hair but little breast development and is not menstruating.

Personality. "Inventory responses suggest that the informant ascribed the most virtuous behaviours and denied minor, commonly occurring problems in the description of this child. This style of responding may have affected other scale elevations such that the resulting profile interpretation minimizes current problems. Children seen for evaluation who obtain similar scores are less likely to be characterized by disobedient, aggressive, or antisocial behaviour."

No hypotheses were generated at the first evaluation, and, despite the potential "minimization," there are no indications of any significant shifting in reporting on the clinical scales.

Intelligence. WISC-R: verbal IQ 115 (107), performance IQ 92 (96), full-scale IQ 105 (102); PPVT IQ 122 (119).

Academic achievement. WRAT: reading 113 (121), spelling 120 (120), arithmetic 88 (102).

Case 12 (NB 10565)

Physical findings. This was another Canadian-Indian child. Only weight was recorded at birth and was at the 50th percentile. Height measurements accelerated from the 50th percentile at 6 months of age to over the 90th percentile at 8 years, 6 months. OFC measurements have consistently been around the 25th percentile. When last examined at age 11 years, 3 months, her growth parameters were essentially unchanged. She remains above the 90th percentile in height, between the 50th and 75th percentiles in weight, and between the 25th and 50th percentiles in OFC. She has fine pubic and axillary hair but is not yet menstruating.

Consistent with her ethnic origin, she showed slightly upslanting palpebral fissures, slight epicanthus, and a broad nasal bridge. In addition, she had mild posterior rotation of the ears, and the helices were thinned in the middle portion. Her joints were lax. Dermatoglyphics were essentially normal with a *t'* axial triradius on the right, I4 patterns bilaterally, and an I3 pattern on the right.

Personality. "Inventory responses suggest that the informant ascribed the most virtuous of behaviours and denied minor, commonly occurring problems in the description of this child. This style of responding may have affected other scale elevations such that the resulting profile interpretation minimizes current problems. Children seen for evaluation who obtain similar scores are less likely to be characterized by disobedient, aggressive, or antisocial behaviour.

Concern about limited intellectual endowment may be raised by a psychological evaluation. Poor school performance may be associated with an inability to retain and apply basic skills taught in previous grades."

As was noted in the previous case (No. 11), both of these hypotheses were evident at the first evaluation, and on follow-up concerns about intellectual ability were less than those previously reported. In addition, concerns regarding general adjustment, academic achievement, and developmental problems that might be manifest in classroom difficulties were not reported.

Intelligence. WISC-R: verbal IQ 62 (72), performance IQ 85 (78), full-scale IQ 71 (73); PPVT IQ 77 (73).

Academic achievement. WRAT: reading 85 (83), spelling 97 (83), arithmetic 75 (83).

45,X/46,XX/47,XXX

Case 13 (NB 9342)

Physical findings. This child was found to be a mosaic at birth with the 45,X line predominating (67:10:20). Since then, there has been an apparent decrease in both abnormal cell lines, although the monosomic line still predominates (60:37:3). At birth, weight was at the 25th percentile, OFC between the 25th and 50th percentiles, and height at the 75th percentile. However, growth has been relatively slow since then and was below the 3rd percentile at 8 years, 9 months. When reexamined at 11 years, 1 month, she remained small but had some relative increase in weight, innercanthal distance, ear length, and chest measurements. Her height is still below the 10th percentile. Her OFC is between the 25th and 50th percentiles and her weight between the 10th and 25th percentiles. She is a healthy and very muscular child with no stigmata of Turner syndrome. She had a medial flare to the eyebrows and a slight strabismus. She is not yet menstruating.

Personality. "There was no indication based on the PIC responses given that further psychologic/psychiatric evaluation was necessary at this time."

This is the same hypothesis generated at previous visit.

Intelligence. WISC-R: verbal IQ 102 (108), performance IQ 100 (101), full-scale IQ 101 (105); PPVT IQ 112 (96).

Academic achievement. WRAT: reading 111 (116), spelling 99 (102), arithmetic 90 (99).

Chromosome Examination of 20,222 Newborn Children: Results From a 7.5-Year Study in Århus, Denmark

Johannes Nielsen, Mogens Wohlert, Jens Faaborg-Andersen, Gunnar Eriksen, Karl Brogaard Hansen, Lone Hvidman, Bente Krag-Olsen, Ib Moulvad, and Poul Videbech

Cytogenetic Laboratory, Århus Psychiatric Hospital, Risskov, and Gynecological-Obstetric Department Y, Århus Municipal Hospital, Denmark

INTRODUCTION

The present chromosome investigation is part of a larger prospective study of the influence of environmental factors on pregnancy, birth, and the fetus. In Denmark, with approximately 51,000 births per year, approximately 400 children with chromosome abnormalities are born yearly, and induced abortion is carried out on 50 fetuses on account of chromosome abnormalities diagnosed prenatally.

Apart from Down syndrome, Edwards syndrome, and Patau syndrome, relatively few children with chromosome abnormalities are diagnosed at birth; in fact, the majority of these children are never diagnosed. A total of 15% of all sex chromosome abnormalities are likely to be diagnosed throughout life. In this chapter, we focus especially on the sex chromosome abnormalities.

Follow-up studies of 203 unselected children with sex chromosome abnormalities in the U.S., Scotland, Canada, and Denmark [1–7] have shown that children with certain sex chromosome abnormalities, especially girls with triple X and boys with the XXY or XYY syndromes have increased risk compared to sibs and children with normal chromosomes to have deviations

Birth Defects: Original Article Series, Volume 22, Number 3, pages 209-219

in their mental development. This is mainly in regard to learning problems and delayed speech and motor development.

On the background of the present studies, we find that there are strong indications that this risk can be eliminated completely or partly by diagnosing these children at birth and giving information, counseling, assistance, and supportive measures of various kinds early during childhood and later on and for certain groups hormone treatment and realistic information and counseling from puberty onwards to the extent that there is a need for it [8].

On the basis of our experience of these conditions in chromosome examination of 11,148 children in Århus from 1969 to 1974 [3,9], we started a new investigation on October 1, 1980, at the Cytogenetic Laboratory in Risskov in close cooperation with the Gynecological-Obstetric Department in Århus, an investigation with the following main goals: 1) to gain further knowledge concerning a number of environmental conditions for pregnancy, birth, fetal development, and de novo chromosome abnormalities; the environmental conditions include parental diseases, working conditions, drug and alcohol consumption, and smoking; 2) to gain sufficient knowledge to advise pregnant women better concerning sick leave at the beginning of or early during pregnancy because of environmental risks at work; 3) to inform and counsel parents of newborn children with chromosome aberrations of any kind; 4) to diagnose at-risk families with inheritable chromosome disorders that in future pregnancies might lead to the birth of children with mental retardation and/or congenital abnormalities; such families can be offered genetic counselling and the possibility of having prenatal chromosome examination in future pregnancies; 5) we also hope to be able to make follow-up studies of children who may have an increased risk of developmental deviation for reasons such as maternal exposure at work; drug treatment during pregnancy with psychotropic, antiepileptic, or other drugs; alcohol abuse; and exposure to organic solvents during pregnancy. We are also interested in follow-up studies of babies who were small for date and who have been exposed to birth complications of various kinds.

METHODS

The investigation is prospective and longitudinal and is carried out using a questionnaire and case history study. In the Århus area, where there are approximately 3,000 births per year, all pregnant women are asked to participate in the study by filling in a questionnaire at week 33 of pregnancy together with the father of the child. The questionnaire is returned to the obstetric department at the time of admission for birth. Approximately 85%

of all parents have filled in the nine-page questionnaire, which requests detailed information about their occupation before and during pregnancy; previous diseases; tobacco, drug and alcohol consumption; and X-ray exposure of both parents.

Information concerning pregnancy, delivery, and the child is recorded by the obstetricians participating in the study in a five-page case record. Information includes age and marital state of the parents; number of previous births, abortions, or stillbirths; disorders and congenital aberrations in previous children; indication for amniocentesis and results if performed; previous disorders in the mother; disorders during pregnancy; and detailed information concerning pregnancy, delivery, and the child.

All data are coded and entered in a computerized register, which, according to the Danish Register Law, does not allow the use of names and person identification numbers but only serial numbers. The register is thus an anonymous research register. At the present time, data from approximately 10,000 pregnancies have been entered in the register.

At the time of admission for birth, the prospective mothers are offered the opportunity to have chromosome analysis from a sample of umbilical cord blood taken immediately after the birth of the child. So far, 91% have accepted this offer, and many consider this an offer that ought to be made to all pregnant women in Denmark.

The results and significance of the chromosome examination are sent to the general practitioner together with information about the birth, usually within 2 weeks of delivery. However, the results of the chromosome examination are available from the Cytogenetic Laboratory 7 days after the birth. If a chromosome aberration is found, firsthand information is given to the parents by the chief of the Department of Obstetrics and Gynecology by telephone, and the parents are offered the opportunity to contact the chief of the Cytogenetic Laboratory in Risskov to get an appointment for further information either immediately or whenever they wish.

All families having a child with a chromosome abnormality are given information, counselled, and offered follow-up care. Practically all accept this offer. Families with a structural chromosome abnormality are offered family chromosome investigation and genetic counselling.

We routinely analyze 5-BUDR-stained metaphases. If we find abnormalities, we analyze a further 30 or more metaphases, and, if necessary, we also use other staining methods.

MATERIAL

As previously mentioned, detailed information about all pregnancies and births in the Århus area is included in the study, that is, approximately 3,000 births per year. We are, at present, preparing an analysis of all data from the first 6,000 pregnancies. We plan to continue the investigation for a number of years, and the chromosome examination of the newborn children will, most probably, continue as a permanent service investigation because of its preventative possibilities.

RESULTS

Postnatal Chromosome Examinations

Including a previous period of investigation from 1969 to 1974, together with the present period from 1980 to 1983, we have so far carried out chromosome examinations on a total of 20,222 newborn children. During the first period from 1969 to 1974, we completed chromosome examinations of 99.7% of all newborns in the area and during the last period of 90.7%.

There was no significant difference in the incidence of chromosome abnormalities between the two periods of investigation, and they are consequently added together in the present short chapter. We identified 58 children with sex chromosome abnormalities and 116 with autosomal abnormalities, giving an incidence of nine per 1,000 for all chromosome abnormalities. The six fetuses aborted as a result of prenatal diagnosis are also included in these figures (Table I).

The incidence of XXY syndrome was 1.5 per 1,000, XYY syndrome 1.1 per 1,000, and triple X syndrome 1.3 per 1,000. The incidence of Turner syndrome was 0.7 per 1,000, or one case per 2,000 girls (Table II).

In Denmark, with approximately 51,000 liveborn children per year, one would expect that 140 children would be born with a sex chromosome abnormality, but induced abortion is, at present, made in 10% of these as a result of prenatal diagnosis. In that there is no significantly increased mortality among children with sex chromosome abnormalities, one can expect that in Denmark, with 5 million inhabitants, there will be approximately 13,000 persons with sex chromosome abnormalities, 12,000 of whom have XXY, XYY, triple X, or Turner syndrome.

TABLE I. Chromosome Abnormalities Among 20,222 Newborn Children Examined in Århus 1969–1974 and 1980–1983

Chromosome abnormalities	Children with chromosome abnormalities	
	Total	Rate/1,000
Sex chromosomal	58	3
Autosomal	116	6
Total	174	9

TABLE II. Incidence of Sex Chromosome Abnormalities in 20,222 Newborn Children

Karyotype	Total	Rate/1,000 newborns	
47,XXY	16	1.53	
47,XYY	11	1.05	2.32
47,XXX	13	1.33	
45,X	7	0.72	
Others ♀	4	0.41	0.54
Others ♂	7	0.67	
Total	58		2.86

TABLE III. Incidence of Autosomal Abnormalities in 20,222 Newborn Children

Karyotype	Total	Rate/1,000 newborns	
+21	25	1.23	
+13, +18	5	0.25	
+8 and ring	3	0.15	0.60
Unbalanced abnormalities	4	0.20	
+mar	11	0.54	
13/14 translocation	22	0.09	
14/21 and 15/21 translocation	7	0.35	3.91
Reciprocal translocation	27	1.34	
Inversions*	5	0.25	
Others	7	0.35	
Total	116		5.74

*Cases of inversion 9, 14.32/1,000, are not included.

The incidence of Down syndrome was 1.23 per 1,000, and the incidence of other aneuploid or unbalanced autosomal chromosome abnormalities was 0.60 per 1,000. The incidence of balanced chromosome abnormalities, that is mainly translocations and inversions, was 3.91 per 1,000 (Table III).

In Denmark, with 51,000 newborns per year, one can expect that 300 children with autosomal abnormalities are born yearly, but induced abortion is, at present, performed for 15% of these after prenatal diagnosis. Adding the sex chromosomally abnormal, 25% of those who have aneuploid or unbalanced structural abnormalities will be aborted, among these appproximately 40% have Down syndrome.

We found as many persons with chromosome abnormalities among relatives as among the newborns. We thus found 18 persons with chromosome abnormalities for each 1,000 liveborn children (Table IV). This figure will continue to increase as more relatives are examined and new children are born to the families with inheritable structural chromosome abnormalities.

In this chapter, we do not deal with chromosome variants, but it might be mentioned that a study of the boys with enlarged Y chromosome [10] has shown that these boys suffered from intrauterine asphyxia leading to emergency caesarean section more frequently than expected.

Prenatal Chromosome Examinations

During the last investigation period from 1980 to 1983, during which chromosome examinations of 9,074 children were performed, prenatal chromosome examinations were made according to the indications used in Denmark [11,12].

Table V shows that, in spite of a comprehensive prenatal cytogenetic service from the Institute of Human Genetics at the University of Århus, the prenatal examinations led to induced abortions of only three of the 16 fetuses that might be expected to become mentally retarded. Amniocentesis was not performed in any of the nine mothers of children with Down syndrome. In Denmark, the incidence of children with aneuploid or unbalanced autosomal chromosome abnormalities has fallen by approximately 15% on account of induced abortion after prenatal diagnosis. There was an indication for amniocentesis in four of the nine mothers of children with Down syndrome, in three because of maternal age 35 years or more and in one because of Down syndrome in the family (Table VI). One of the four mothers did not want to have amniocentesis because under no circumstances did she want an induced abortion. In the other cases, the parents had not received sufficient information about the possibility of prenatal chromosome examination.

There is still a need for more information for the population in general and for physicians concerning the indications for prenatal chromosome ex-

TABLE IV. Chromosome Abnormalities Among 20,222 Newborns and Their Relatives Examined in Århus, 1969–1974 and 1980–1983

Newborns and their relatives	Individuals with chromosome abnormalities	
	Total	Rate/1,000
Newborns	174	9
Relatives	180	9
Total	354	18

TABLE V. Incidence of Chromosomal Abnormalities in 9,074 Children 1980–1983 and Induced Abortion After Prenatal Diagnosis*

Karyotype	Total	Induced abortions
+21	9†	—
+18	3	2
Del., deriv., ring	4	1
+mar	5	—
Translocations	24	—
Inversions	4	—
Others	2	—
Total	51	3

*Prenatal examinations = 453 (5% of all pregnancies).Induced abortions = six. This includes three fetuses with sex chromosome abnormalities (see Table VII; 1.3% of all prenatal examinations).
†Amniocentesis not done (see text).

TABLE VI. Parental Age Distribution of the Nine Children With Down Syndrome

Age		Other indications for prenatal examination
Maternal	Paternal	
26	30	—
20	30	—
31	27	—
30	36	—
24	25	—
34*	35	Down syndrome in the family
41†	43	—
39*	40	—
36*	37	—

*Were not given sufficient information about prenatal exam.
†Did not want prenatal examination.

amination, especially in the event of advanced maternal age. Table VII shows that induced abortion was made in as many fetuses with sex chromosome abnormalities as in those with autosomal abnormalities, that is, three children with XXY syndrome, triple X syndrome, and Turner syndrome out of a total of seven children with sex chromosome abnormalities who were diagnosed prenatally.

In Denmark, the incidence of children with sex chromosome abnormalities has decreased by 10% on account of induced abortion after prenatal examination. As a result of the diagnosis at birth of all children with chromosome abnormalities in the Århus area and the possibility of information and coun-

TABLE VII. Sex Chromosome Abnormalities in Liveborns and in Induced Abortions, Århus, 1980–1983

Karyotype	Total	Induced abortions
47,XXY	8	1
47,XYY	6	—
47,XXX	5	1
45,X	4	1
Others	5	—
Total	28	3*

*A total of seven sex chromosome abnormalities were diagnosed prenatally.

TABLE VIII. Fetuses With Sex Chromosome Abnormalities Diagnosed at Prenatal Chromosome Examination in Denmark From 1970 to 1982

	1970–1980		1981–1982		Total		
Karyotype	Diagnosed	Induced abortion	Diagnosed	Induced abortion	Diagnosed	Induced abortion	%
XXY	13	10	6	5	19	15	79
XYY	8	8	5	3	13	11	85
XXX	15	10	9	8	24	18	75
X	10	8	11	8	21	16	76
Total	46	36	31	24	77	60	78
Percent		78		77		78	

selling of the parents, and follow-up studies carried out at the Cytogenetic Laboratory in Risskov, half of the fetuses aborted, ie, those with sex chromosome abnormalities, would most probably have developed relatively normally and had a life of reasonable quality, as has been the case for practically all children with sex chromosome abnormalities we have followed in the Arhus area since we started chromosome examination of newborn children in 1969.

Table VIII shows that in Denmark induced abortion is made of approximately 80% of the fetuses with sex chromosome abnormalities diagnosed prenatally. To the authors, this appears to be discrimination against the approximately 12,000 people with sex chromosome abnormalities in Denmark, most of whom live virtually normal lives. It is our impression that fewer parents would choose to have an induced abortion of a fetus with sex chromosome abnormalities if they received more comprehensive information about the developmental possibilities for these children together with the offer of professional information, counseling, and support for their child with a sex chromosome abnormality if they chose not to have an induced abortion.

During the last few years, increasing resources have been used to offer prenatal chromosome examination in Denmark and elsewhere. At the same time, in our opinion, governments should allocate as much, or even more, resources to look for environmental factors that may lead to congenital disorders, including de novo chromosome abnormalities, as well as to deviations in the mental development of children in general.

Especially during the last part of pregnancy, exposure to drugs, alcohol, organic solvents and other agents may lead to quite severe deviations in behavior and learning ability and to a decrease in intelligence level. Resources have been readily allocated for prenatal examinations and abortion of children who, if they are born, proved to be a financial burden to society, but it is very difficult to get financial support for the work we are carrying out that examines the environmental influences on pregnancy, birth, and the fetus, an examination that we believe has a number of preventive aspects.

ETHICAL CONSIDERATIONS CONCERNING CHROMOSOME EXAMINATION OF NEWBORN CHILDREN

There is general agreement that the diagnosis of aneuploid chromosome abnormalities at birth is of great value, and similarly that inherited structural autosomal chromosome abnormalities with their increased risk for carriers of giving birth to children with mental retardation and congenital disorders should be diagnosed. The diagnosing of sex chromosome abnormalities at birth and transmission of this information to parents has, however, been considered unethical by some. On account of our experience with the follow-up studies since 1969 of children with sex chromosome abnormalities and with the cooperation with groups who make similar studies in the U.S., Canada, and Britain, we have become convinced that information, counselling, and support for parents of children with sex chromosome abnormalities is very important for the development of these children. Such support includes assistance in getting these children early into nursery classes and in beginning speech therapy if needed and remedial teaching if there are educational difficulties; the knowledge that the parents can obtain guidance whenever they need it, is, to a great extent, of preventive value for the otherwise increased risk of deviation in the mental development of these children.

On the basis of the increased use of prenatal chromosome examinations in Denmark from eight cases in 1970 to 5,000 in 1983, followed by induced abortion of approximately 75% of fetuses with sex chromosome abnormalities diagnosed prenatally, it is also very important through investigations like this to obtain information concerning the development of children with sex chromosome abnormalities for geneticists and physicians who must advise the parents as a result of prenatal examination. It is furthermore our experience that if parents are prepared to give and capable of giving their child with a sex chromosome abnormality good and stable conditions, and if they are ready to accept and act on information and counselling, and if social resources are available in day institutions and schools, then they can expect that their child will develop within the normal range.

It is very important that those who counsel parents of a fetus with a sex chromosome abnormality do not forget to give information about the positive aspects of the development of the child in question and about all that can be done in order that the development of this child may be within the normal range. It is also important that, if the parents choose not to have induced abortion, they are offered skilled assistance and counselling after the birth of the child.

CONCLUSIONS

We have, so far, made chromosome examinations of 20,222 newborn children in the Århus area. The last 9,074 examinations were made as part of a comprehensive prospective study of the influence of environmental conditions on pregnancy, birth, and the fetus, a study that will continue for a number of years.

We identified 174 children with chromosome abnormalities, that is, nine per 1,000 newborns, and 180 relatives with chromosome abnormalities, that is, also, nine per 1,000 newborns. There was thus a total of 354 individuals with chromosome abnormalities corresponding to 18 per 1,000 newborn children. Chromosome examination has turned out to be valuable in the information and counselling that can be given to parents of children with chromosome abnormalities whether autosomal or affecting the sex chromosomes. The results of the investigation have allowed increased knowledge to be acquired, especially about the mental development of children with sex chromosome abnormalities; this may lead to a decrease in the frequency of induced abortions of these fetuses. Chromosome examination of relatives of newborn children with structural autosomal chromosome abnormalities and the consequent genetic counselling have also been one of the important results of the investigation.

It is our impression at the Cytogenetic Laboratory in Risskov that, as a result of the present study, there are good reasons to offer chromosome examination of all newborn children as a general service provided that there are sufficient resources for chromosome examination, counselling, and follow-up studies. We consider that it would be unethical to perform chromosome examination without subsequent counselling and follow-up. Furthermore, at the present time, we have a number of interesting results concerning the influence of the work environment on pregnancy, birth, and the fetus. In the long run, we hope that our investigations will be able to increase present knowledge concerning the etiology of certain congenital disorders, including de novo chromosome abnormalities, and perhaps also identify certain environmental conditions operating prenatally that may cause deviations in the mental development of children.

REFERENCES

1. Evans JA, Flindt R de von, Greenberg C, Ramsay S, Hamerton JL: A cytogenetic survey of 14,069 newborn infants. IV. Further follow-up on the children with sex chromosome anomalies. Birth Defects 18(4):169–184, 1982.
2. Leonard MF, Sparrow S, Schowalter JE: A prospective study of development of children with sex chromosome anomalies—New Haven study III. The middle years. Birth Defects 18(4):193–218, 1982.
3. Nielsen J, Sørensen AM, Sørensen K: Follow-up until age 7 to 11 of 25 unselected children with sex chromosome abnormalities. Birth Defects 18(4):61–97, 1982.
4. Ratcliffe SG, Tierney I, Nshaho J, Smith L, Springbett A, Callan S: The Edinburgh study of growth and development of children with sex chromosome abnormalities. Birth Defects 18(4):41–60, 1982.
5. Robinson A, Bender B, Borelli J, Puck M, Salbenblatt J, Webber ML: Sex chromosomal abnormalities (SCA): A prospective and longitudinal study of newborns identified in an unbiased manner. Birth Defects 18(4):7–39, 1982.
6. Stewart DA, Netley CT, Park E: Summary of clinical findings of children with 47,XXY, 47,XYY and 47,XXX karyotypes. Birth Defects 18(4):1–5, 1982.
7. Walzer S, Graham Jr JM, Bashir AS, Silbert AR: Preliminary observations on language and learning in XXY boys. Birth Defects 18(4):185–192, 1982.
8. Nielsen J: "Fejludvikling af kromosomerne. En orientering." Århus: Arkona, 1979.
9. Nielsen J, Sørensen AM, Sørensen K: Mental development of unselected children with sex chromosome abnormalities. Hum Genet 59:324–332, 1981.
10. Videbech P, Nielsen J, Wohlert M, Eriksen G, Hansen KB, Hvidman L, Krag-Olsen B: The impact of large Y chromosome on pregnancy, fetus and birth. Clin Genet 26:281–288, 1984.
11. Friedrich U. Personlig meddelelse, 1983.
12. Forebyggende fosterkromosomundersøgelser i Danmark. Rapport fra cytogenetisk centralregister. Bidragydere: Friedrich U, Horn N, Jensen PKA, Lundsteen C, Mikkelsen M, Mosbech J, Nielsen J, Nielsen KB, Peterson GB, Philip J, Rasmussen K, Therkelsen AJ, Tommerup N, Videbech P. ugeskr Laeger 145:1760–1762, 1983.

Prospective Study of Development of Children With Sex Chromosome Anomalies: New Haven Study IV. Adolescence

Martha F. Leonard, MD, and Sara Sparrow, PhD

Yale Child Study Center and Department of Pediatrics, Yale University Medical School (M.F.L.), and Yale Child Study Center (Psychology) and Department of Psychology (S.S.), Yale University, New Haven, Connecticut 06510

INTRODUCTION

This is the fourth report of a prospective study of children with sex chromosome anomalies identified at birth in New Haven, Connecticut. Previous reports in 1974 [1], 1978 [2], and 1982 [3] summarized data from the first 2.5 years, the first 9 years, and the preadolescent years, respectively. This chapter focuses on progress and evaluations to the age of 17–18 years.

Thanks to grants from The National Foundation—March of Dimes, we have had the opportunity to meet about every 3 years with investigators from medical centers in Scotland, Denmark, Japan, Canada, and other parts of the U.S. to share our findings in comparable studies. Reports from two earlier sessions have been published by The National Foundation—March of Dimes [Birth Defects: Original Article Series, Volumes 15 and 18]. This paper was presented in part at the last collaborative meeting in Edinburgh in June, 1984, with addition of data from more recent contacts with our subjects.

Until the mid-1960s knowledge about persons with sex chromosome anomalies was based entirely on subjects who had clinical manifestations, such as Turner or Klinefelter syndromes, or on subjects already identified as emotionally disturbed (from inpatient or outpatient mental health facilities) or delinquent (prison inmates). These early studies gave no indications of the

Birth Defects: Original Article Series, Volume 22, Number 3, pages 221-249

incidence of such aneuploidies in the general population, nor of the physical, cognitive, and emotional conditions of persons with the same aneuploidies who did not show typical syndromes or who were not in groups already characterized by personality disorders or delinquency. Could there be many such persons who function so well that their genetic anomaly never has been suspected?

Large-scale screening of newborn infants in the mid-1960s led to identification of approximately one baby in 400 with a sex chromosome variation. Accumulation of prospective developmental data from such children has been important in genetic counseling. Because of the rarity of sex chromosome aneuploidy, no one center sees enough of these children to feel confident about the validity of conclusions based on its own experience. The joint publication of results by investigators in several centers has provided large enough numbers of children with each karyotype for valid conclusions about the range of expected growth and development.

Now we know that aneuploid children can have normal intelligence but that many function in the low-average to borderline range, especially if they have complicating biologic or environmental handicaps. Many have delayed language development and difficulty with some aspects of school learning, especially speech, language, and reading. No specific personality or character attributes have been found at least until age 18 years. Boys and girls with extra X or Y chromosomes have tended to be tall. Those with an extra X chromosome also have had some evidence of hypogonadism and delayed pubertal changes. Unless they have other complications, these aneuploid young men and women are functioning reasonably well in their families, schools, social groups, and beginning employment.

SUBJECTS

When large-scale genetic screening became feasible, Herbert Lubs, MD, and Frank Ruddle, PhD, human geneticists at Yale University performed cytogenetic studies on 4,500 live infants born consecutively at the Yale—New Haven Hospital from October 1, 1967, to September 30, 1968 [4]. They identified 11 who had anomalies of the sex chromosomes: four with 47,XXY, three with 47,XYY, three with 47,XXX, and one with 45,X. A comparison group was selected consisting of an infant of the same sex, race, and approximate birth weight, born just before or after each infant with a sex chromosome anomaly. For the 47,XYY boy, who was a dizygous twin, his unaffected twin brother was a control. Details of the cytogenetic studies have been reported previously [2,4].

In addition, one more subject has been added to our study. For several years at the Yale—New Haven Hospital, pregnant mothers with reasons to be considered at risk have been offered amniocentesis and genetic consultation during pregnancy. One author (M.F.L.) has volunteered to the geneticists to consult with any parents prenatally if our research results would be helpful to them. From time to time, a diagnosis of a sex chromosome anomaly has been made by study of cells obtained at amniocentesis. Most parents, when informed of the sex chromosome aneuploidy of their fetus, have elected a therapeutic abortion. However, in 1981 a couple was referred to me after determination that their third child, still unborn, had an extra Y chromosome. The parents already had decided to continue the pregnancy but wanted further information about prognosis that I could offer on the basis of the combined observations of this group of investigators. The child was born in November, 1981, weighing 6 pounds, 5 ounces. He was in good condition and had no congenital anomalies. I have seen and tested him three times, at 8 months, 27 months, and 4 years of age.

SUMMARY OF OUR PREVIOUS STUDIES

All infants were examined thoroughly at birth. Although some had mild anomalies, such as clinodactyly, no infant had any abnormalities that suggested a need for chromosome analysis. The most common congenital anomaly was inguinal hernia, which occurred in two of the 11 boys. The parents of the affected children were tested cytogenetically, and all had normal karyotypes. When the children were about 1 year old, the parents of both the aneuploid and the comparison infants were approached with an invitation to participate in a study of child development without any explanation of the underlying genetic basis. All but one set of the 22 couples agreed. At age 1 year Gail Landy, MD, and at 2 years Martha Leonard, MD, performed developmental evaluations on the sex aneuploid children and the comparison group without knowledge of the nature of the study or which child belonged to which group. The children were tested with the Yale developmental schedules.* These schedules also were used for children up to age 5 years at

*The Yale developmental schedules were developed by Dr. Sally Provence at the Yale Child Study Center and have been used there for about 25 years. They consist of a number of items from several standardized tests, including the Gesell, the Viennese Scale, the Merrill Palmer, the Bayley, and the Stanford-Binet. They are organized into a composite, by which children from birth to age 6 years can be tested.

the third evaluation, whereas those over 5 years were tested with the Stanford-Binet test (SB). The Goodenough draw-a-person test, Peabody picture vocabulary test, Beery-Buktenica test of visual motor integration, and Leiter scales were administered to many children. The 5 year evaluation also included one or two play interviews conducted by John Schowalter, MD, a child psychiatrist, who was blind as to which children had the sex chromosome variation. The developmental examiners already had broken the code by this time, and the basis of the study and the individual genetic conditions of each child then were discussed with the parents.

When the children were about 9 years old, the parents of the index children were contacted again. They were interviewed as to the child's progress; the child had a brief play interview with Dr. Schowalter; measurements were made of height, weight, and head circumference; and permission was obtained to request records of school achievement. No psychometric testing was done at this time. The comparison children were not seen again, but measurements and school records were requested by letter.

In the summer of 1980, when the children were 12–13 years old, a more thorough evaluation was offered to each family. This consisted of a parent interview by Drs. Leonard and Schowalter; a psychometric test by Sara Sparrow, PhD, child psychologist; a physical examination and appropriate endocrine studies by Myron Genel, MD, pediatric endocrinologist; a genetic evaluation and recheck of karyotype with banding by Margretta Seashore, MD, pediatric geneticist; a brief psychiatric interview by Dr. Schowalter; and a conference to discuss all the findings with the parents by Drs. Leonard and Schowalter. Dr. Sparrow's testing included the revised Wechsler intelligence scale for children (WISC-R), and for some children (depending on indications and available time), the Beery-Buktenica test of visual motor integration (VMI), the wide range achievement tests (WRAT), and the Peabody picture vocabulary test (PPVT). The parents of the comparison children were sent a questionnaire.

Results of the evaluations at ages 1, 2, 5, 9, and 12 years have been reported in detail in three previous publications [1–3]. In summary, none of the children had any physical abnormality that suggested a genetic variation, and all looked quite normal except the girl with Turner syndrome (45,X), who was very short. There was a trend toward increased stature in the children with an extra X or Y chromosome, eight of ten being in the 50th to above the 95th percentile for height. Cognitive ability tended to be slightly less in the affected children by the time of the second evaluation, at the age of 2 years, with language skills being the most impaired. At the third evaluation, at 5 years, only one affected child had an IQ above 100, whereas six of the nine comparison children tested had scores at or above 100. At 12.5–13 years, the seven children tested with the WISC-R had full-scale IQ scores of 55–98.

Figures 1–3 from our previous report [3] show the scores of serial developmental and psychometric tests for the individual children. Nearly all the aneuploid children have shown a downward trend. In contrast, the comparison children had fluctuations in scores but no consistent trends. We believe that the tests reflect a relative decrease in cognitive ability of these aneuploid children during the early school years. The only control child who was tested again was the twin brother of N0384, 46,XY (N0383). His full-scale IQ score on the WISC-R at 13 years was 103, with superiority of his verbal score over performance.

Achievement tests at age 12–13 years also showed impairment in most of the aneuploid children. Five of the affected children showed generalized impairment of 1.5–3 grade levels. Four boys scored close to grade level, but three of them had repeated a grade and were one grade below their agemates. The control twin achieved a little below age level in the reading and spelling subtests of the WRAT. The other two comparison children for whom we have data were given achievement tests in school because they had some learning difficulty, and the test confirmed it.

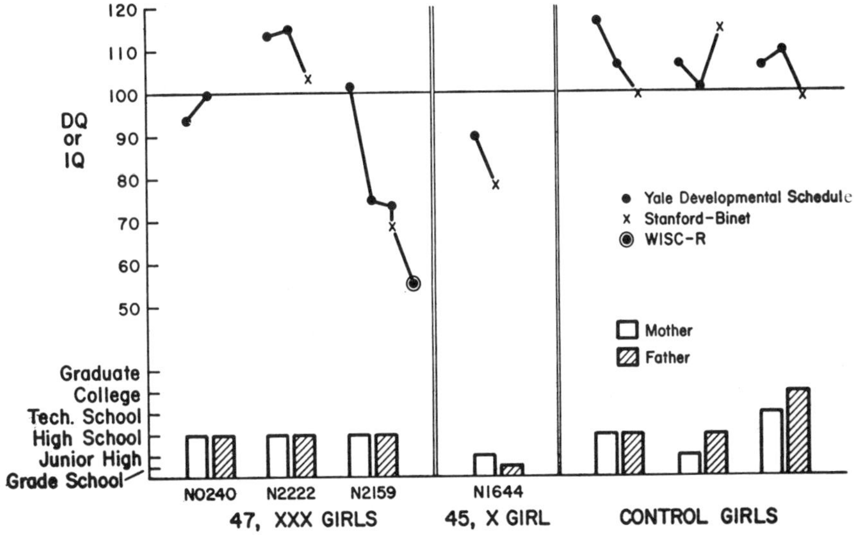

Fig. 1. Serial developmental and psychometric tests for individual girls correlated with educational levels of their parents. From Leonard et al [3], with permission.

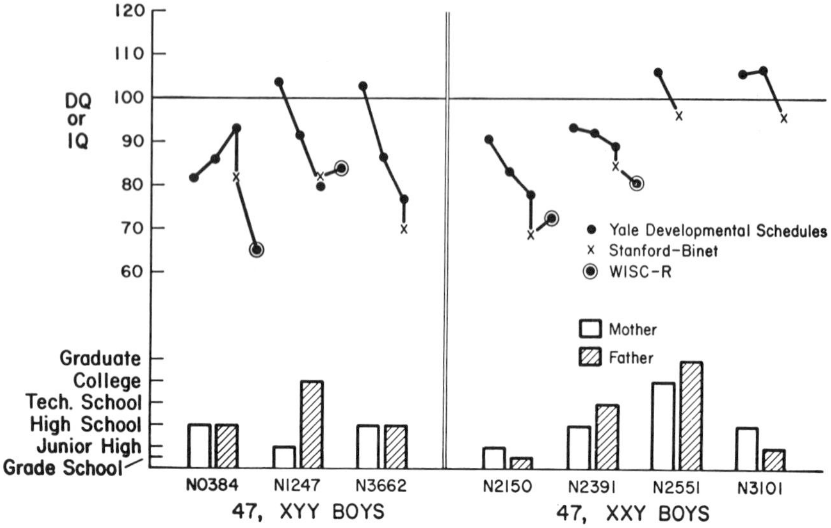

Fig. 2. Serial developmental and psychometric tests for individual boys correlated with educational levels of their parents. From Leonard et al [3], with permission.

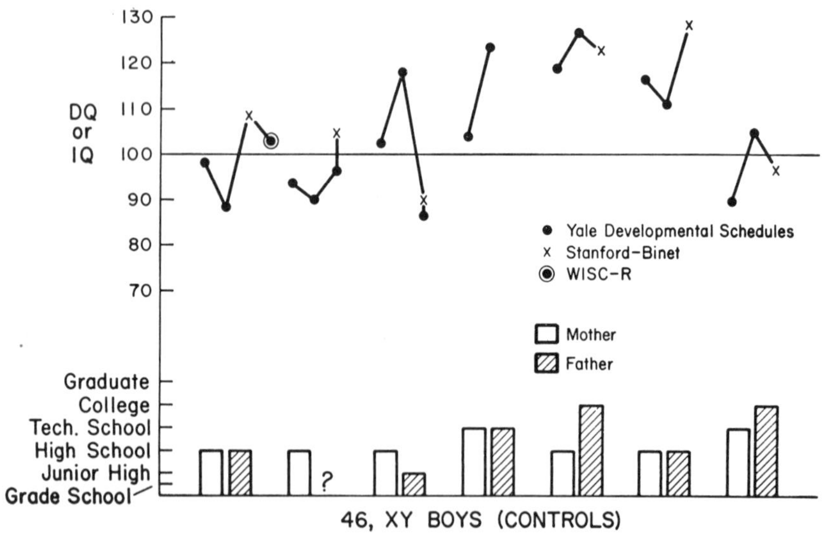

Fig. 3. Serial developmental and psychometric tests of control boys correlated with educational levels of their parents. From Leonard et al [3], with permission.

Every one of the 11 aneuploid children had difficulty in some aspect of academic learning. Eight had repeated a grade. Four had attended special classes, and one was continuing in a class for educable retarded children at the time of the third report. Most of the children had delayed onset of language development. All were reported to have difficulty in some aspect of speech or language development or reading.

The aneuploid children were not distinguishable from the controls by their personality traits. Both the aneuploid and the comparison children showed a range of characteristics appropriate to their life experiences. For instance, those who were struggling in school tended to be anxious about doing well and low in self-esteem. All parents described their children in positive terms—lovable, outgoing, thoughtful, polite, respectful, and friendly. The XYY boys were not distinguishable by any specific personality characteristics.

METHOD OF CURRENT EVALUATION AT AGE 17–18 YEARS

In 1984, an effort was begun to find all the subjects for an update of their progress. The plan was for a parent interview, a physical examination and anthropometric measurements, a brief psychiatric interview, a Vineland adaptive behavior scale assessment, and a conference with parents and subject to summarize findings and answer any questions. Questionnaires were sent to the parents of the subjects who had moved away and to the parents of the comparison children. Some data were obtained from physicians who now are caring for these young people.

RESULTS

As before, most of the families have been very cooperative with this study when and if they could be located. To date, we have located all 11 aneuploid subjects and most of the comparison group. The girl with Turner syndrome cannot be reached directly, but her endocrinologist has furnished recent information about her. The parents of one XXX girl have declined further participation. The mother of one of the others has been interviewed, and information about the third has been obtained by telephone. One XXY boy has been seen, and his endocrinologist has shared details of his growth and pubertal development. The other three young men with this karyotype have been located, and a few details have been obtained by telephone. Two boys with the XYY karyotype have been seen and the control twin brother of one. The questionnaire has been returned by the mother of the third. Questionnaires also were returned by the parents of three of the comparison subjects, and parents of seven additional subjects have been interviewed by phone.

Cognitive Development and Academic Progress

Most of our aneuploid subjects are functioning in the low-average or borderline range according to psychometric testing at the age of about 13. We have elected not to repeat psychometric examinations, because we anticipated that little more would be learned than was demonstrated by the last tests. The exceptions were one XXY boy with an average score and our young XYY boy, who at age 4 years is functioning in the high-average range. Most of the subjects have had to struggle in school. Many have repeated one grade, and all have needed and received some remedial help.

The girl with *Turner syndrome* (N1644) is considered learning-disabled, probably more from deprivation than because of her missing chromosome. According to testing in school with the WISC-R, her full-scale IQ was in the mid-low-average range. She showed relative weakness in short-term memory and visual-motor coordination and better ability in reasoning, common sense, and perceptual organization. At 16 years, she was receiving special services in school, including vocational training. At 18, she has dropped out of school and is said to be staying at home and helping there and in the neighborhood.

Of the *girls with one extra X chromosome*, one (N0240) was doing well in school, in tenth grade at age 16½. She has had no psychometric studies. The other two were tested by WISC-R at 12 and 13½ years. They had full-scale IQ scores of 55 (N2159) and 88 (N2222), with performance subtests better than verbal. One of these (N2159) spent some time in a residential school for mildly retarded children. She is believed to have fetal alcohol syndrome, as her mother drank heavily during pregnancy and for the first 10 years of the girl's life. At 17, she was in a regular eleventh grade, but her attendance was erratic. She dropped out after a few months and has left home. The other (N2222) had a tutor earlier, but at nearly 18 is a senior in high school and making fairly good grades, although she is said to care little for studying.

Three of the boys with the *47,XXY* karyotype have been tested with the WISC-R at the ages of 8½, 12½, and 13½. Their full-scale IQ scores were 73, 81, and 98 (Table I). In all, the performance IQ was greater than the verbal, in two by significant amounts, 25 and 16 points. This was consistent with their earlier delay and impairment in language development. There was no other consistent pattern in the subtests. One (N2150) was in special education during elementary school. He is now a sophomore in high school making barely passing grades. Another (N2391) has needed tutorial help and dropped out of school after repeating ninth grade twice. He now is going to night school and working in a fast-food restaurant. The third (N2551) is in the twelfth grade, earning Bs and Cs and hoping to go to college. The fourth (N3101) dropped out of school after his sophomore year of high school. He

TABLE I. Latest Psychometric Tests

		WISC-R															PPVT IQ	VMI IQ
					Subtests scaled scores													
Subject	Age (years, months)	Full-scale IQ	Verbal IQ	Performance IQ	Inf.	Sim.	Arith.	Voc.	Comp.	Dig.	Pict. comp.	Pict. arr.	Block des.	Obj.	Coding	Maze		
45,X																		
N1644	12	Mid-low aver.																
47,XXX																		
N2159	12,3	55	52	66	1	2	5	1	3	2	4	1	5	6	8	6	55	50
N2222	13,7	88	84	95	8	8	8	6	7	—	7	7	6	11	16	—	92	—
47,XXY																		
N2150	12,6	73	68	84	5	6	5	2	6	—	9	9	6	6	7	—	77	81
N2391	8,8	81	70	95	8	1	9	Ref.	3	2	9	8	11	9	10	8	81	99
N3101	13,7	98	97	100	10	11	10	7	10	—	9	10	10	11	—	—	101	68
47,XYY																		
N1247	13,2	84	91	78	9	9	11	7	7	—	9	5	7	—	6	—	—	—
N0384	12,4	65	70	64	9	5	6	3	3	4	4	3	1	5	9	5	86	—
N3662	13,10	78	82	75														
N4001	4,0	DQ* 115	125*	122*														
46,XY																		
N0383	13,0	103	115	90	12	16	13	12	10	—	9	7	9	7	11	—	98	—

*Yale developmental schedules: DQ = developmental quotient, language, and non-verbal quotients, respectively.
Ref. = refused.

had achieved poorly in spite of average ability according to tests. He has just been fired from a part-time job because of a fight. He has started a correspondence course to work for his high school equivalency certificate and then hopes to join the Navy.

We have two more sets of test results in *47,XYY* boys to add to those in our previous report (see Table I for the latest test results). One 47,XYY boy (N3662), now living outside Connecticut, was tested by a school psychologist at the age of 13 years, 10 months. The full-scale IQ on the WISC-R was reported to be 78, with a verbal IQ of 82 and a performance IQ of 75. This boy performed best in information, similarities, and picture arrangement, lowest in picture completion, vocabulary, and comprehension. In the eighth grade at that time, his achievement test scores in reading, spelling, and arithmetic were at the fifth through seventh grade levels. The psychologist considered this a minimal estimate and believed that the boy's potential was somewhat higher. The other two boys with XYY karyotype also had slightly low scores on the WISC-R at 12 and 13 years, with full-scale IQ scores of 84 and 65. All three boys had verbal scores higher than performance scores. Our new young XYY boy (N4001) has been tested three times with the Yale developmental schedules–at 8 months, 27 months, and 4 years of age. His developmental quotients were 99, 126,* and 115, respectively. His language quotient at the 4 year examination* was 125, and his nonverbal cognitive skills quotient 122 (with average assumed to be 100). He is functioning very well in all areas, with an excellent vocabulary, fluent speech, and good syntax for his age.

All three older XYY boys have had to struggle in school. All have repeated one grade and have had remedial help. At 15½ to 16½ years, all were in ninth grade but not accomplishing well. All were described as having poor study habits and inadequate attention, motivation, or interest in academic subjects. At 18 years, one (N0384) is in his senior year of high school in a special education class. Another (N1247) at 17 is a junior in high school with art and mathematics his best subjects. He says he hates school and feels dumb. The third (N3662) has quit school for a while; he is now in a vocational technical school studying automobile mechanics.

The karyotypically normal twin of N0384 is a freshman in college at the age of 18. The other six comparison boys and three comparison girls we have heard from are seniors in high school at the ages of 17 to 17½.

*Motor and personal skills were not recorded, so this represents only language and adaptive sectors, which were in the same range as in the subsequent evaluation.

Behavior

Our behavioral data come from parental reports, some school reports, and an interview of some subjects by John Schowalter, MD, child psychiatrist. We also have begun to use the Vineland adaptive behavior scale as a way to assess and compare adaptive development.

The girl with *Turner syndrome* (N1644) is described as adapting well to her life situation. She always has been friendly, outgoing, and cheerful in spite of considerable deprivation.

The *triple X* girl (N2159), who is mildly retarded, has had some delinquent behavior and is of considerable concern to her mother. She was a frequent truant from school and was caught shoplifting with a group of girls. After breaking probation, she was sent to a residential treatment center for 1.5 years. Since she has been home, she has rebelled against her mother's rules. She left home 3 months ago and claims in a recent telephone call to have an apartment of her own and a job. According to her mother's report on the Vineland adaptive behavior scales, she is functioning in the average range in communication, daily living skills, and socialization. However, the mother listed a significant number of maladaptive behaviors, such as stubbornness, truancy, anxiety, dependency, and running away. The other XXX girl we still can reach (N2222) is functioning well. She is friendly, well liked, and responsible in her part-time job although still somewhat immature in her handling of money.

The children with the most difficulty from aggression and acting out are *not* the boys with an extra Y chromosome, but the one girl described above with 47,XXX and two boys with 47,XXY karyotypes. All three have had severe disturbances in their family lives and relationships so that it is not clear how much, if any, of their symptoms can be attributed to their genetic constitution. One *XXY boy* (N2391) showed a pervasive developmental disorder during his 5 years and had both individual and group psychotherapy in the latency years. At 17½ years he was described by his pediatrician as having major emotional problems. He rebelled against his mother's rules, used alcohol and marijuana, and chose friends who often got him into trouble. He now is on probation following a minor theft. Another (N3101) is described by his mother at the age of 17 as beloved, sensitive, caring, and nonconforming at home usually but sometimes upset, angry, and verbally aggressive to her. At school he was said to be belligerent, defiant, and fearless, with a quick temper. At the age of 14 years, he was considered by Dr. Schowalter to have an oppositional disorder. Another XXY boy (N2551) is described as forthright, considerate, and likable. Although quiet and passive earlier, he has learned to be assertive. The fourth XXY boy has had no behavioral problems according to his mother.

The *47,XYY* twin (N0384) has continued to have a little immaturity of speech and to be less well coordinated than his twin brother. He always has been considered the kinder and more generous of the two. At 16½ years, he scored low or moderately low in all sectors of the Vineland adaptive behavior scales and had an intermediate level of maladaptive behavior. Poor concentration and swearing were his main maladaptive characteristics, with some sullenness, lying, anxiety, impulsiveness, and defiance. At 18 years, he is reported by his mother to have a quick temper but not much different from his twin brother. He does not fight except with his brother. He is active in sports, coaches a basketball team, and never has had any delinquent behavior. Another 47,XYY boy (N1247) experiences anger at times and reacts by withdrawal. He and his mother agree that his anger is less and better controlled than that of his father, who rules the family autocratically, has abused the mother occasionally, and still used physical punishment on this boy up to the age of 16. The boy admitted brief trials of alcohol and amphetamines but gave them up because they interfered with his athletic prowess. He and his parents have had some sessions with the school psychologist beginning at that time. On the Vineland adaptive behavior scales he scored in the adequate range for communication and daily living skills but low in some aspects of socialization—leisure-time activities and coping skills. He also scored significantly high in maladaptive behavior, including such items as impulsiveness, defiance, stubbornness, and poor concentration. Our third boy with double Y (N3662) was reported by his mother to be friendly, helpful, even tempered, and cooperative. He does not like to fight and gets nervous if someone tries to start something with him. He loses his temper only with his sister. A report from his school said that he was unhappy that year, attributed at least in part to the recent split in his family, which resulted in his living with his father and his sister with their mother. He was described by the school psychologist as experiencing some difficulty in forming meaningful relationships with peers and having low self-esteem. He was more comfortable with adults. None of these boys has had unusual tendencies to aggression or antisocial behavior. The fourth boy with the extra Y chromosome (N4001) was described and observed at age 4 years to be delightful, happy, helpful, and friendly. On the Vineland adaptive behavior scale he scored adequate or moderately high in all domains—communication, daily living skills, socialization, and motor. He had no significant maladaptive behaviors.

Growth (Tables II and III)

The trends in height and weight that were recorded previously have continued in our subjects. Our one girl with Turner syndrome (N1644) has

TABLE II. Physical Growth

Identity	Age (years, months)	Height		Weight		Head circumference		Segment			Arm span (cm)
		cm	Percentile	kg	Percentile	cm	Percentile	Upper cm	Lower cm	Upper/lower	
45,X											
N1644	14,6	135.9	<3	26	<3						
	16,2	139.7	<3	34.5	<3						
47,XXX											
N0240	13,9	155	25	42	25						
	16,8	160.3	30	46.4	10						
N2222	12,4	157	75								
	17,10	173	95	55	45						
N2159	9,6	137.2	60	27.5	25	52	25				
	17,9	167.6	75								
47,XXY											
N2150	13,4	157	50	35.7	10	52	2	72.5	86		160
	17,10	172	20								
N2391	12,6	164	90	67	>97						
	17,8	183.4	85	77.3	77						
N2551	13,2	160	60	44	45						
	17,8	180.3	75	65.5	45						
N3101	15,7			58.5	45						
	17,7	184	85	67	50			82	92		168
47,XYY											
N0384	13,8	163.3	50	43	25	55.2		75.3	88.0	0.86	165.0
	18,0	180.3	75	65.9	50						
N1247	13,4	168.8	85	56.5	80			76.8	92.0	0.83	
	16,5	189.7	90	79.8	90						
	18,0	193	>95	88	85	57.1					193.3
N3662	17,3	168–9	95	61.4	25						
N4001	4,1	106	75	17.5	75						

TABLE III. Physical Growth (comparison subjects)

Identity	Age (years, months)	Height		Weight	
		cm	Percentile	kg	Percentile
46,XX					
1	13, 0	156	50	45.5	45
	17, 10	160	25		
2	17, 10	158	20	53	25
3	9, 6	131	25	24.5	10
46,XY					
1	18, 0	170.2	15	64.6	25
2	13, 0	152	30		
	17, 9	176	45	66	40
3	13, 1	152	25		
	17, 3	164.4	<5	64.0	40
4	13, 0	168	93	56.5	80
	17, 8	185	90	91	90
5	12, 8			42.8	40
	17, 10	170	15	54	10
6	13, 0	163	75	45.5	50
	17, 8	179	70	72	55
7	17, 10	173	25	73	60

continued to be short and small. Her bone age at 14 years, 11 months was 11–12 years. At 16 years, 2 months, she was 139 cm tall and weighed 34.5 kg, both well below the 3rd percentile. She has grown with the help of hormone-replacement therapy, but is still well below the 3rd percentile in both height and weight.

Our three girls with the triple X karyotype had heights at the 30th to 95th percentiles at their last observations, at 16–17½ years. Their weights were relatively less, in the 10th–45th percentiles. Three of the comparison girls had heights at the 20th, 25th, and 50th percentiles when last contacted.

The boys, both XXY and XYY, also have tended to be tall, ranging from the 75th to over the 97th percentile in height, except for one boy, 47,XXY (N2150), who was at the 50th percentile when last seen at 13 years, 4 months. At 17 years, 10 months, he is at the 20th percentile. The XYY boy who is a fraternal twin was 11.9 cm taller than his twin brother at 16½ years. Our new patient with the XYY karyotype (N4001) was 106 cm tall (75th percentile) and weighed 17.5 kg (70th percentile) at 4 years, 1 month. The other XXY and XYY boys have most recent weights in the 25th–85th percentiles. The seven comparison boys had a normal distribution of heights–90th, 70th, 45th, 25th, 15th, 15th, and <5th percentiles at 17 to nearly 18 years.

Health

All the children essentially have been well except for the new child with XYY (N4001), who has had recurrent asthma attacks. The girl with Turner syndrome (N1644) recently has had a thorough evaluation for the first time. She was found to have a horseshoe kidney, ureteral duplication, a 5–7 thoracolumbar scoliosis, and persistent serous otitis media with mild hearing impairment.

Sexual Development

The girl with Turner syndrome (N1644) had minimal sexual development before treatment. At the age of about 15, when she finally began to receive good medical care, she was started on androgens in an attempt to increase her height. After 9 months with little effect, her treatment was changed to estrogens, premarin 2.5 mg three times daily. Her Tanner staging of 1 at age 15 years, 10 months, had increased to stage III–IV by 16 years, 3 months, and she had begun to have some vaginal spotting. At that time, she was changed to premarin for 3 weeks of the month and provera the fourth week. Now, at nearly 18, she is said to have normal secondary sex characteristics and regular menstrual periods.

None of the *triple X* girls has been seen since the last report. According to their parents, they are developing normally and all have achieved menarche, two at the ages of 12–13 years and one a little later. One was reported by her mother to be possibly sexually active and has been taking a contraceptive pill since the age of 16.

One young man with *47,XXY* (N2551), who no longer lives in Connecticut, has been receiving hormone therapy since the age of about 14. His mother reported that he still is worried about his appearance. Details of his hormone levels and therapy have been sought from his endocrinologist.

Only one of the boys with the XXY karyotype has been seen since the age of 13. He (N3101) had some upper lip hair and a little axillary hair at 15 years, 9 months. His phallic length was approximately 13 cm, and his gonads were 3–5 cm in diameter. At 17 years, he was considered Tanner IV–V in sexual development. His beard was coming in, and he had axillary hair and foul-smelling sweat. He had a Tanner IV escutcheon. His gonads were 5 cm in length. His hormone levels are listed in Table IV. No hormone therapy had been considered indicated up to that time. When interviewed at 17 years, 8 months, he showed great sensitivity about the size of his genitalia and expressed a wish for treatment that would help him feel more normal. Studies on the other two XXY boys still are pending. The mother of one (N2391) reported that at 17½ he has a deep voice but does not need to shave. He is

TABLE IV. Pubertal Development and Hormones*

Identity	Age at visit (yr, mo)	Pubic hair	Acne	Axillary hair	Penis length (cm)	Testicle volume R	Testicle volume L	Testosterone (ng/ml)	LH (MIU/ml)	FSH (MIU/ml)	Estradiol (pg/ml)	Estrogen
47,XYY N 0384	13, 8	Tanner III	None	Tanner III	10	14 ml	12 ml	Total 5.5, free 1.7	1	5	14	
	16, 7	Tanner V	Mild-mod.	Tanner V	14	17 ml	15 ml					
N 1247	13, 0	Tanner II–III		None	10	16 ml	17 ml	Total 2.24, free 1.12	1	4	23	
	16, 5	Tanner IV	Mod.	Tanner IV	13	20 ml	20 ml					
47,XXY N 2150	13, 4	Tanner III			12	4.5 ml		Total 7.6, 31% free	14	16		28
N 3101	14, 1								3	0		
	14, 9	Mod.			Normal	Sl. small		Total 2.88, free 1.09	2	1		
	15, 9				13	3 × 4 × 5 cm						
	17, 0			Present		5 cm long		Total 5.94, free 2.28	7	9		

*Normals for this laboratory in adult males: testosterone, total 2–8 ng/ml, free 1–4 ng/ml; LH, 2–20 MIU/ml; FSH, 2–10 MIU/ml.

sexually active. When he consulted a new physician recently, the physician raised the question of Klinefelter syndrome because of the appearance of his genitalia.

The boys with the XYY karyotype have developed normally in secondary sex characteristics. The hormone levels of the two who have been measured are listed in Table IV.

SUBJECTS' AWARENESS OF ANOMALY

By this time, most of the boy subjects and the girl with Turner syndrome have become aware of their aneuploidy. Two of the boys with XXY karyotype and the 45,X girl have been seen in endocrine clinics, where they have had opportunities to discuss their anomalies and implications for sexual development. Only one has been reported to be distressed by the knowledge of his anomaly. He (N2551) was very angry when he learned the reason for his having to receive hormone injections. He said that he should have been told much earlier. The mother of one 47,XYY boy and the mothers of two XXX girls have preferred not to tell their children about their genetic constitution.

DISCUSSION

This study, like the other prospective studies described in this book, was motivated by the need to know the incidence of sex chromosome anomalies in the general population and the impact on development in an unbiased sample. Now we know that about one in 400 liveborn infants will have a sex chromosome aneuploidy. Except for Turner syndrome, they will have no distinguishing physical features and will not be diagnosed unless or until delayed pubertal changes suggest a sex chromosome anomaly or unless prenatal amniocentesis and genetic studies provide knowledge of a fetus's karyotype.

Our group is too small to provide statistically significant differences between the adolescents with the various karyotypes, but one can make a few generalizations about the group. As before, the most distinguishing features have been in growth, pubertal development, and learning. Lack of one X chromosome has been associated with short stature and delayed puberty. An extra chromosome, either X or Y has predisposed to tallness. These young people are not giants, but most are between the 50th and 97th percentiles for height and taller than most of their family members. The comparison children have a more usual distribution of heights. Girls with Turner syndrome have minimal pubertal changes until they receive hormone therapy. Boys with 47,XXY karyotype may have delayed pubertal changes. One of our four in this group has received hormone therapy. Another has been examined and tested annually, and up to the age of 17 is considered to have normal pubertal development. However, both these young men are sensitive about their genitalia. A third has not yet been examined, but his voice has changed, and he has not spoken of any concern. It begins to seem that not all males with the 47,XXY karyotype will show typical Klinefelter syndrome at least by the age of 18. Possibly, testosterone levels decrease after the early pubertal changes, or perhaps some boys would be found to have mosaicism if fibroblasts or cells from other tissues could be studied.

All of our aneuploid subjects have shown impairment in some aspects of learning. Psychometric testing resulted in scores ranging from average to mild mental retardation, with most in the low-average to borderline range. Many of the subjects have repeated one grade, and nearly all have had some special education or remediation. The area of most difficulty has been language development—delayed onset, limited vocabulary and syntax, difficulty with concepts, and lack of fluency. These learning difficulties are not specific to the genetic differences and can be helped with good educational techniques. The sex chromosome anomaly alone has not had a devastating effect on the intelligence or learning of any child. It has not caused retardation without other complications. Some of the compounding variables have been prematurity, neonatal hypoxia, possible arrested hydrocephalus, fetal alcohol syndrome, and maternal deprivation. Some of the biologic and environmental hazards that have complicated these children's development are listed in Table V. Only two of the subjects did not have perinatal complications, and many had severe and traumatic experiences. We believe that the chromosome aneuploidy increases a child's vulnerability. When other biologic or environmental complications are added, the resulting impairment may be worse than it would be for a child with a normal karyotype. One subject in each group (47,XXX, 47,XXY, and 47,XYY) has shown intelligence in the average range even with added complications.

Until the age of 17–18, we still find no specific personality or character abnormalities in the boys with the XYY karyotype compared to the XXY boys or the comparison group. Some in each group have quick tempers, and some in each group have had enough emotional disturbance to seek psychiatric help. However, antisocial, delinquent, or criminal tendencies have occurred in none of the young men with XYY karyotype. One young woman (N2159) has been in trouble with the law because of shoplifting, and one young man with XXY karyotype (N2391) for minor theft.

All these young people seem to be well accepted by their families and all but one by their peer groups. The families apparently do not view their children as seriously impaired. They seem to be able to cope with their child's learning problems and to ask for and receive appropriate help.

The findings from this and the other similar prospective studies already have proved useful in genetic counseling. Now that we have followed a large number of boys and girls with sex chromosome anomalies into adolescence, we can share our findings with parents who learn a prenatal diagnosis of sex chromosome aneuploidy. Some parents, faced with the prospect of any abnormality or learning difficulty in their child, find it intolerable and a reason for an elective abortion. However, others, such as the parents of N4001 (47,XYY), find the outcomes of these children and others in the accompanying reports reassuring enough to warrant continuation of the pregnancy. Each couple will make their own decision, but sharing the findings of these prospective studies can assist them. We can say that at least into adolescence these young people are failing to confirm the dire consequences suggested by early surveys on selected populations [5].

This study has raised many ethical questions. We began in 1967 before strict human investigation regulations and initiated the study without awareness by parents or child development investigators of the nature of the study or the genetic identity of subjects and comparison children. The cytogenetic analyses were done at the time of each child's birth. One year later, the parents of the sex chromosome aneuploid children and of the comparison children were invited to have their children participate in a study of child development with no indication as to why they were chosen. Such a study no longer could be done in this way, ie, withholding information from parents about their children. However, we believe it was fortunate that the parents did not have to be made aware of their children's anomalies at birth, since publicity at that time in the lay, as well as the scientific, literature equated the XYY karyotype with aggression, criminal tendencies, and mental illness. In fact, this early bias has continued to be printed in some nonmedical media. It would have been hard or even impossible for parents to treat their children normally if they had anticipated such a prognosis from the time the child was born. The genetic diagnoses were shared with all parents at the first contact after the children were five years old. By that time, the children were functioning reasonably well and were well accepted in their families. The news about their karyotypes was accepted without apparent serious concern. Most parents said that they were glad not to have known sooner, for it would have made them very anxious, and they probably would have treated their children differently.

TABLE V. Biologic and Experiential Factors That Could Affect Development

Identity	Birth weight (gm)	Apgar score (1 min/5 min)	Perinatal complications	Newborn intensive care	Other factors
45,X					
N1644	2,675	5/6	Preclampsia Tetanic contractions Vacuum extraction Cord around neck Baby depressed	Yes	Single teenage mother Emotional and social deprivation Poverty Conductive hearing loss
47,XXX					
N0240	3,300	8/9	Maternal hyperthyroidism, treated	No	None known
N2159	2,160	9/9	Maternal alcoholism Maternal anemia Toxemia Premature rupture of membranes Cord around neck, compressed	Yes	Maternal alcoholism, severe until patient 10 years old Parents divorced
N2222	3,560	⩽2/5	Pitocin drip Low-midforceps	Yes	Parents divorced
47,XXY					
N2150	2,745		Anemia Slow labor Pitocin	No	Twelfth of 13 children

N2391	2,410	9/9	Bleeding in first trimester Premature rupture of membranes Maternal infection Small placenta Breech delivery	No	Father and stepfather alcoholic Parental discord Two divorces
N2551	2,560	9/9	Maternal virus infection in second or third trimester	No	
N3101	3,360	9/9	Maternal herpes infection Toxemia Anemia		Parental discord Two divorces Father alcoholic
47,XYY					
N0384	1,170	7/7	Maternal virus infection Cord compression Twin Breech delivery	Yes	Neonatal pneumonia Rapidly enlarging head—possibly arrested hydrocephalus
N1247	3,465	8/9	None	No	Severe marital discord with violence in early years
N3662	2,385	<2/3	Premature rupture of membranes Placental insufficiency Dysmaturity Delayed respirations Seizures Hyponatremia	Yes	Parents divorced
N4001	2,870		None	No	

Our follow-up has been very low-key. We have asked to meet with the families about every 3 years. We believed that annual evaluations, as in some longitudinal studies, would be too intrusive and would tend to remind the parents and children of a genetic condition for which there is no specific treatment. We think our approach has minimized the feeling in these young people and their parents that the chromosome variant makes them terribly different and a source of severe pain and anxiety.

Whether to continue this type of genetic screening is another ethical question. We have chosen not to do so, for we believe that it is unethical to screen persons for a condition that cannot be changed and has no treatment. For this reason, we have not added to our original cohort until one more subject was made available by amniocentesis and prenatal screening.

CONCLUSIONS

The most significant effects of a sex chromosome anomaly of the sort studied here (one missing X or one extra X or Y) are on growth, sexual development, and cognitive development. Lack of an X chromosome leads to short stature, and an extra X or Y usually is accompanied by tallness. A girl with 45,X karyotype shows the typical phenotype of Turner syndrome, with webbed neck, delayed or absent secondary sex characteristics unless treated, and probable sterility even after hormone treatment. She also may have congenital anomalies in other organs, such as heart or kidneys. Our study, as well as those of our colleagues reported in this volume, has shown variable pictures in the boys with the XXY karyotype. Although one of our subjects received hormone treatment early, another has developed normally up to the age of 17, including serial serum hormone levels. Therefore, it is not clear yet whether all males with this karyotype will show Klinefelter syndrome of hypogonadism and sterility. Girls with 47,XXX and boys with 47,XYY do not show any specific syndrome.

The effect on cognitive development is variable. One subject with each karyotype (47,XXX, 47,XXY, and 47,XYY) has normal intelligence by psychometric testing, whereas the rest are mostly in the low-normal to borderline range. Many had a selective learning difficulty, especially in the area of language development. The only subjects of ours who were seriously impaired in cognitive development had other contributing causes. It may well be that the genetic abnormality is in the form of vulnerability that enhances the effects of other biologic, environmental, or experiential factors [6]. No consistent effect on personality was observed. In particular, the boys with XYY karyotype have shown no severe or unusual aggressive or delinquent tendencies up to the age of 17–18, compared to their siblings, the XXY boys, or the comparison group.

By combining our data with those of the other studies in this volume, one can learn the range of expected development for children with each of the sex chromosome anomalies up until the age of 17–18. This can be useful for genetic counseling of couples whose fetus has been identified as having a sex chromosome aneuploidy. We believe that it offers considerable reassurance to those parents who wish to continue the pregnancy.

ACKNOWLEDGMENTS

We acknowledge with appreciation the consultations for these subjects supplied by Margretta R. Seashore, MD, pediatric geneticist, and Myron Genel, MD, pediatric endocrinologist, at the Yale—New Haven Medical Center; by John Schowalter, MD, child psychiatrist at the Yale Child Study Center; and especially by Sally Provence, MD, at the Yale Child Study Center, who formulated the original plans for this study and nurtured it with her interest and encouragement. This study was supported in part by the Children's Clinical Research Center (NIH grant RR00125).

REFERENCES

1. Leonard MF, Landy G, Ruddle FH, Lubs HA: Early development of children with abnormalities of the sex chromosomes: A prospective study. Pediatrics 54:208–212, 1974.
2. Leonard MF, Schowalter JE, Landy G, Ruddle FH, Lubs HA: Chromosomal abnormalities in the New Haven Newborn study: A prospective study of development of children with sex chromosome anomalies. In Robinson A, Lubs HA, Bergsma C (eds): "Sex Chromosome Aneuploidy: Prospective Studies on Children." New York: Alan R. Liss, Inc., for The National Foundation—March of Dimes, BD:OAS XV(1):115–152, 1978.
3. Leonard MF, Sparrow S, Schowalter JE: A prospective study of development of children with sex chromosome anomalies—New Haven study III. The middle years. In Stewart DA (ed): "Children With Sex Chromosome Aneuploidy: Follow-up Studies." New York: Alan R. Liss, Inc., for The National Foundation—March of Dimes, BD:OAS 18(4):193–218, 1982.
4. Lubs HA, Ruddle FH: Applications of quantitative karyotype to chromosome variation in 4,400 consecutive newborns. In Jacobs P, Price WH, Law P (eds): "Human Population Genetics." Edinburgh: Edinburgh Press, 1970.
5. Hook EB, Kim D: Prevalence of XYY and XXY karyotypes in 337 nonretarded young offenders. N Engl J Med 283:410, 1970.
6. Leonard MF, Rosenberg, L: Interaction of environmental and genetic factors in Klinefelter's syndrome. Pediatrics 52:141–143, 1973.

APPENDIX A. SHORT HISTORIES OF EACH OF THE SUBJECTS

45,X (N1644)

Ellen* was the first child of a 19-year-old unmarried mother, who had preeclampsia and tetanic contractions that caused fetal bradycardia in labor. Ellen was delivered by vacuum extraction and was depressed and hypotonic. She was cared for in the newborn intensive care unit. Her early development was slow—smiling at 6–7 months, sitting alone at 11 months. These delays may have been caused by understimulation, for she improved rapidly in the care of an aunt and uncle. Ellen and her mother left Connecticut and were not located again until she was 9½ years old. She had started school at age 7 and was succeeding fairly well. She was a cheerful, friendly girl, considerably shorter than her 7-year-old half-sister. On psychometric testing, she was in the borderline range. She never was able to receive adequate medical attention until friends of the family took her in as a foster child at the age of 15. She was then very short and had only minimal pubertal development. With hormone therapy she has grown a little faster but still is well below the 5th percentile in height and weight. She has made satisfactory gains in pubertal development. On psychometric testing at 12 years, she scored in the mid-low-average range, although the psychologist believed that her ability was higher. She left school at age 18 to return to her family and help in household duties.

47,XXX (N0240)

Alice was a full-term baby with no perinatal complications. She was considered a little slower than her sister, but her developmental evaluations at 1½ and 2½ years showed her development to be in the average range. Her parents decided to discontinue the study thereafter, because reminders of a chromosome variation in their daughter caused them considerable anxiety. However, they have provided further details by telephone. At 10 years, Alice was in fourth grade, an average student. She had spent 1 year in a reading readiness class between kindergarten and first grade. She was well coordinated. She was somewhat shy, but her personality development was considered normal. At 13½ she was going into eighth grade. Her best subjects were reading and languages. She was described as quiet, respectful, caring, and sensitive. At 16½ she was finishing tenth grade with low-average marks and planned to go to summer school. She had good study habits, and

*Not real names.

her best subject was English. She was popular with her friends and had a part-time job.

47,XXX (N2159)

Angela's story up to age 12 years was included in the last report [3]. Premature by 5–6 weeks, she weighed 2,160 gm at birth and remained in the hospital for 2.5 weeks. Her development was complicated further by the severe alcoholism of her mother during the pregnancy and Angela's first 10 years. Angela also had fluctuating conductive hearing loss. Her speech was delayed and immature. She repeated kindergarten and thereafter was in special classes for learning-disabled children. On her psychometric test at 12 years, 3 months, she scored in the mildly retarded range with her performance IQ considerably higher than her verbal score. At 15 she was arrested twice for minor thefts and was placed in a residential center for retarded children for 1.5 years. She returned home at 17 and attended eleventh grade with some regular and some special classes. She said that she liked school, but often did not go and left both school and home at 17½. She now is said to be living alone and working. Angela's mother considers her immature, impulsive, and still in need of protection. Angela can sew and cook and gets along well with people. She still has temper tantrums when she cannot get her own way.

47,XXX (N2222)

Marcy was delivered after 43 weeks of gestation by low-mid forceps and needed intubation and suction before spontaneous respirations began. She was a sober baby at first but developed well. She had no health or coordination problems. Her development was above average at the 2 year evaluation, including excellent language development. Throughout the elementary grades, she worked hard and got along well but lacked self-confidence. She had some difficulty with composition or telling a story fluently. When she was 11 years old, she had to struggle in school, and she had a tutor one summer. Her psychometric test at age 13½ placed her in the low-average range, with performance a little better than verbal subtests. At 16 she finished tenth grade with below-average marks and went to summer school again. She qualified for help in reading in her first or second year of high school. She is now nearly 18, a senior in high school, earning fair grades, and not planning further education. She does not read much and tends to let her work slide.

Marcy always has been tall, along the 90th percentile for height. Her pubertal development has been normal, with menarche at about 13 years. She always has been popular and friendly. She is emotional and moody at times.

She was upset by her parents' divorce when she was 12, but has adjusted well to her mother's remarriage. She now has a part-time job. She has been told about her aneuploidy, but seems to have paid little attention.

47,XXY (N2150)

Daniel's early course was described in our previous report [3]. He was the twelfth of 13 children of hard-working parents. All but one of his sibs have achieved well in school, and several have college educations. Daniel's language development was delayed. He had a year of reading readiness between kindergarten and first grade and special help each year in elementary school, where he was placed in slow classes. On his psychometric testing at 13 years he scored in the borderline range. He was then in seventh grade, but his achievement test scores were at the third to fourth grade level. Now, at 17 years, 10 months, he is in tenth grade. He has been active in sports, especially football.

At 13½ he was in the 50th percentile for height. He had mild acne. He has axillary hair and Tanner III pubic hair development. His testicular volume was just below the 50th percentile for age, and his penis was 12 cm long, above the 90th percentile. His FSH was a little high, and his serum testosterone level was within the average range. No therapy was recommended. He is to be reevaluated soon.

47,XXY (N2391)

Paul's history to the age of 12½ years was described in our previous report [3], but by typographic error he was listed as 47,XYY. His karyotype is *47,XXY*. His mother's pregnancy was complicated by bleeding and hormone injections, and he was born at 37 weeks by breech. He had bilateral inguinal hernias repaired at 22 months. His walking and talking were delayed. Although his cognitive ability was in the average range, he was handicapped by extreme anxiety, elective mutism, and constriction of affect and play. He had marked difficulty with separation when he started nursery school. When he was seen at 5 years, he was considered to have a pervasive developmental disorder, and psychotherapy was recommended. He has had special help in school and both individual and group psychotherapy. By psychometric testing at 8½ years he placed in the low-average range with a problem in language use and understanding. He seemed unhappy, tense, and inhibited and became immobilized and passive with frustration. At 12½, Paul was considered helpful at home, but self-conscious and sensitive with others. His mother called him vulnerable and anticipated that he always would need help. He sometimes stole from her. At 16 he was suspended from junior high school. He had been using marijuana and alcohol and had been caught with a group stealing beer. His mother reports that he is happier and more stable now. He has a job in a fast-food restaurant and is attending night school. He has many friends. Paul's problems probably have been enhanced by the severe maritial difficulties of his mother in two marriages and the alcoholism of his father.

Paul is tall, just above or just below the 90th percentile on serial measurements. He has not had consistent medical care and has not yet received an endocrine evaluation. His pediatrician says he has a severe emotional disturbance.

47,XXY (N2551)

Mark's history to age 13 years was included in our last report [3]. He had no perinatal complications and has had good health. A heart murmur has been described, which has been determined to be functional. Mark's last psychometric examination was at the age of 8, when he scored an IQ of 97 on the Stanford-Binet scales. He had some difficulty in the early grades with reading comprehension, reversed letters, and spatial relationships, with which he was helped by a learning-assistance center. He also had poor coordination in gross and fine motor skills. At 17 years, 8 months, he is a senior in high school, making Bs and Cs and hoping to go to college.

Because of his distant location he has not been seen again. He is described by his mother at this time as likeable, considerate, friendly, and active. He has changed from his early shyness to appropriate assertiveness. He has good relationships with people. His voice has changed, and he shaves daily. He has received hormone injections since the age of about 12 years. His height was close to the 50th percentile until the last few years and at age 17½ is above the 75th percentile.

47,XXY (N3101)

Edward was the result of his parents' ninth pregnancy. There had been four miscarriages and one baby who lived only 32 minutes. Edward was a healthy baby with satisfactory early development, including his language. By psychometric testing his intelligence is in the average range, but he has not achieved well in school. He repeated third grade and had special help for reading. He was thought to have ability, but to lack motivation for learning, and was functionally learning-disabled. Although beloved by family and friends, he was unhappy in school, belligerent, and nonconforming. He dropped out of school at 17 years rather than repeat tenth grade. He did not care about learning and felt that the teacher was against him all year. Now, at 17½, he has started a correspondence course to work for high school equivalency and hopes to join the Navy. He had a part-time job in a restaurant but recently was fired because of a fight.

Edward is close to the 90th percentile for height and still is growing. He has been seen yearly in the endocrine clinic, where his pubertal changes and hormonal levels have been considered normal for his age, so no therapy has been recommended for him yet. However, he is sensitive about his appearance and now would like help if it would make his genitalia look more like those of other men. He gets along well with peers and with children. His relationship with his mother vacillates between loving and considerate and angry and belligerent. He has lived through two unhappy marriages of his mother.

47,XYY (N0384)

Harvey's early history was described in our previous summary [3]. He had early complications of prematurity and possibly arrested hydrocephalus. His language development was delayed, and his distortion of articulation persisted through elementary school. His motor incoordination and developmental profile suggested a mild diffuse neurologic abnormality. He has struggled through school, most of the time in small "developmental" classes. He had difficulty with concentration and did not do homework. At 18 he is a senior in high school, performing fairly well in a special class and hoping to go on to technical school. He has many friends and enjoys sports, excelling in golf and bowling. He drinks occasionally but denies use of drugs. He is described as generally happy and independent, but always resentful that his twin brother has been a year ahead of him in school. He has a quick temper, but no more so than his brother, and expresses it by slamming doors. He fights only with his brother. He has had no antisocial or delinquent behavior according to his mother.

He surpassed his twin brother in height by the age of 3 years, and is in the 75th percentile at 18, about 6 inches taller than his twin. Harvey's mother has not shared with him the knowledge of his extra chromosome, and does not plan to do so, for she feels that "he has enough problems."

47,XYY (N1247)

Robert's history to 13½ years was described in our previous report [3]. He had no prenatal or neonatal complications. By psychometric testing he placed in the low-average range. His language development was a little slower than his other achievements. He repeated first grade and has struggled since in school, especially with language skills and reading. He had speech therapy in school for articulation problems. At 18 he is a junior in high school, achieving best in art and mathematics, worst in history. He and his mother agree that he is disinterested in learning and has poor study habits. He now works part time with audiovisual equipment in the store his father manages and thinks he would like to specialize in that work.

He is well physically and is a good athlete, enjoying weightlifting, basketball, and baseball. He has experimented with alcohol and dexedrine, but stopped because it interfered with his sports. He has been tall at least since 5 years, and now at 18 years is 6 feet, 4 inches, above the 95th percentile. He has had some emotional difficulties, partly rebellion against the strict and punitive discipline of his father. He has had treatment from a clinical psychologist several times, more recently for depression and sleep disturbance after break-up of a relationship with a girlfriend. He is friendly and gets along well with everyone. He gets angry but controls his anger better than his father, usually by going to his room and not by fighting. He has had no antisocial or delinquent behavior.

47,XYY (N3662)

Jonathan was a first child who was born postmaturely after 44 weeks of gestation and premature rupture of the membranes. He weighed only 2,370 gm, and the placenta was described as small and hard. Respirations were started with difficulty. His Apgar scores were less than 2 in 1 minute and 3 in 5 minutes. He was kept in the newborn special care unit for 8 days, and his course was complicated by seizures, hypoproteinemia, hyponatremia, and questions of intracranial bleeding and sepsis. His only physical anomalies were bilateral inguinal hernias, corrected surgically when he was 2 years old. His expressive language developed slowly, and he had some distortion of articulation in the early years, but otherwise his development was satisfactory. Although he seemed well motivated, academic subjects were hard for him. At his last psychometric test his full-score IQ on the WISC-R was 78, with verbal slightly better than performance. The psychologist indicated that his potential probably was higher than this. Jonathan had one year in a learning disability class between kindergarten and first grade. He was in ninth grade at 15 but left school after that. His study habits and concentration were described as poor. At 17 he is in a vocational technical school studying automobile mechanics.

He has not been seen here since he was 8 years, 9 months, but his mother reported by mail that he likes people and gets along well with them. He is very helpful and cooperative and likes to work. He loves the outdoors. Rarely, he displays temper, especially with his sister, but he has good control. He does not like to fight. His height always has been above the 75th percentile, and at 17 is at the 95th percentile.

Growth, Psychologic Characteristics, and Sleep-Wakefulness Cycle of Children With Sex Chromosomal Abnormalities

Makoto Higurashi, MD, Hisashi Kawai, Masaya Segawa, Kumiko Iijima, Yukie Ikeda, DHS, Fumihiko Tanaka, Shinzo Egi, MD, and Shigehiko Kamashita

Department of Health Sciences, Yamanashi Medical College, Yamanashi (M.H.), Department of Psychology, Psychiatric Research Institute of Tokyo, Tokyo (H.K.), Segawa Neurological Clinic for Children, Tokyo (M.S.), Departments of Maternal and Child Health (K.I.), Pathology (F.T.), and Pediatrics (S.K., S.E.), University of Tokyo, Tokyo, and Department of Education for Handicapped Children, Tsukuba University, Ibaragi (Y.I.), Japan

INTRODUCTION

The term *growth* has commonly been used for those aspects of maturation which can be reduced to a measurement of size. The term *development* refers to changes in the function of the organism. Because these two aspects cannot be sharply differentiated, the term *growth and development* is generally used without discrimination of either the magnitude or quality of maturational changes. Changes in size, psychologic findings, and circadian rhythms in patients with sex chromosomal abnormalities are described in this chapter.

This study on girls with Turner syndrome and boys with XXY and XYY sex chromosome constitutions was designed to follow the longitudinal growth and to analyze the psychologic findings. The pathophysiology of behavioral disorders is difficult to detect. However, abnormalities of synchronization of physiologic rhythm are known to have a causal relation to this disorder. To detect rhythm desynchrony, it is useful to examine sleep parameters. In this chapter we examine the sleep-wakefulness cycle in XXY and XYY boys.

Birth Defects: Original Article Series, Volume 22, Number 3, pages 251-275

MATERIALS AND METHODS

The subjects were 16 girls with Turner syndrome, two boys with XXY, and eight with additional Y chromosome material who lived in Tokyo and whose ages ranged from 2 months to 14 years when measurements were begun (Table I). Chromosome analyses were done on peripheral blood, and, when mosaicism was suspected, two tissues (skin and bone marrow) were examined in almost all cases. Six of the 16 cases with Turner syndrome were 45,X, seven were 45,X/46,XX mosaics; three other mosaics were one 46,X,i(Xq)/46,XX; one 45,X/46,XXp−/46,XX; and one 46,XXp−/46,XX.

TABLE I. Case Material

	Parental age (years)				
Case No.	F	M	Birth weight (gm)	Birth length (cm)	Karyotype
Turner syndrome					
1	31	27	2,580	45.8	45, X
2	28	27	2,480	45.0	45, X/46, XXp−/46, XX
3	31	30	2,400	43.5	45, X
4	32	25	2,500	49.0	45, X/46, XX
5	31	31	2,480	44.0	45, X/46, XX
6	30	29	2,600	48.0	45, X/46, XX
7	33	29	2,980	46.0	45, X/46, XX
8	45	37	2,500	45.0	45, X
9	39	30	3,250	48.0	46, X, i(Xq)/46, XX
10	35	34	3,050	50.3	45, X/46, XX
11	32	25	2,500	—	45, X
12	27	27	2,680	43.5	45, X/46, XX
13	31	25	2,990	48.0	45, X
14	33	30	2,010	42.5	45, X/46, XX
15	30	26	2,960	49.5	45, X
16	31	32	2,690	48.1	46, XXp−/46, XX
Klinefelter syndrome					
1	31	29	3,120	49.8	47, XXY
2	33	32	3,465	50.0	47, XXY
XYY constitution					
1	31	24	4,100	52.2	47, XYY
2	24	23	2,350	45.6	47, XYY
3	35	27	3,350	48.0	47, XYY
4	24	22	3,500	53.7	46, XY, −15, +t(15:Y)
5	28	27	2,490	44.7	47, XYY
6	30	30	2,800	48.9	47, XYY
7	25	24	2,460	45.7	46, XY/47, XYY
8	35	34	4,210	52.6	46, XY/47, XYY

All the anthropometric measurements were made using the Martin and Saller [1] anthropometric method. The measurements were performed at a time close to the date of birth of each child and 6 months later. The longitudinal data were collected over a period of 11 years from 1974 to 1984.

The measurements were of weight, height (or length), lower-limb length, biacromial diameter, biiliac diameter, transverse chest diameter, chest depth, upper limb length, chest girth, hand length, hand breadth, foot length, cephalic length, cephalic breadth, cephalic circumference, bizygomatic breadth, internal biocular breadth, and morphologic face height. Until the age of 2 years, length was measured using the horizontal calibrated board in the supine position, and thereafter standing height was measured.

Standard deviation score (SDS) was used to compare the normal control measurements with those of the patients. SDS is calculated from the following formula: $SDS = (X + \overline{X})/SD$ where SDS is standard deviation of the normal control, $\overline{X}$ is the mean of the normal control, and $\overline{X}$ is the measurement of the patient. Control values for 5-year-old Japanese girls were reported by Ikeda et al [2], Kida et al [3], and Okada [4] for girls older than or equal to 6 years of age. The measurements taken for the 47,XYY boys were of stature and body weight.

Five cases with Turner syndrome, one mosaic XYY individual, one Y-15 translocation and one XXY boy were examined on clinical psychologic tests [Rorschach test, picture-frustration study, Wechsler intelligence scale for children, and Bender Gestalt test (Table II)].

Sleep-wakefulness cycles for an infant with XXY and one with XYY were recorded on a day-to-day plot method (Figs. 1, 2). As for sleep measures, total day sleep time and total night sleep time were assessed. The day sleep

TABLE II. Psychologic Testing*

Case	Age (years, months)	Diagnosis	Psychological testing (years, months)
OY	11, 4	Turner	Ro test, P-F, BGT, WISC
SM	11, 4	Turner	Ro test, P-F, WISC (6, 5; 11, 4), BGT
SA	12, 4	Turner	WISC (6, 5; 7, 2), BGT (7, 2), mother's report
KT	15, 5	Turner	Ro test, P-F, BGT, WISC
YK	16, 4	Turner	Ro test, P-F, BGT, WISC, mother's report
KK	1, 5	XY/XYY	JPDST, behavior observation, mother's report
TR	7, 3	XY/15-Y	Ro test, BGT, P-F, WISC, mother's report
AT	3, 4	Klinefelter	JPDST, behavior observation, mother's report

*JPDST, Japanese edition of Denver developmental screening test; Ro test; Rorschach test; P-F, picture-frustration study; BGT, Bender Gestalt test; WISC, Wechsler intelligence scale for children.

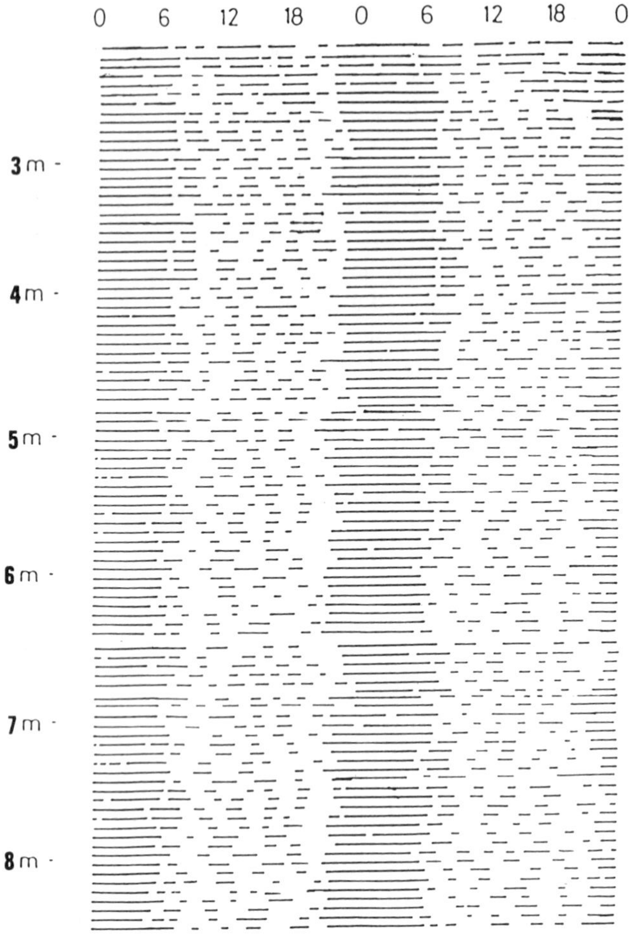

Fig. 1. Circadian oscillation in an XXY child. Top, hours; left months of age.

time indicated the total night sleeping hours observed in 12 hr between 8:00 and 20:00. Night sleep time indicated the total sleeping time that occurred in 12 hr between 20:00 and 8:00.

RESULTS

Growth in Children With Turner Syndrome and XXY and XYY Constitutions

Turner syndrome. Until the age of 5 years all measurements on the 16 cases are compared with those from normal control girls using the SDS.

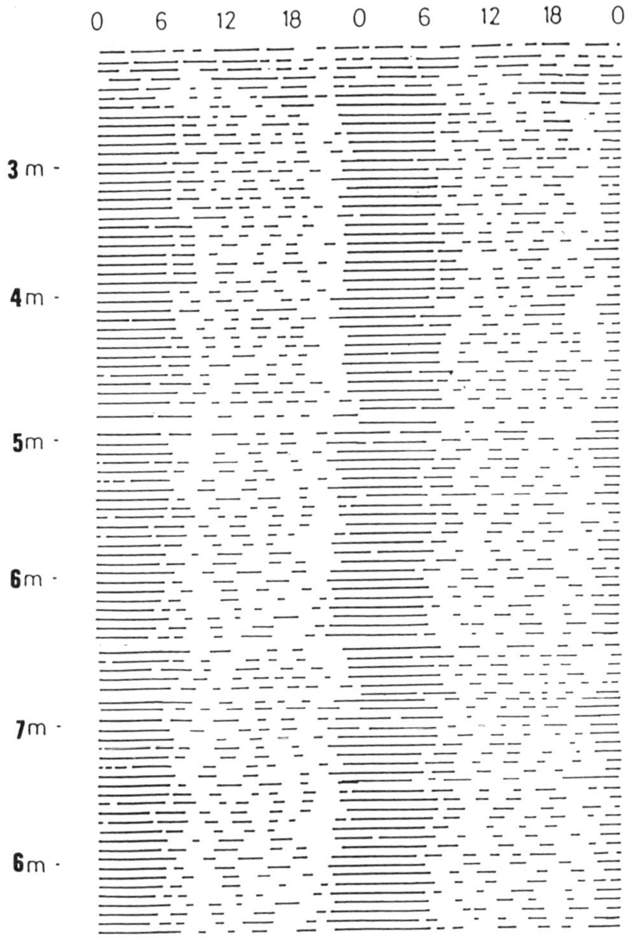

Fig. 2. Circadian oscillation in an XYY child. Top, hours; left, months of age.

However, in the girls older than age 6 years, control data were not available for eight of the 16 measurements (upper limb length, hand length, hand breadth, foot length, foot breadth, bizygomatic breadth, internal biocular breadth, morphologic face height). SDSs are presented in Figure 3 for the eight girls less than 5 years old and in Figure 4 for the eight girls over 6 years old. Table III shows the means of SDS anthropometric measurements of 16 patients.

The height of all the patients was less than that of the control girls; in 13 patients it was below −2 SDS and in seven patients below −3 SDS. The

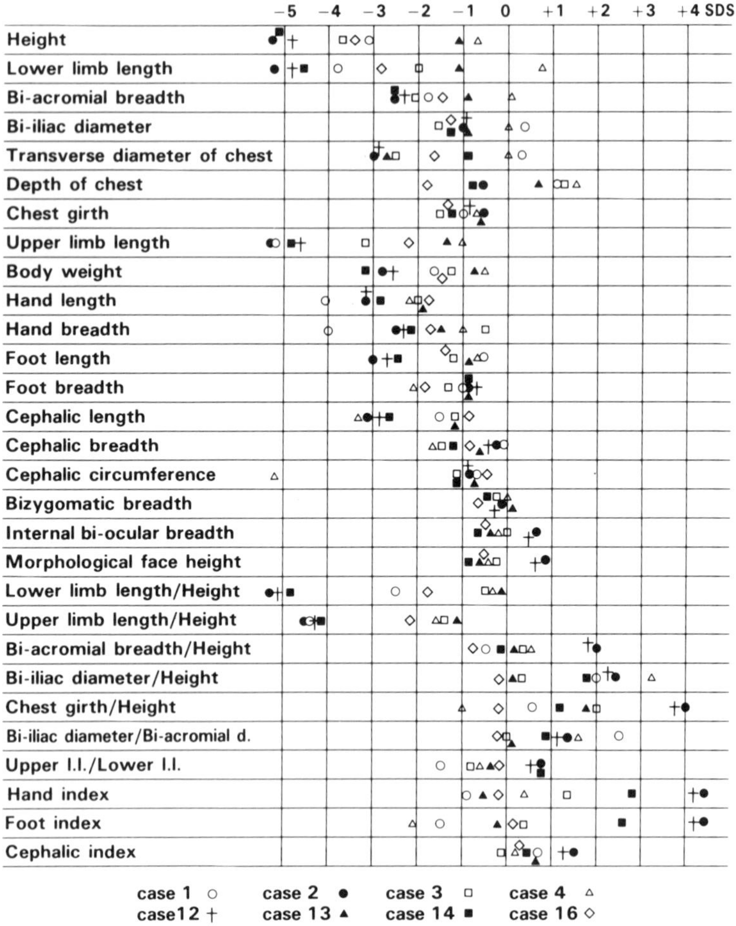

Fig. 3. Standard deviation score vs normal controls of the eight patients with Turner syndrome less than 5 years old.

mean of SDS of height in girls less than 5 years old was −2.38; in cases older than age 6 years old it was −2.40.

The lower limb length of all the patients except one was shorter than that of normal controls. The mean of SDS of lower limb length in cases less than 5 years old was −2.90, that in cases older than 6 years old −1.80.

The upper limb length of all the patients who could be compared with normal control girls was less than that of the control girls. The mean SDS of upper limb length in eight patients was −3.43. The mean SDS of biacromial

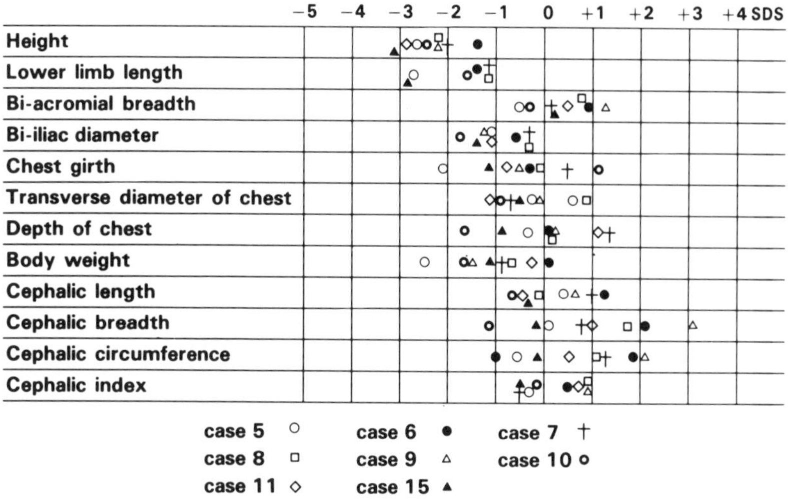

Fig. 4. Standard deviation score vs normal controls of the eight patients with Turner syndrome older than 6 years old.

TABLE III. Means of SDS in 16 Patients With Turner Syndrome

	Means of SDS	
Measurement	⩽5 Years old	⩾6 Years old
Height	−2.38	−2.40
Lower limb length	−2.90	−1.80
Biacromial breadth	−1.68	+0.40
Biiliac diameter	−0.78	−0.98
Transverse diameter of chest	−1.65	−0.30
Depth of chest	+0.20	+0.03
Weight	−1.73	−1.04
Cephalic length	−2.08	+0.23
Cephalic breadth	−0.76	+0.68
Cephalic circumference	−1.34	+0.45
Girth of chest	−0.93	−0.45

breadth in cases less than 5 years old was −1.68, but that in cases older than 6 years old was +0.40. The mean SDS of biiliac diameter of the patients was −0.78 in the younger group and −0.98 in the older group. Chest girth, chest transverse diameter, and depth of chest in the patients was almost equal to that of control girls. Weight was less than that of control girls, but only two patients were less than −3 SDS. Results of the patients' head and face measurements were similar to those of normal girls, although the cephalic circumference of one patient (case 4) was −5 SDS. The head size in the older group tended to be relatively larger than that in the younger group. The patients tended to have small hands and feet, and the hand and foot indices were greater than those of normal girls. Relative lower limb length and relative upper limb length to height ratios were less than those of normal control girls. On the other hand, the relative biacromial breadth and the relative biiliac diameter to height in the younger group were greater than those of normal controls.

Longitudinal anthropometric measurements were also obtained in this part of the study. In Figure 5, all the measurements of height and lower limb length of the several patients are plotted; these were below the mean of the controls girls. In Figure 6, chest girth and chest depth are shown; these were mostly above the mean of controls. Figure 7 shows the growth curves for height and weight of five girls with the chromosome constitution 45,X; Figure 8 shows similar curves for four girls with 45,X/46,XX mosaicism in comparison with the growth curves of normal Japanese girls established by

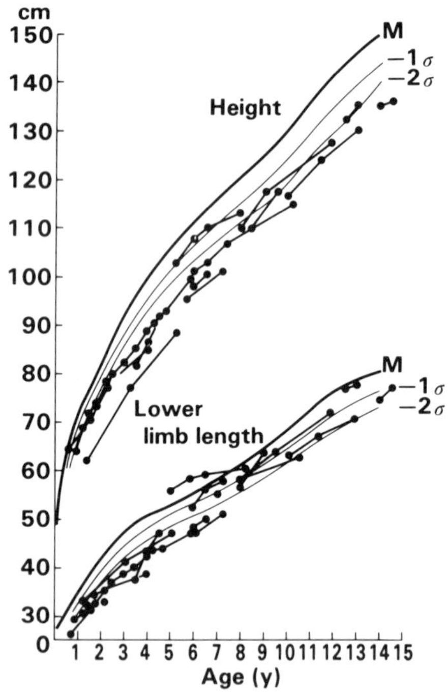

Fig. 5. Height and lower limb length of girls with Turner syndrome at each measurement plotted against chronologic age.

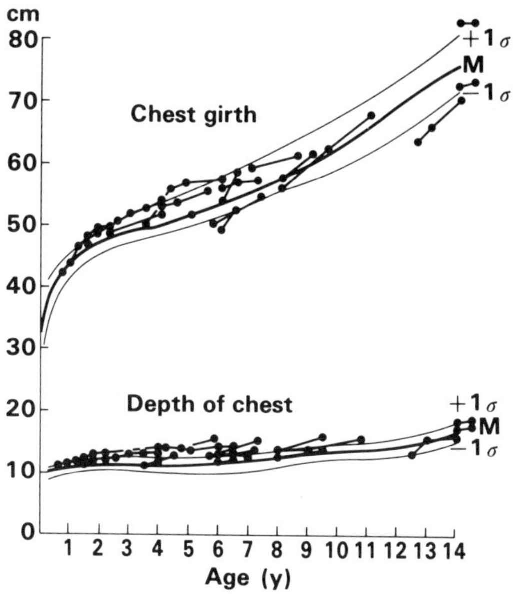

Fig. 6. Chest girth and depth of chest of girls with Turner syndrome at each measurement.

Takaishi et al [5]. The height of all nine of these girls was less than the 3rd centile at almost all ages. On the other hand, weight was between the 3rd and 25th centiles. It seems that there were no differences in the growth curves of patients with 45,X and those with mosaic chromosome constitutions.

XXY. The measurements done before 5 years of age on the two boys are compared with those from normal control boys using SDS and are presented in Figure 9. All measurements on both XXY boys were less than those of the control boys, but the degree of reduction was less marked than in Turner syndrome.

XYY, XYY mosaics, and Y-15 translocation. The birth length of these eight boys was compared with normal Japanese infants. Two boys had above normal length for gestational age, and five with 47,XYY had lengths below the mean normal for gestational age. On the other hand, with the exception of one case, they were within the normal range of weights for gestational age for normal Japanese infants.

Comparison of birth weights of 47,XYY cases with normal Japanese standards for gestational age indicated that 47,XYY babies, with one exception, were not small or large for dates. Neither the size at birth nor

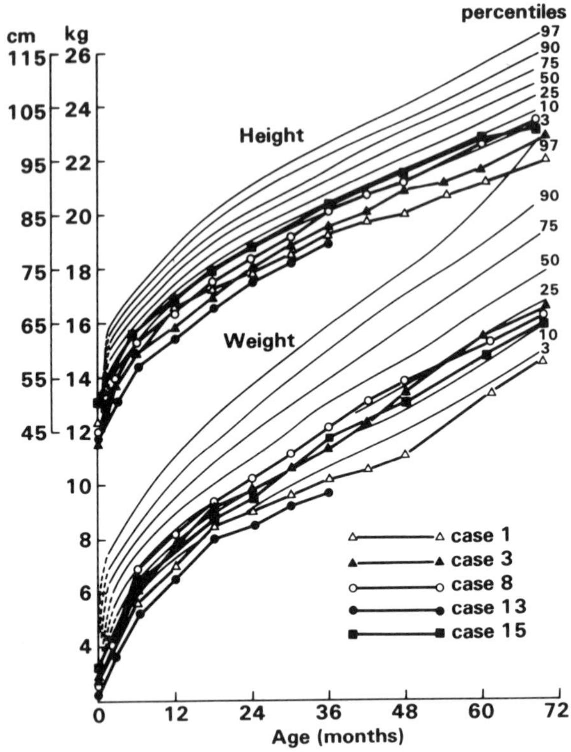

Fig. 7. Height and weight of five girls with 45,X chromosome constitution.

growth curves from infancy to 9 years of age for 47,XYY boys show an abnormal pattern.

Eight individuals were followed for up to 9 years (five with 47,XYY; one with 46,XY-15, +t(15p−:Yq+), and two with 46,XY/47,XYY). The growth curves for stature in five boys older than 3 years of age are shown in Figure 10 in comparison with the growth curve of normal Japanese boys established by Takaishi et al [6]. The height of 47,XYY boys from infancy to 7 years of age was between the 10th and 90th percentiles with one exception in which height rose above the 90th after 8 years of age.

Psychologic Characteristics in Turner Syndrome and XXY and XYY Constitution

Because the numbers of patients were limited and the intellectual levels of the patients were of wide range, generalizations from the findings are difficult. However, relatively common features were found.

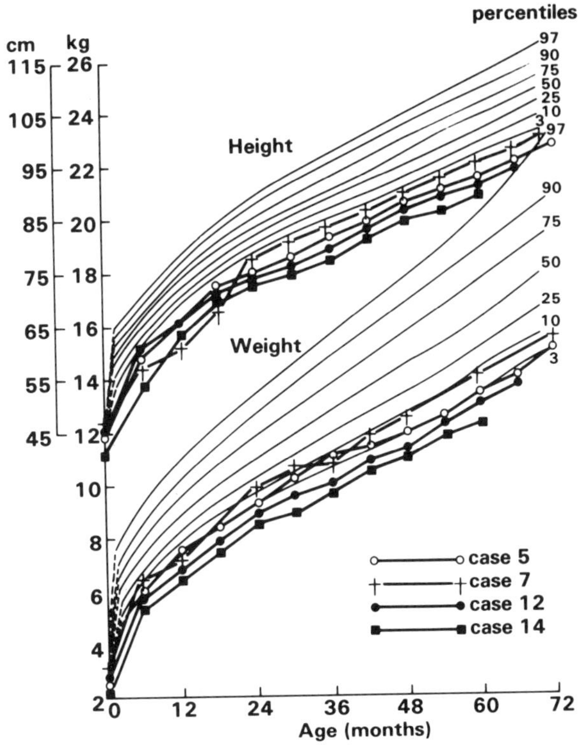

Fig. 8. Height and weight of four girls with karyotype 45,X/46,XX.

Turner syndrome (Tables IV–VI). 1) Low responsiveness to emotional stimuli (patients OY, SM, and KT were embarrassed by the emotional stimuli), 2) poor emotional expression (lacking variety of emotional response resulting in the appearance of quiet stability), 3) passive, negative mental attitudes, 4) low mental activity (including psychologic) and low productivity, 5) narrow and limited interests, 6) passive and common sense adaptability (patients OY and SM show somewhat low adaptability related to their lower IQs), 7) lower impulsiveness and aggressiveness, 8) low aggression in response to frustration, not easily discouraged (except for patient SM), 9) little aggressiveness, does not respond appropriately when criticized, 10) passivity in personal relations, lack of interactions (only patient YK has continued, interactive relations), 11) wide range of IQ from subnormal to normal, 12) adaptation according to IQ levels, and 13) no psychopathologic findings.

In addition, two patients, KT and SA, whose IQs are relatively high, showed other psychologic characteristics. First, there was a stubbornness, or

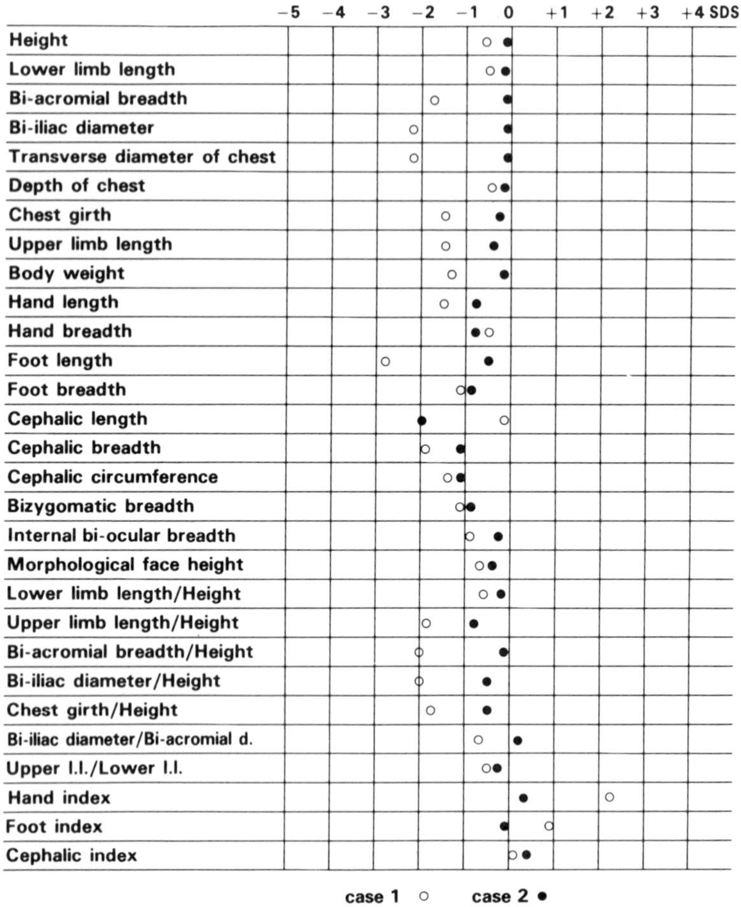

Fig. 9. Standard deviation score vs normal control of the two patients with XYY constitution less than 5 years old.

lack of flexibility, that made it difficult to change behavior or viewpoint according to conditions. However, whether this characteristic is related to an intellectual factor is not clear at this stage. Patient YK had another characteristic, that she was not influenced by standards outside herself; being an inner-life-dominant type, she thinks and feels in her mind. She can cure herself when some troubles occur. She is somewhat passive in her personal relations but can interact with others.

XXY and Y-15 translocation. It is not possible to make generalizations in that there are only two cases with psychologic results. Patient KK, the

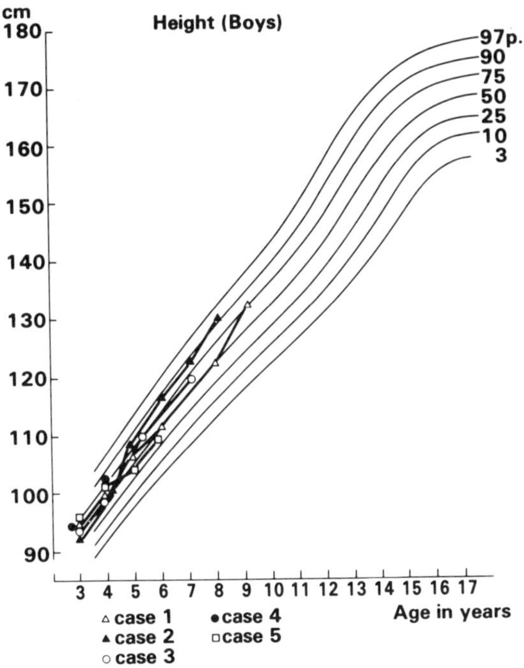

Fig. 10. Height curves for five boys with XYY chromosome constitution.

mosaic XYY individual, was 1 year, 4 months, of age and hence too young for us to test his psychologic characteristics. He was well both psychologically and physically, having many positive aspects emotionally and developmentally.

Patient TR, with the Y-15 translocation, had good general adaptability, but he also had rigidity making it difficult for him to change his viewpoint. He was also emotionally suppressed, being very sensitive but unable to express his feelings. His mood was rather negative, and his mental activity was low. He was somewhat sensitive, but also passive in his personal relations. Impulsiveness and aggressiveness were not evident. However, his control might be poor when he acts out. His IQ was within normal limits, but there was a marked discrepancy of 19 points between VIQ and PIQ. The reason for the discrepancy was his poor verbalization of his thinking and feelings. He did not show direct aggression to others or to himself but had a tendency to work by himself.

The boy with an XXY sex chromosome constitution, was 3 years, 4 months, of age. His development was within normal limits, and no adaptation

TABLE IV. Rorschach Summary Scoring Table

	QY	SM	KT	YK	TR
R	25	9	9	13	16
Rej		5			
R,T	20.3	24.8	44	12.3	10.9
R,T(NC/CC)	23.7, 17.6	24.8/Fail	32.5/53.2	14/11	8.8/13.0
W%	28	78	44	62	44
D%	64	22	55	38	56
Dd%	8				
F%	80	56	67		62.5
F+%	52	68	33		70
R%	60	68	56	92	80
M		1		4	1
FM	4	2	3	6	4
m					
k(k.kF.Fk)					
K(K.KF.FK)	1				
Fc				1	
cF.c					
c′		1			
FC				1	
CF				1	1
C					
ΣC:M		0:1		1.75:4	1:1
Fc + c + c′:FM + m	0:4	1:2	0:3	1:6.5	0:4
8,9,10/R%	56	11.1	33.3	46.2	50
FC:CF + C				1.5:1.5	0:1
FC + CF + C:Fc + c + c′		0:1		1:1	1:0
FM:M	4:0	2:1	3:0	6.5:4	4:1
A%	80	56	100	54	67
H%	8	11.1		31	(1)
P%	8 (2)	33 (3)	22.2 (2)	39 (5)	25 (4)
DR	53	4	2	4	4
CR	5	4	1	3	4

problem had been found. He was stable in his emotionality, but he had some stubbornness. These characteristics were not necessarily related to this syndrome but are common in young children. His attachment to his mother was also well established.

Sleep-Wakefulness Cycles of One XXY Infant and One XYY Infant

Figure 11 shows the sleep-wakefulness cycle of the XXY infant via a double-track method. Circadian rhythm was observed until 3 years of age with shifting of the sleeping time to the night period, after which day sleep gradually tended to decrease with age. Age variation of day and night sleep

TABLE V. Results of WISC

Measure	OY	SM		SA		KT	YK	TR
Verbal IQ	53	93	86	105	125	115	101	104
Performance IQ	89	74	93	100	128	108	102	122
Total IQ	67	82	88	103	130	114	102	115
Verbal test								
Information	3	11	12	12	17	13	11	11
Comprehension	3	7	8	4	10	13	12	11
Arithmetic	1	9	7	11	13	11	4	8
Similarities	2	9	4	13	15	13	13	12
Vocabulary	4		9	11		13	12	12
Digit span	4		7	14		11	9	10
Performance Test								
Picture completion	7	6	8	9	15	10	8	8
Picture arrangement	1	5	10	13	12	10	10	17
Block design	7	9	7	12	17	13	9	16
Object assembly	11	5	9	7	13	9	10	16
Digit symbol	10		9	6		11	12	9
Maze	14		11	13		14	13	14

in an XXY child compared with that of a normal child is also shown. The XXY infant showed normal variation in early childhood, although in infancy there was more sleeping time than normal. Figure 12 shows the comparisons of both day and night sleeping times of the XXY boy with the average values of normal children. The XXY boy slept more both day and night than normal, although in early childhood he slept similarly. Similar studies were performed in a child with the chromosomal abnormality XYY and revealed that his sleep parameters were quite normal in infancy and childhood (Figs. 13, 14).

These studies revealed there was no delay in the development of circadian rhythm in either disorder, suggesting normal development in the mechanism for modulation of the circadian rhythm in both disorders. However, in the XXY child there might be a delay in the development of regulation of the state of waking; in early infancy he had prolonged sleeping time in both day and night instead of apparent circadian modulation.

DISCUSSION

Chromosome mutations have a profound influence on development. Their presence in survivors at birth generally causes obvious, and often severe, sublethal, deviations in physical development, intelligence, or psychosocial adaptation. There is, however, another aspect to the developmental influence

TABLE VI. Table for P-F Study Scoring

Score	OY	SM	KT	YK	TR
GCR%	40↓*	60	79↑	83↑	68↑
E%	65↑	50	23↓	48	30↓
I%	25	29	40↑	21	20
M%	10	21	38	31	50↑
O-P%	10	24	13	17	9
E-D%	30↓	36↓	54	63↑	64
N-P%	60↑	40↑	33	21↓	27
E′	0↓	3↑	0↓	1.5	0↓
E	2↓	3↓	2.5↓	7.5↑	5.5
e	11↑	4.5	3	2.5	1↓
I′	1	0.5	1	1.5↑	1
I	4	4	7↑	3.5	1.5
i	0↓	1.5	1.5	0↓	2↑
M′	1	1.5	2	1	1
M	0↓	0.5↓	3.5	4	7↑
m	1↓	2.5	3.5	2.5	3
E%	5	0↓	4	4	9
I%	0↓	14↑	2↓	2↓	2
E + *I%*	5	14	6↓	6↓	11
E − *E%*	5↓	14	6↓	27	16
I − *I%*	20↑	5	27↓	13	5
M + *I%*	10↓	36	40	33	52↑

*↓, −ISD; ↑, +ISD.

of chromosomal abnormalities. Only about half the oocytes that are fertilized survive long enough to embed and lead to a recognizable pregnancy, and one can deduce that the chromosome abnormality rate is over 10% of these pregnancies, if taken from the fourth week on, so that more than 40% of all spontaneous abortions are chromosomally abnormal [7]. However, the proportion of abnormality in the abortion falls from 50% or higher at 4–8 weeks to a low point of 5% or less at 20 weeks of gestation, which is also the approximate frequency of perinatal deaths [8]. According to Polani [7], about 97% of autosomally abnormal fetuses are spontaneously aborted, but this is the fate of only 83% of those with sex chromosome anomalies, and these are almost always the sex chromosome monosomies (45,X). When the autosomes are invloved, development is more seriously affected and postnatal existence more threatened. Hence, it is instructive to study the development in sex chromosomal abnormalities in order to compare it with that in autosomal abnormalities.

Anthropometric Measurements

Longitudinal anthropometric measurements on 16 cases with Turner syndrome were made in this part of the study. The growth curves for stature of

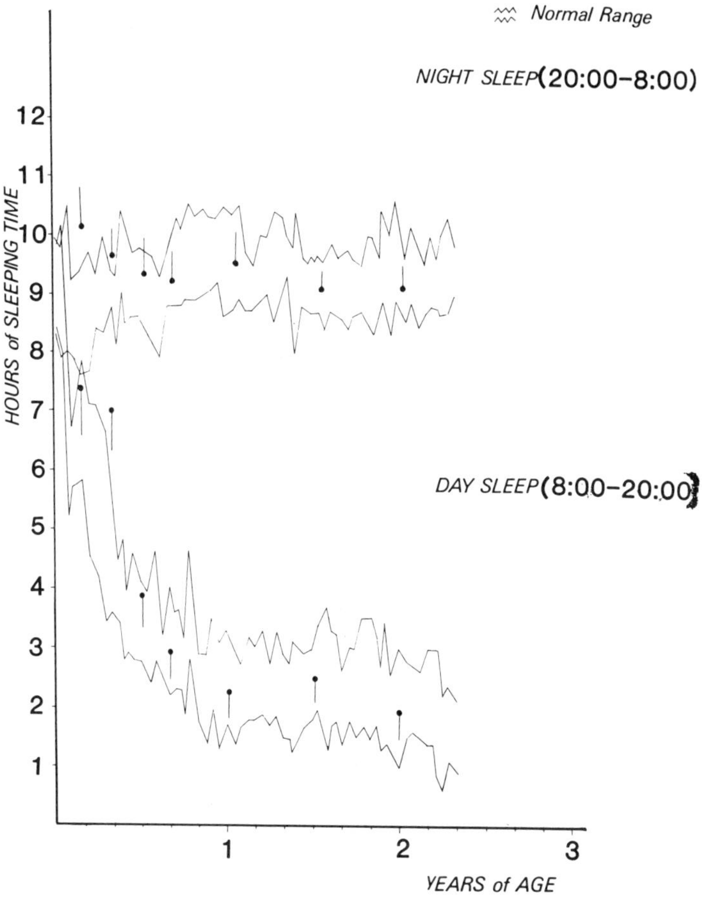

Fig. 11. Age variation in day and night sleep in a child with XXY constitution.

almost all cases of Turner syndrome were below the 3rd precentile at all ages. On the other hand, the growth curves of weight were not so markedly lower than those of the normal controls. As is shown in Figures 7 and 8, there were no differences in the growth curves of patients with 45,X and those with mosaicism.

Although many reports have pointed out that girls with Turner syndrome are characterized by short stature, few studies on longitudinal, anthropometric growth processes in infants with this condition have been reported. Nielsen et al [9] studied the development of Turner syndrome during childhood and the longitudinal findings of stature and weight of patients from 7 to

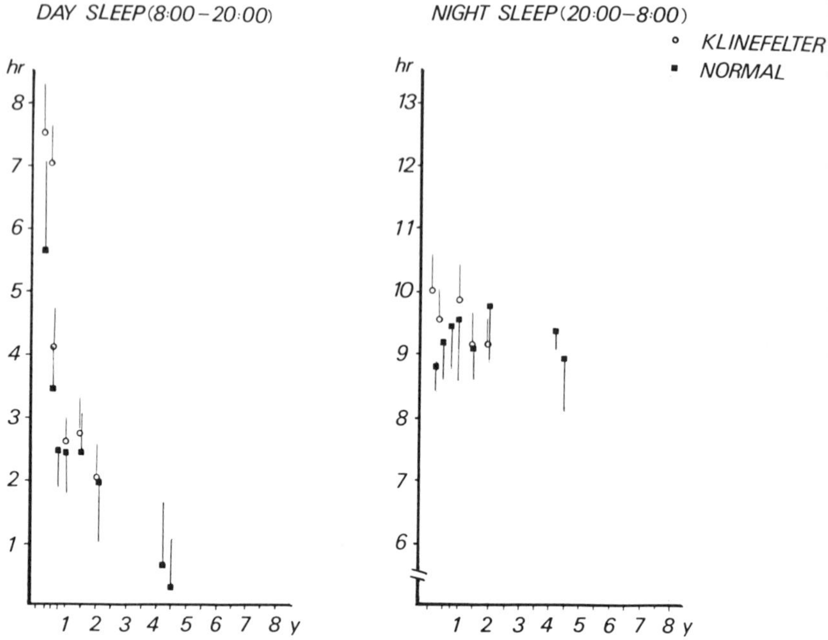

Fig. 12. Age variation in day and night sleep in a child with XXY constitution.

18 years of age. According to their results, stature of the majority of girls with Turner syndrome aged 7–18 is below the 3rd percentile. In the present study, the stature curves of girls with Turner syndrome from birth to age 14 years were also below the 3rd percentile, confirming that patients with Turner syndrome have a significantly shorter stature from birth onwards.

The findings of note in longitudinal anthropometric measurements in Turner syndrome are those for biacromial breadth, biiliac diameter, and head size. The mean SDS of biacromial breadth in cases less than 5 years old was −1.68, but that in cases older than age 6 years was +0.40. The mean SDS of biiliac diameter of the patients was −0.78 in the younger group and −0.98 in the older group.

Earlier descriptions of Turner syndrome include mention of a shield-shaped chest and widely spaced hypoplastic nipples as common features. However, these features are not noticeable before 5 years of age but gradually develop after 6 years of age.

The shoulders of Turner syndrome girls show a tendency to broaden with age, but this is not the case with the development of the pelvis. It is also

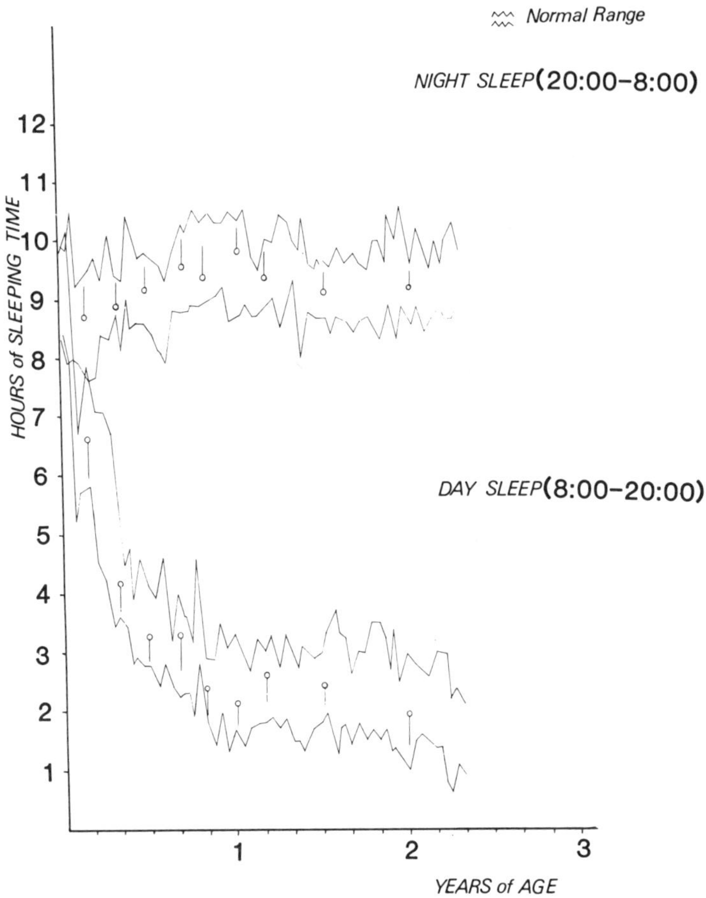

Fig. 13. Age variation in day and night sleep in an XYY child.

interesting that the head size in the older group tended to be relatively larger than in the younger group.

We cannot make any generalizations from the present data on the anthropometric measurements of XXY boys, as we measured only two cases. The growth curves for height in each 47,XYY boy were within the normal range between early infancy and 9 years of age, as were all growth curves for weight. Physical malformations were hardly ever observed, and intellectual development was within the normal range in all cases. However, follow-up is necessary; it might well be that a spurt in growth will occur later.

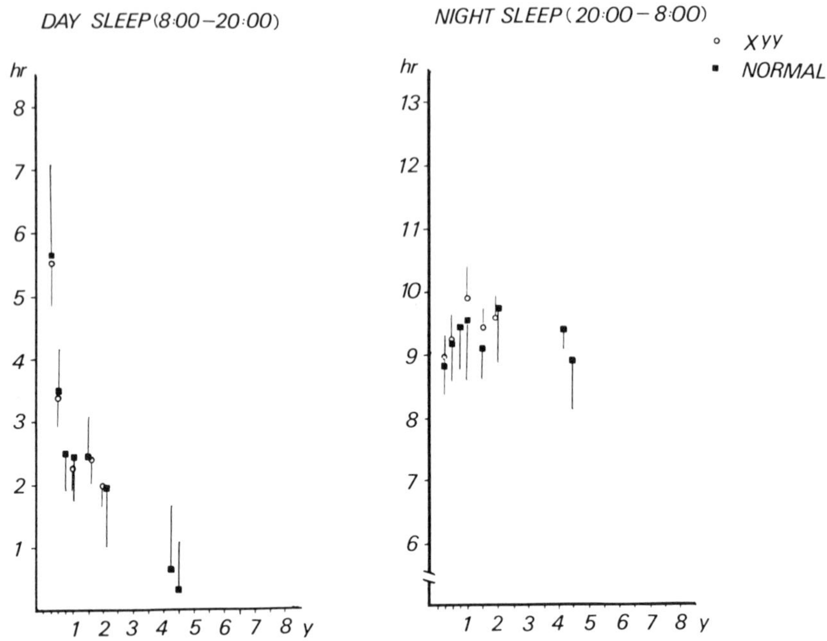

Fig. 14. Age variation in day and night sleep in an XYY child.

Psychologic Study

Intellectual aspects in Turner syndrome. IQs of our patients were widely distributed with scores from 67 to 130, as has been described in other reports [10–12]. It is said that the discrepancies between verbal IQ (VIQ) and performance IQ (PIQ) have clinical significance if the differences are 15 points or more [13]. Among our cases, only one (OY) shows marked discrepancy with PIQ higher than VIQ; three cases show a somewhat higher PIQ than VIQ. Therefore, a total of four of five cases showed higher PIQ. These results do not correspond with those of Money [14]. The findings of the other researchers have indicated such characteristics of perceptual-cognitive dysfunction as space-form blindness [14], poor visual organization [15], field-dependence [16], and cognitive deficit [17] in Turner syndrome. Because our study is based only on the WISC and Bender-Gestalt, our findings do not examine all these areas. However, our results were discordant with the Buckley [12] and Bekker and Van Gemund [15] reports, which indicated a lower score in "object assembly" compared with other tests and poor visual organization. Scores for picture completion, block design, picture arrangement, and object assembly, the factors in visual organization, are not low in our subjects, and no distortions are found in the Bender. There was no evidence of perceptual-cognitive dysfunction in our study. Nielsen [18] suggested that cognitive-perceptual function matured slowly and might be distorted when there is some delay. Further research is needed on the cognitive-perceptual dysfunctions.

Psychologic characteristics in Turner syndrome. No psychopathologic findings were found in our data. Reviews of other studies have characteristically shown no psychopathologic findings characteristic of Turner syndrome. For example, Christodorescu et al [19] investigated three patients admitted to a psychiatric hospital and found that one patient had an immature personality because of mental retardation. Of two patients with borderline levels of IQ, one had an infantile personality and the other abnormal behavior. Money and Mittenthal [20] followed 73 patients and found that three of them had severe mental disorders and that seven showed lack of ability in social life both within and outside their homes. Nielsen [18] found among 45 patients that one had psychopathologic symptoms and four had anxiety, immaturity, or separation anxiety because of overprotection by their parents. Thus review of the literature reveals that some parents show mental disorders with variable levels of adaptation. These, however, are not characteristic of Turner syndrome. It is our impression from the psychologic characteristics described above that the patients develop problems when excessive amounts of adaptation are forced, resulting in embarrassment from emotional stimuli, passiveness, and inability to protect themselves fully.

We found 13 psychologic characteristics of Turner syndrome, which are not in contradiction to the findings of Money, Nielsen, and other researchers. For example, the "inertia of emotional arousal" Money describes is the same as our "poor emotional responsiveness"; "equability" is based on poor emotional expressions; and "acceptance," "compliancy," and "resignedness" are based on a passive attitude, difficulty in self-defense, lack of aggressiveness, and passive personal relations. "Low activity" in Nielsen's report is in accordance with our low mental activity and productivity and limited interests. The tolerance of frustration in these patients corresponds to their lower aggressiveness and rare expressions of discouragement, but this tendency also results in passiveness and lack of ability to cope with adaptation. We should emphasize that these tendencies do not imply maladaptation; similarly, emotional stability and lower impulsiveness also have their positive aspects. The good social competence described by Nielsen seems to reflect their common sense as well as their passive and stereotyped adaptations.

Nielsen's finding of low extraversion in these patients is supported by our data on the Rorschach test, which show experience, balance, and case-

introversive type. Four cases of contractive type are characteristic of the poor experience processes, emotional flatness and passive attitude, low activity-productivity, limited interests, and poor imagination. Both Money and Nielsen describe the development of these psychologic characteristics, but we have not sufficient data to discuss these issues. (Our findings are on basic psychologic characteristics rather than descriptions by other researchers such as Money and Nielsen.) It is important that we make clear the situations in which these patients' characteristics are to be used and to what they can be applied.

Psychologic characteristics of the XYY and XXY individuals. It is not possible to describe general characteristics of both these syndromes because of the limited number of cases. Patient KK was well both psychologically and physically. He had many positive aspects emotionally and developmentally.

Patient TR, with a Y-15 translocation, had good general adaptability, but he also had rigidity making it difficult for him to change his viewpoint. He was also suppressed emotionally, being very sensitive but unable to express his feelings. His mood was rather negative, and his mental activity was low. He was somewhat sensitive in his personal relations, but his passiveness appeared also in this dimension. Impulsiveness and aggressiveness was not evident. However, his control might be poor when he acts out. His IQ was within normal limits but there was a marked discrepancy between VIQ and PIQ (PIQ > VIQ). The reason for the discrepancy was his poor verbalization of his thoughts and feelings. He did not show direct aggression to others or to himself but had a tendency to work by himself.

As Valentine et al [21] suggested, the development of XYY boys may be normal at least during infancy and early childhood.

The development of the boy with XYY constitution was within normal limits, and no problem in adaptation was found. He was stable in his emotionality, but he had some stubbornness. These characteristics were not restricted to this syndrome and are found in other young children. Attachment to his mother was also well established. Forssman and Hambert [23] found an incidence of 1.97% of individuals with this syndrome among criminals or hard-to-manage males with subnormal intelligence. Sperber et al [23] suggested that some patients had abnormal behavior. These characteristics often appeared during early puberty and seemed to relate to physical symptoms and immaturity of the genital organs. Furthermore, it is reported that they are unambitious, dependent, and obedient and that 25% of them have mild mental retardation or subnormality of sexual behavior or potency. However, until puberty, the only general characteristic of these patients may be mild intellectual impairment. Further study based on large samples is needed to clarify the nature of the psychologic characteristics of these syndromes.

Sleep-Wake Cycle

The remarkable characteristic of circadian rhythms is that they do not represent learned or habitual responses to daily changes in the environment. Rather, these rhythms are innate and are subject to genetic manipulation [24–27]. It has been known for over 40 years that the rest-activity cycle and, therefore, the closely related sleep-wake rhythm in mammals occurs even in the absence of periodic events in the environment.

Circadian rhythm of sleep develops after birth; it can be observed as early as 1 month of age and becomes established by 4 months of age. After that, daytime sleep has the character of naps, which decrease with age, and physiologic naps disappear around 4–5 years of age. REM sleep appears around 2.5 months of age and length of REM sleep tends to increase around 3 months of age.

A correlation between growth hormone (GH) secretion and the sleep-wake cycle is found after the third month of life, and it is from childhood that a circadian periodicity in GH secretion develops in man [28], showing spikes of secretion synchronizing with sleep after 4–5 years of age.

With this information in mind, we started this study, which has revealed no delay in the development of circadian rhythms in any of these disorders. However, in the XXY infants, there may be a delay in the development of regulation of the state of waking, because in early infancy they had prolonged sleeping time both day and night instead of the usual circadian modulation.

From this preliminary study, the following ideas are suggested. 1) There is no abnormality in that part of the CNS that is concerned with establishment of the circadian rhythm of sleep-wakefulness cycles in these diorders. 2) Prolonged daytime sleep suggests dysfunction or abnormality of the part of the CNS concerned with the shortening or diminishing of day sleep, which has a critical age of development around this period of childhood. 3) Considering the maturational process of sleep and sleep-related parameters, there might be disorders of the CNS modulating the rest-activity cycle and NREM-REM cycle, or hypothalamus-pituitary system, and/or related structures. 4) These disorders could cause desynchronization between sleep and hypothalamic pituitary function and be related to abnormal behavior after maturation. 5) Although further study is necessary, medical and psychotherapy around early to late childhood might alleviate or prevent the behavioral abnormalities around adolescence.

SUMMARY

This study of girls with Turner syndrome (16 cases) and boys with XXY (two cases) and XYY (eight cases) constitutions was designed to follow the longitudinal growth and to analyze the psychologic findings and the sleeping patterns in them. The growth curves of stature in Turner syndrome were below the 3rd percentile at all ages. The mean of SDS of height in cases less than 5 years old was −2.38, that in cases older than 6 years old −2.40. The shoulders in Turner syndrome have a tendency to broaden with time, in contrast to the development of the pelvis, which is not so good.

The height of 47,XYY boys from infancy to 9 years of age was between the 10th and 90th percentiles, with one exception in which the height was above the 90th after 8 years of age.

Relatively common psychologic findings in cases with Turner syndrome were as follows: low responsiveness to emotional stimuli, poor emotional expressions, passive and negative mental attitudes, low activity and low productivity, narrow and limited interests, passiveness and common sense adaptability, lower impulsiveness and aggressiveness, etc.

The circadian oscillation was observed up to 3 years of age with shifting of the sleeping time to the night period. After that, day sleep gradually tends to decrease with age. Age variation in day and night sleep of an XXY boy was compared with that of a normal child. This showed the normal variation in circadian rhythms in early childhood, although the infant with this syndrome slept more, both day and night, than normal. In early childhood, no difference was noted. Similar studies were performed on a child with the chromosomal aberration XYY and revealed the sleep parameters to be quite normal in infancy and early childhood.

ACKNOWLEDGMENTS

This work was aided in part by a grant from the Ministry of Education and the Ministry of Health and Welfare, Japan, and The Mitsubishi Fund.

REFERENCES

1. Martin R, Saller K: "Lehrbuch der Anthropologie. Bd. I., 3. Aufl." Berlin: Fischer, 1957.
2. Ikeda Y, Higurashi M, Hirayama M, Ishikawa N: A longitudinal study on the anthropometric measurements of stature, sitting height, lower limb length and upper limb length in Japanese children with Down's syndrome. Jpn J Hum Genet 22:197–198, 1977.
3. Kida N, Sakamoto S, Ishii Y, Katsuki S: Physical growth of children and influence of various factors upon it. Part 1. Brief summary of the results of the observation. Jpn Sci Labor 33:686–697, 1957.

4. Okada N: The age changes in body proportion of the Japanese. J Anthropol Soc Nippon 79:139–150, 1970.
5. Takaishi M, Fujimara K, Ohmura S: Studies on evaluation of the physical growth of infants and preschool children. 1. Growth centile chart. J Child Health 34:340–345, 1976.
6. Takaishi M, Fujimara K, Ohmori S: Studies on evaluation of physical growth of preschool children and school children. J Proc Jpn Soc School Health 24:305, 1982.
7. Polani PE: Chromosomal and other genetic influences on birth weight variation. In: "Size at Birth." Amsterdam: Elsevier North-Holland, 1976, pp 127–164.
8. Machin GA: Chromosome abnormality and perinatal death. Lancet 1:549–551, 1974.
9. Nielsen J, Nyborg H, Dahl G: Turner syndrome. Acta Jutlandica, XLV (Medical Series 21), 1977.
10. Kihlbom M: Psychopathology of Turner's syndrome. Acta Paedopsychiat 36:75–81, 1969.
11. Money J and Granoff D: IQ and the somatic stigmata of Turner's syndrome. Am J Ment Defic 70:69–77, 1965.
12. Buckley F: Preliminary report on intelligence quotient scores of patients with Turner's syndrome: A replication study. Br J Psychiatr 119:513–514, 1971.
13. Field JG: Two types of tables for use with Wechsler's Intelligence Scales. J Clin Psychol 16:1–7, 1960.
14. Money J: Two cytogenetic syndromes: psychologic comparisons. I. Intelligence and specific-factor quotients. J Psychiatr Res 2:223–231, 1964.
15. Bekker FJ, van Gemund J: Mental retardation and cognitive defects in XO-Turner's syndrome. Maandschr Kindergeneesk 36:148–156, 1968.
16. Shaffer JW: A specific cognitive deficit observed in gonadal aplasia (Turner's syndrome). J Clin Psychol 18:403–406, 1962.
17. Alexander D, Ehrhardt AA, Money J: Defective figure drawing, geometric and human in Turner's syndrome. J Nerv Ment Dis 142:161–167, 1966.
18. Nielsen J: Turner's syndrome. Acta Jutlandica, XLV (Medicine Series 21), 1977.
19. Christodorescu D, Collino S, Zellingher R, Tautu C: Psychiatric disturbances in Turner's syndrome. Psychiatr Clin 3:114–124, 1970.
20. Money J, Mittenthal S: Lack of personality pathology in Turner's syndrome: Relation to cytogenetics, hormones and physique. Behav Genet 1:43–56, 1970.
21. Valentine GH, et al: The growth and development of four XYY infants. Pediatrics 48:583, 1971.
22. Forssman H, Hambert G: Incidence of Klinefelter's syndromes among mental patients. Lancet 1:1327, 1963.
23. Sperber MA, Salomon L, Collins MH and Stambler M: Childhood schizophrenia and 47,XXY Klinefelter's syndrome. Am J Psychiatr 128:1400, 1972.
24. Davis FC: Ontogeny of circadian rhythms. In Aschoff (ed): "Handbook of behavioral neurobiology." New York: Plenum Press, 1981, Vol 4, pp 257–274.
25. Konopka RJ: Genetic dissection of the Drosophila circadian system. Fed Proc 38:2602–2605, 1979.
26. Possidente B, Hegmann JP: Circadian complexes: Circadian rhythms under common gene control. J Comp Physiol 139:121–126, 1980.
27. Richter CP: Inborn nature of the rat's 24 hour clock. J Comp Physiol Psychol 75:1–4, 1971.
28. D'Agata R, Vigneri R, Polosa P: Chronobiological study of growth hormone secretion in man: Its relation to sleep-wake cycles and to increasing age. In Scheving LE, Halberg F, Pauly JE (eds): "Chronobiology." Tokyo: Igaku Shoin, 1974, pp 81–87.

Psychologic Study of XYY and XXY Men

Alice Theilgaard, DMSc

Department of Psychiatry, University Hospital, Copenhagen, Denmark

INTRODUCTION

This investigation is part of an extensive cytogenetic, endocrinologic, neurologic, psychologic, psychophysiologic, psychosocial, and criminologic study of noninstitutionalized males with sex chromosome aberrations. The purpose of the investigation is twofold: first, to analyze the intellectual organization of the probands and to assess whether they show consistent and characteristic patterns of cognitive functioning and, second, to give an integrated and global picture of the dimensions of personality in XYY and XXY men. More specifically, the aim is to test the following hypotheses. 1) Are there distinctive XYY and XXY syndromes of psychologic features? 2) Do XXY men show more "feminine" and XYY men more "masculine" characteristics? 3) Does the extra chromosome have an adverse effect on development? 4) Finally a general hypothesis—and one not necessarily contrary to the developmental hypothesis—has to be examined. It deals with the question of neurologic impairment and neuropsychologic dysfunction possibly related to the extra chromosome, manifested in the form of a minimal brain damage syndrome.

METHODS

Sampling Methods

The birth cohort consisted of all males born in the municipality of Copenhagen from January 1, 1944, to December 31, 1947, amounting to 31,438 men. The tallest 15% were selected for examination, since XYY and XXY

Birth Defects: Original Article Series, Volume 22, Number 3, pages 277-292

tend to be more frequent among tall men [1,2]. Information about height could be obtained in 28,884 cases. The losses were due mainly to death or emigration. A total of 4,591 men over 184 cm in height constituted 15.9% of the height distribution in the population of 28,884. The fact that the material was selected for height calls for a comment. Height data from XYY men ascertained in studies of unselected male populations show that XYY men tend to be very tall [3–12]. As a way of maximizing the chances of obtaining a sufficient sample of XYY men for intensive individual study, chromosomal analyses were made of all men in the top 15% of the height distribution of the Danish male population. A sampling of men in the remaining 85% would also have been desirable, but the probability of finding XYY men among shorter men is so small as to make an effective study of possible shorter XYY men too expensive to conduct.

It is more likely that the XXY height distribution has a wider range than that of the XYY. However, Raboch [10], examining 100 XXY men found that they were significantly taller than their controls. A recent Danish study [11] of Klinefelter syndrome in childhood, adolescence, and youth (a sample unselected for height) shows that the XXY males were significantly taller in all age groups than the XY.

The height restriction of the present sample should not then imply serious modifications of the findings. Buccal smears and blood samples were collected from 4,139 (90.2%). (For a detailed description of the case-finding procedure, see Witkin et al [2].) The karyotyping procedures resulted in the identification of 12 XYY and 16 XXY men [13]. All the 12 XYY men, 14 of the 16 XXY men, and 52 matched controls (XY) agreed to participate in the individual case study.

Selection of Controls

Two types of controls were assigned, designated *NoBPP* and *BPP*. The NoBPP control was matched individually to the sex anomaly case for age, height, and social class. In addition to these three variables, the BPP control was also matched according to educational level and to scores on the army selection test: the BPP—the Børge Prien Prøve—a Danish group intelligence test described in Rasch [14].

The reasons for applying two groups of controls were 1) to check the effects of the level of intellectual functioning (the BPP group) and 2) to keep problems of overmatching under control (the NoBPP group).

Basic Data

The data comprise the results of 16 hr of individual psychologic and psychophysiologic testing, the neurologic examination, the two types of interviews conducted by the psychologist and the social worker, and data from social, school, medical, and forensic records.

This vast array of information stemming from the total project has been rated, scored, and coded by staff members, who had no knowledge of the genotypes of the subjects from whom the data were obtained. The exact number of XYY and XXY subjects in the sample was not known to them.

The scoring and evaluation of the data from the projective part and the preparation of the case reports were made independently of the data from the cognitive tests and the interviews.

During the 2 days of examination, all tests were presented in a standardized sequence and administered according to conventional procedure. Meals and coffee breaks were also included at predetermined levels.

Selection of Psychologic Tests

The *cognitive* part of the test battery comprises 23 tests intended to measure the following functions: 1) intelligence—Wechsler adult intelligence scale (WAIS) [15]; 2) memory—visual gestalts [16], paired associates [17], facial recognition test [18]; 3) concentration—Stroop color-word naming test [19], matching familiar figures test [20]; 4) field-dependence—embedded figure test [21], rod and frame test (portable apparatus) [22]; and 5) hemispheric lateralization—dichotic listening (syllables and digits) [23], hand preference scale [24], finger tapping (left and right) [17], and hand tapping (left and right) [17]. Furthermore, subtests from the French kit [25] were included. These represent six factorially discrete cognitive dimensions.

The Projective Tests

The personality part of the test battery consisted of four tests—the Rorschach test, the thematic apperception test [26], the draw-a-person test [27], and the word association test [28].

Psychologic Interview

Independent information from a 2 hr semistructured interview (constructed by Helen Ziegler and the author) that terminated the test period on the second day was added to the material obtained through the projective techniques on personality dimensions. The essence of the answers could be rated quantitatively, and the questions were precoded. The ratings, made independently by two psychologists (Helen Ziegler and the author), were concordant to a high degree.

Social Worker Interview

The 59 precoded questions in this interview concern educational and occupational development, social class, marital status, parenting of both the subject and of the biologic (and possible substitute) parents. Details concerning the number of residences, institutional placement, causes of job changes, course of military service, and standard of present residence were also obtained for each subject, and finally 28 questions regarding present and previous health conditions were answered.

Statistical Procedure

The total statistical material comprising approximately 61,000 data points was analyzed (by Dr. David R. Owen) by the following tests: Cochran's Q test, simple measure matched-groups ANOVA, and correlated measures matched-groups ANOVA.

RESULTS

Intelligence Level

An analysis of variance on Wechsler adult intelligence scale (WAIS) full-scale IQ shows that the XYY group scores are significantly lower ($P < 0.001$) than their NoBBP controls. As was expected from the nature of the selection criteria, there was no statistically significant difference between the group scores of the XYY and those of their BPP controls. Similarly, the XXY group scored significantly below ($P < 0.001$) their NoBPP controls, whereas there was no statistically significant differences between the XXY men and their BPP controls. The XYY group did not differ significantly from the XXY group. The range of IQ in the XYY group was 77–124, in the XXY group 88–128. (It should be noted that the Danish version of the WAIS has been standardized for a few age groups only. Undoubtedly, a translation bias of a positive character [29] [estimated to be about 10 IQ points] was thereby introduced. Therefore, without modification, the values of the Danish version of the WAIS cannot be considered equivalent to IQ scores obtained from the original version.)

Qualitatively, the outcome of the intelligence test gives no evidence of different cognitive styles in the two proband groups. No differential pattern of cognitive scores on the WAIS was revealed in the analysis of the subtest scores, neither in the verbal IQ, performance IQ pattern nor in the factor analysis according to Goodenough and Karp [30]. The differences between the proband groups are nonsignificant on the WAIS masculinity-femininity score. Neither the verbal nor the nonverbal memory test reveals differences between the XYY men and the XXY men.

Two tests were more specifically aimed at evaluating concentration/attention: the color-word naming test and the matching familiar figures. In the former, the XYY men have the smallest number and the XXY men the highest number of corrected errors ($P < 0.08$). In the latter, there is likewise no convincing evidence of differences between the two proband groups.

Field-Dependence/Independence

There are no statistically significant differences between the two proband groups regarding solution times and errors in the embedded figures test. Likewise, the analysis of variance of the portable rod and frame test does not show any statistically significant differences between the XYY and the XXY men.

French Kit

In accordance with the pattern analysis of the WAIS, the results of the analysis of variance show that the patterns of XYY and XXY men do not differ. Furthermore, a direct comparison of the individual test results shows nonsignificant differences.

Hemispheric Lateralization

In dichotic listening, no differences are revealed between the proband groups in any of the measures. In the hand preference scale, the XYY men have the highest scores compared to controls, but the difference is not significant. Ninety-three percent of the XYY group are right-handed. This figure is closer to the 88% characteristic of the general population, calculated according to Annett's genetic model [31–33], compared to the case for the XXY group, in which the proportion of right-handed persons constitutes only 79%.

It is well known that there are more extreme right-handed or left-handed men than women, and consequently more ambidextrous women than men. The distribution of the hand preference scores shows that the ambidextrous (scores +74 to −75) constitute only 25% of the XYY group, but 71% of the XXY group.

Both proband groups are significantly slower in finger tapping as well as hand tapping than their controls. Again, there are no significant differences between the sex aneuploids.

Summarizing the findings of the comparisons of the total amount of psychologic data obtained from the XYY men and the XXY men, it appears that the two groups are more alike than different. Thus the cognitive part of the study does not give evidence of different cognitive styles. Both the XYY group and XXY group show a slight general deficit in global intelligence, but a wide spectrum of IQ scores is possible in both groups.

With regard to personality, the two groups of sex aneuploids seem to have more rather than less characteristics in common. There is a slight difference concerning the defensive pattern, the XYY men tending to be more evasive, the XXY men more inclined to use denial.

The XYYs also seem more rigid in their thinking, the XXYs being more indecisive. The latter group tends to be more submissive and dependent than the former; they are inclined to show less aggression against others, and criminal acts committed by XXYs tend to be less impulsive than is the case with XYYs.

The most distinct differences appearing between the two groups are related to the sexual domain. The XYYs are younger at first intercourse, have a higher frequency of masturbation in childhood and as adults (with less guilt feelings), report more libido and unconventional sex, and have more partners than the XXYs. Also, they show signs of a more differentiated sex role and are judged to be more masculine and to have less problems with their masculine role than XXYs. On the whole, however, the XYYs show a lower self-acceptance and their mood tends to be more pessimistic than that of the XXY men.

Concerning the major psychopathologic issues—"psychotic," "borderline psychotic," "organicity" (here used in the sense of signs of focal brain damage), "somatization" (tendency to react with psychosomatic symptoms), "anxiety," and "depression"—the ratings of the two proband groups do not differ. None of these sex chromosome aberrations seems to present a specific or an exaggeratedly morbid picture.

In conclusion, it can be stated as an answer to hypothesis 1 that the pictures emerging from this study of the XYY and XXY conditions present more similarities than differences among the groups with regard to cognitive and personality aspects. The diversities manifesting themselves belong mainly to the sexual domain: the gender role and the sexual behavior. To a lesser extent, dissimilarity with regard to mediation of aggression is seen.

DISCUSSION

If the term *syndrome* is applied in the strict sense, in signifying a characteristic dynamic combination or pattern of behavior, then, in the author's opinion, the degree and nature of the described disparities between the two karyotypic anomalies are not so essential that they warrant the description "distinctive psychologic syndromes."

The hypothetical question *Do XXY men show more "feminine" and XYY men more "masculine" characteristics?* naturally implies considerations concerning the psychologic nature of man and woman. Recent decades have presented an intensive debate colored by one-sided viewpoints. They seem to be dictated by emotion and prejudice rather than knowledge. This leads to a discussion characterized by polarization and polemic. It is an everyday experience—and it is supported by numerous examples from experimental psychology—that our conceptual framework influences our observations. This is especially true when it concerns interpretations of complicated phenomena of which we have only an incomplete knowledge. From the middle of the 1960s, investigations of biologic as well as psychologic sex differences made rapid strides and the experimentally verified sex differences of a cognitive and emotional nature limit the questions to the following five. 1) Are XXY superior to XYY men in the verbal field, and do the proband groups differ from their controls? 2) Are XYY superior to XXY men on visuospatial tasks, and do the proband groups differ from their controls? 3) Are XYY more field-independent compared to XXY men or to their controls? Are the XXY men more field-dependent than their controls? Are there any differences with regard to lateralization? 4) Are XYY more aggressive than XXY men? Do the proband groups show any differences compared to the controls? 5) Are there any differences in the acceptance and quality of the sex role when the sex aneuploids are compared among themselves and to their controls?

The factor analysis of the subtests of the WAIS, particularly the verbal vs the perceptual organization factors, did not reveal any differences in the comparisons of XYY, XXY, or control, nor does an analysis of variance show any significant differences among the groups on the masculinity-femininity scale.

A superiority in verbal vs spatial tasks should also manifest itself in a battery of memory tests differentiated with regard to material- and modality-specific factors. This calls for the coding of easy-to-verbalize vs not-easy-to-verbalize material [17]. Checking the results from these tests does not lend support to any such difference between XXY, XYY, and control.

The segregation of functions as a result of psychologic differentiation is likely to be reflected in the neuropsychologic domain and vice versa. Furthermore, to the extent that psychologic functions are rooted in a neurophysiologic substratum, specialization of functions at a neurophysiologic level might actually be an important determinant of specialization of psychologic functions. From this it follows that field-independent persons, compared to field-dependent, will show greater lateralization of verbal processing in the left hemisphere and of processing of visuospatial material in the right hemisphere. The XXY men show a very slight tendency to be more field-dependent and less lateralized than the XYY, but most striking are the similarities of the sex aneuploid groups compared to the controls concerning these two dimensions. Certainly, whether they are compared to XXY men or to their controls, the XYY men do not tend to follow a more definite masculine pattern with regard to field-dependence or lateralization.

When considering the various aspects of aggression as reflected in the projective battery, and in the psychologic interview, the differences between the groups are not of an impressive magnitude. Most noteworthy are variations in ways of manifesting aggression rather than differences with regard to strength of aggression.

Thus the fantasy productions of the entire projective battery only yield two significant differences, both pertaining to the thematic apperception test (TAT). 1) The XYY men have a higher aggressive-weighted sum than their controls ($P < 0.05$), but this is valid only with the reservation that the results are unadjusted for vocabulary score. 2) The XYY men have fewer antiaggressive themes than the controls ($P < 0.01$). There are no differences between the proband groups.

In the interview, several questions address various aspects of aggression. The self-reporting of the subjects discloses that, in the instances when manifest outwardly directed aggression is concerned (as in the question of having been a bully in school or showing violence against others at the present time), the XYY men do not diverge significantly from either the controls or the XXY men. An exception is "aggression toward wife," where the XYYs differ markedly ($P < 0.001$) from their controls, but barely significant from the XXY men ($P < 0.07$). The information gained from questions pertaining to "relationship to partner" indicates that the XYY men generally have a less satisfactory alliance—also including sexual relations—with their partners ($P < 0.05$). It is obviously difficult to tell cause from effect, but it does not seem farfetched to relate the aggressive attitude toward partners with sexual frustration and lack of self-acceptance, where XYYs also clearly differ from controls ($P < 0.05$ and $P < 0.01$, for the BPP and NoBPP controls, respectively). The XYY men do not differ from their controls when it is a case of verbal aggression, but they are in this respect different from the XXY men, who are significantly less sarcastic and teasing ($P < 0.05$). The latter group seems markedly ($P < 0.001$) more dependent and submissive both in comparison with controls and the XYY group, which does not diverge from its control group. A pronounced characteristic of the XXY men is their antiaggressiveness, which is reflected in the outcome of the TAT as well as in the answers to the interview. Also, a correlation analysis of the several variables pertaining to aggression shows that, although the XYY men follow the general trends indicated by the NoBPP controls with regard to covariation of parameters, the XXY men in quite a few cases do not. This supports the impression gained from the individual case histories that in the XXY men the aggressiveness takes on more disguised and defended forms, whereas in the XYY it has a more direct, unpremeditated outlet. It is consistent that the XYY men report more impulsivity concerning the circumstances of the offensive acts than either their controls ($P > 0.05$) or the XXY men ($P > 0.01$). The XYY men admit to having been more frequently arrested (but not convicted) than their controls ($P > 0.05$). The comparison with the XXY men yields nonsignificant differences. Analyses of the relation between hormonal levels and records of criminal conviction reveal that the controls with convictions have significantly higher testosterone levels than the controls with no convictions. Among XYY men, however, testosterone levels are almost identical for subjects with and without convictions. The complexity of the matter requires that the question of aggression not be viewed in isolation. For example, the fact that the XYY men report the quality of their childhood as significantly poorer ($P < 0.05$) compared both to the controls and to the XXY men underlines the intricate interactions of experiential, environmental, and genetic factors.

Sexuality

The sexual differentiation in the draw-a-person test increases as a function of age [34]. Rating of this variable in human figure drawings provides a reliable measure of human behavior according to Cutter [35], who also found that, the more overt the sexual deviation, the better the differentiation between the drawn figures.

In the present study, the XYY men produce the most differentiated drawings, the XXY the least, the difference being statistically significant ($P < 0.05$). Qualitatively, the drawings of XXY men are generally less detailed and have inadequate proportions and poor integration. Sexual characteristics are either minimal or absent. The drawings of the XYY men show the opposite tendency, being elaborated with overdimensionalized sexual attributes. This seems to mirror the fact that the XXY men report significantly less libido than the XYY, who also show precocious development and heightened sexual interest and activity in comparison with their controls. A reflection of this difference in libidinal strength is also seen in the Rorschach test.

In the TAT, the XXY men indicated more problems with their masculine role than the XYY ($P < 0.01$). The subjective aspects of sex roles as expressed by the subjects in the interview are also in agreement with these findings. The XXY men rate themselves lower than their controls on items pertaining to the masculine gender in childhood as well as in adulthood. This does not imply that the XXY men have a feminine attitude. In fact, none appear to be homosexually orientated. However, they do seem to be sexually more undifferentiated than both the controls and the XYYs. The lack of a crystallized sense of masculinity in the probands naturally also influences the attitudes of parents, sibs, and peers.

That the XXY men have difficulties in living up to the masculine role might be expected; it seems perhaps more surprising that the XYY also show a weakness of the sense of masculinity compared to the controls. In this proband group, it is not so much a case of lack of sexual polarization toward the other sex, as a lowered self-acceptance and assertiveness. Another difference between the proband groups seems to be that the less developed sense of gender is marked in the XXY men already from childhood, whereas the information from the interview suggests that a lack of a strong sense of masculinity in the XYY men first appears at puberty.

The results indicate that the attitude of masculinity—the gender—is a subtle and complicated phenomenon, and its development seems not as dependent on an extra Y chromosome constitution or an elevated testosterone level as on personality and environmental factors. However, it is clear that the activating and organizing role of the sex hormones should not be overlooked. The difference of the plasma concentrations of testosterone between the XYY and XXY men [36] undoubtedly contributes to the demonstrated dissimilarities in the libidinal strength and manifestations of sexuality between the two sex aneuploid groups.

The role of gonadal hormones in the case of aggressiveness is much more obscure, and evidence of the part played by the postnatal androgens in human aggression is still meager [37].

It could be held that the findings on hormone analysis showing an elevated testosterone level in criminals [38] make it seem not unlikely that aggressiveness has a biologic component associated with male sex hormones. Again, it should be underlined that a correlational relationship does not necessarily reflect a causal one. More direct experimental proof is required before the significance of the hormonal factor in the interplay with behavior and environmental factors can be delineated and more definite conclusions drawn.

Does the Extra Chromosome Have an Adverse Effect on Development?

A developmental lag may account for eventual differences in cognitive style between the two proband groups and the controls. Subsidiary issues such as intellectual dysfunction, of a general or specific nature, and delayed or arrested hemispheric specialization also have to be taken into account. This hypothesis of developmental lag also implies emotional immaturity and subordinate questions connected with instability and impulsivity.

Of the hemispheric lateralization tests, the dichotic listening shows a difference between the XYY men and the controls, indicating that left hemisphere is more incompletely lateralized in the former group. For the XXY men, this is true to an even higher degree when it concerns syllables, the processing of which is also more faulty in the right hemisphere compared to their controls. As already described, the hand preference scale accordingly shows more ambidexterity in the XXY group than in any of the other groups. The hand proficiency as measured by the hand tapping tests shows greater discrepancy between the proband groups and their respective controls than does the hand preference test. This can be explained by the fact that the hand preference is more subject to environmental influence.

Individuals who are less than fully right-handed have been found to be relatively more field-dependent than clear right-handers [39,40]. This represents another link between the two theoretic frames of reference: field-dependence/independence and hemispheric lateralization.

As pointed out by Witkin and Goodenough [41], the characteristic sequence in individual development is from a *field-dependent to a field-independent* mode of functioning. A developmental lag would thus imply that the field-dependent aspects of the cognitive style were dominating.

The results from the embedded figure test and the rod and frame test indicate that the trends are in the direction of greater field-dependence in both proband groups, but the differences are statistically significant only in comparison with the nonintelligence-matched controls and solely in the embedded figure test.

Emotional immaturity can manifest itself in a number of ways: by more intense and undifferentiated affects, by impulsivity and lability, by low threshold of frustration, by dependency and lack of responsibility and social skills, by retarded psychosexual development, and by insufficient or more primitive defense mechanisms reflecting an immature ego organization. The outcome of the projective tests as well as the subjects' own reporting bear witness of immaturity showing less differentiated and more labile affects compared to their controls. They also show less situative feeling and poorer judgment, and their personality integration is assessed to be more disharmonious than in their controls. These findings raise a question as to whether an analysis of the child-environment interactions provides better material for appropriate professional guidance not only in relation to behavior problems but also with reference to the cognitive potentialities.

In conclusion, it can be stated that these findings support the hypothesis that the extra chromosome has an adverse effect on development, and thus developmental lag is manifested in a more incomplete lateralization of the hemispheres. That the discrepancy between probands and controls generally is greatest with regard to left hemisphere processing can be interpreted as an indication of maturational arrest. The statement of Semmes [42] claiming that the left hemisphere becomes more specialized during development supports this view.

The final and more general hypothesis deals with the question of neurologic impairment and neuropsychologic dysfunction. This may be related to the extra chromosome even to the extent of considering minimal brain damage syndromes.

In a report on the subjects in the present study, Volavka and coworkers [43,44] focused on the electrocortical parameter and found that the XYY men have a significant lower average frequency of occipital alpha activity than their controls. Both proband groups showed significantly more theta activity than controls, but it was largely within normal limits. It should be noted that the differences found were of a nonspecific nature. The possible influence of drugs was taken into consideration. Volavka and coworkers [43,44] suggest that differences indicate retardation of the development of brain function. The fact that both the proband groups show significantly more theta activity than controls can be taken as an indication of a comparatively lower arousal. This is also in line with the finding of S.A. Mednick and J. Sergeant (personal communication, 1978) that a preliminary analysis of the autonomic nervous system reactions of the XYY men in our study shows a low responsiveness and a slow rate of recovery of the skin conductance response.

The concentration capacity as measured by the color-word naming test and the dichotic listening test as reported by the subjects themselves in the interview is generally poorer in the proband groups than in the controls. The WAIS factor analysis shows results in the same direction, but nonsignificantly, on the attention/concentration variable.

In that the reduced inhibitory control over sensory function could be expected to result in easy distractability, insufficient inhibitory control over motor function may result in restlessness and/or imprecise motor activity. The XYY men report in the interview that they have more difficulties ($P < 0.05$) in relaxing than the controls. They are also more excitable and take more sedative drugs. Both proband groups show a highly significant lowering of motor performance compared to the controls in the hand tapping test.

The impression that inhibitory control is less developed in the probands than in the XY males also gains support from other data on the same population. Thus Boisen's and Rasmussen's tremor study [45] shows that physiologic and intention tremor are seen more often in XYY than in controls and that the mean tremor amplitude is significantly different ($P < 0.01$). In the XXY men, the same tendency is reported although not to the same extent.

The neuropsychologic findings in this study do not present evidence for the assumption that the XYY and the XXY conditions cause focal brain damage. This is consistent with the theoretic position that early alterations in brain development have generalized rather than localized effects on brain functions in later life [46].

The analysis of the data from a neuropsychologic frame of reference reveals in the XYY and XXY men a number of the signs generally attributed to the minimal brain dysfunction syndrome. It does not permit definite conclusions as to causes but suggests that it is a case of minor irreversible impairment as well as of a maturational lag. It should be underlined that it is not a question of severe brain damage but of subtle disturbances, which may lead to reduced tolerance of psychologic and environmental strain.

The significantly different hormonal patterns of XYY and XXY men from those of the controls suggests that the alteration in brain development is due to the combined effect of neurologic and neuroendocrinologic factors. However, our understanding of the fine mechanisms by which altered endocrine conditions affect the metabolism of the brain, and further the way such chemical changes in certain brain cells influence the psychologic pattern of such functions as arousal, activity, excitability, impulsivity, etc, remain to be elucidated. Consequently, the interpretation of their role in the causal chain of the neuroendocrinologic findings are, in our present state of knowledge, highly speculative. The attempt to see "aggression" and "sex life" in the light of hormonal patterns should not be taken as an indication of a sociobiologic reductionistic viewpoint. The endocrine factors present only one of the potentials for influencing behavior. The making of direct straight-line cause-and-effect connections, when the matter is as complex as is the case here, does not enhance scientific thinking.

As a corollary to this, I have postulated [47] that psychosocial experiences affect both the neural substrates and the consequent behavior, which these neural structures subserve, given the particular state of development. For example, the information that XYY men have spent more time in institutions and thus possibly been exposed to more emotional deprivation should also be taken into account in the analysis of causal factors. The XYY and XXY men are, as are all other human beings, the result of an interplay of genetic, environmental, and psychologic processes. As Kessler and Moos [48] state, "To ascribe to a single chromosome specific behavioural effect fails to take account of a broad range of pertinent environmental factors, and ignores the fact that the phenotype is the result of the total integrative properties of the genome interacting with the environment."

Past and present views of these conditions have presented contradictory statements reflecting the enthusiasm with which abnormalities have been sought and recorded in studies of newly discovered conditions—not always with due respect for the validity of the methods applied. Methodologic flaws include sampling bias, generalizations unacceptably made on single cases, lack of control groups, lack of a blind evaluation procedure, and, finally, lack of sufficiently broad spectrum of data. Thus an impetus to reassess the sex aneuploid group presents itself.

Although four hypotheses have been put forward, it has not necessarily been in search of one universally appropriate theory. In matters as complicated as these concerning interrelations of biologic, psychologic, and social factors, it is important to attack the question from different perspectives, and one theory does not make another redundant. This procedure has not elicited such a cacophony of ideas from so many diverse sources that it is difficult to distinguish and understand the nature of the issues under discussion. On the contrary, in the attempt to examine the four hypotheses, a picture has emerged that does not show any inconsistencies or controversies, regardless of the perspective from which it is seen, thus manifesting the force of inner coherence.

REFERENCES

1. Noël B, DuPont JP, Revil D, Dussuyer I, Quack B: The XYY syndrome: Reality or myth? Clin Genet 5:387–394, 1974.
2. Witkin HA, Sarnoff AM, Schulsinger F, Bakkestrøm E, Christiansen KD, Goodenough DR, Hirschhorn K, Lundsteen C, Owen D, Philip J, Rubin DB, Stocking M: Criminality in XYY and XXY men. Science 193:547–555, 1976.
3. Bartlett DJ, Hurley WP, Brand CR, Poole EW: Chromosomes of male patients in a security prison. Nature 219:351, 1968.
4. Crandall BF, Carrel RE, Sparkes RS: Chromosome findings in 700 children referred to a psychiatric clinic. J Pediatr 80:62, 1972.
5. DeBault LE, Johnston E, Loeffelholz P: Incidence of XYY and XXY individuals in a security hospital population. Dis Nerv Syst 33:590, 1972.
6. Hook EG, Kim D-S: Height and anti-social behaviour in XY and XYY boys. Science 172:284, 1971.
7. Jacobs PA, Brunton M, Melville MM, Brittain RP, McClemont WF: Aggressive behaviour, mental sub-normality and the XYY male. Nature 208:1351, 1965.
8. Jacobs PA, Price WH, Richmond S, Ratcliff RAW: Chromosome surveys in penal institutions and approved schools. J Med Genet 8:49, 1971.
9. Nielsen J: Prevalence and 2.5 years incidence of chromosome abnormalities among all males in a forensic psychiatric clinic. Br J Psychiatr 119:503–512, 1971.

10. Raboch J: A clinical study of 100 chromatin-positive men. Fertil Steril 15:331-337, 1964.
11. Sørensen K: Klinefelter's syndrome in childhood, adolescence and youth. In preparation.
12. Vianna AM, Frota-Pessoa O, Lion MF, Decourt L: Searching for XYY males through electrocardiograms. J Med Genet 9:165, 1972.
13. Philip J, Lundstrøm C, Owen D, Hirschhorn K: The frequency of chromosome aberrations in tall men with special reference to 47,XYY and 47,XXY. Am J Hum Genet 28:404–411, 1976.
14. Rasch G: "Probabilistic Models for Some Intelligence and Attainment Tests. Copenhagen: Danish Institute for Educational Research, 1960.
15. Wechsler D: "Wechsler Adult Intelligence Scale." New York: The Psychological Corporation, 1955.
16. Anderson R: Verbal and visuospatial memory. Scand J Psychol 17:198–204, 1976.
17. Heshe J, Röder E, Theilgaard A: Unilateral and bilateral ECT: A psychiatric and psychological study of therapeutic effect and side effects. Acta Psychiatr Scand 275 [Suppl]:1-180, 1978.
18. Benton AL, van Allen MW: "Manual: Test of Facial Recognition," Iowa City: Department of Neurology. Univ. of Iowa Hospital, 1973.
19. Reisby N, Theilgaard A: The interaction of alcohol and meprobamate in man. Acta Psychiatr Scand 208[Suppl]:5–204, 1969.
20. Kagan J, Rosman BL, Day D, Albert J, Phillips W: Information processing in the child: Significance of analytic and reflective attitudes. Psychol Monogr 78:1, 1964.
21. Witkin HA: Individual differences in ease of perception of embedded figures. J Pers 19:1–15, 1950.
22. Oltmann PK: A portable rod-and-frame apparatus. Percept Motor Skills 26:503–506, 1968.
23. Kimura D: Right temporal-lobe damage. Perception of unfamiliar stimuli after damage. Arch Neurol 8:264–271, 1963.
24. Oldfield RC: The assessment and analysis of handedness: The Edinburgh Inventory. Neuropsychologia 9:97–113, 1971.
25. French JW, Eksrom RB, Price LA: Manual for kit of reference tests for cognitive factors. Princeton, NJ: Educational Testing Service, 1963.
26. Murray H: "Thematic Apperception Test: Manual." Cambridge: Harvard University Press, 1943.
27. Machover K: "Personality Projection in the Drawing of the Human Figure: A Method of Personality Investigation." Springfield, IL: Charles C Thomas, 1949.
28. Rapaport D, Gill MM, Schafer R: "Diagnostic Psychological Testing." New York: International Universities Press, 1968.
29. Hess, G: "WAIS anvendt på 698 50 årige." Copenhagen: Akademisk Forlag, 1974.
30. Goodenough DR, Karp SA: Field dependence and intellectual functioning. J Abnorm Soc Psychol 63:241–246, 1961.
31. Annett M: A model of the inheritance of handedness and cerebral dominance. Nature 204:59–60, 1964.
32. Annett M: The binominal distribution of right, mixed and left handedness. Q J Exp Psychol 19:327–333, 1967.
33. Annett M: The distribution of manual asymmetry. Br J Psychol 63:343–348, 1972.
34. Swensen C, Newton KR: The development of sexual differentiation on the draw-a-person test. J Clin Psychol 11:37–41, 1955.
35. Cutter F: Sexual differentiation in figure drawings and overt deviation. J Clin Psychol 12:369–372, 1956.

36. Schiavi RC, Theilgaard A, Owen D, White D: Sex chromosome anomalies, hormones and aggressivity. Arch Gen Psychiatr 41:93–99, 1984.
37. Meyer-Bahlburg HF: Aggression, androgens and the XYY syndrome. In Friedman RC, Richart RM, Vande Wiels GD (eds): "Sex Differences in Behaviour". New York: John Wiley & Sons, 1974, pp 433–453.
38. Ehrenkranz J, Bliss E, Sheard MH: Plasma testosterone: Correlation with aggressive behaviour and social dominance in man. Psychosom Med 36:469–475, 1974.
39. Silverman AJ, Adevai G, McGough WE: Some relationships between handedness and perception. J Psychosomatic Res 10:151–158, 1966.
40. Pizzamiglio L: Handedness, ear preference, and field dependence. Percept Motor Skills 38:700–702, 1974.
41. Witkin HA, Goodenough DR: "Cognitive Styles: Essence and Origins: Field Dependence and Field Independence." New York: International Universities Press, 1981.
42. Semmes J: Hemispheric specialization. A possible clue to mechanism. Neuropsychologia 6:11–26, 1968.
43. Volavka J, Mednick SA, Sergeant J, Rasmussen L: Electroencephalograms of XYY and XXY men. Br J Psychiatr 130:43–47, 1977.
44. Volavka J, Mednick SA, Rasmussen L, Teasdale T: EEG spectra in XYY and XXY men. Electroencephalogr Clin Neurophysiol 73:798–801, 1977.
45. Boisen E, Rasmussen L: Tremor in XYY and XXY men. Acta Neurol Scand 58:66–73, 1978.
46. Luria AR: "The Working Brain: Introduction to Neuropsychology. London: Penguin Press, 1973.
47. Theilgaard A: Psykologiske funktioners repraesentation i hjernen. Nord psykiat. Tidsskrift 7:484–494, 1973.
48. Kessler S, Moos RH: The XYY karyotype and criminality: A review. J Psychiatr Res 7:153–170, 1970.

Summary Overview of Behavioural Development in Individuals With Neonatally Identified X and Y Aneuploidy

Charles T. Netley, PhD

Department of Psychology, The Hospital for Sick Children, Toronto, Ontario M5G 1X8, Canada

47,XXY

It was agreed on by the participants at the Edinburgh meeting that an attempt should be made to collate the behavioural data from each centre involved in studying neonatally identified X and Y aneuploid individuals. To this end, the questionnaire shown in Table I was designed and distributed to all conference participants. It is recognized that the procedure of simply summing data from different centres is a relatively crude procedure, ignoring as it does, the methodologically and culturally unique features of each study. However, other chapters deal with such specifics while this presents the broad trends in behavioural development of X and Y aneuploids as revealed by the force of large numbers.

The following is based on the material submitted by investigators in Boston, Denver, Edinburgh, Japan, Toronto, and Winnipeg. In general, the control observations were obtained from assessments of family members of probands although in the Boston series the controls were unrelated individuals with familial balanced autosomal re-arrangements. The Edinburgh control groups, also composed of unrelated individuals, were not included in the analyses because of their large size and predominantly middle or upper-middle class composition. In the sections describing each karyotype the numbers of cases and controls sometimes vary with the kinds of behaviour

Birth Defects: Original Article Series, Volume 22, Number 3, pages 293-306

TABLE I. Data Sheet Distributed to Edinburgh Conference Participants

Case No.	Karyo-type	Age at test	Verbal IQ	Performance IQ	Full-scale IQ	Educational problems	Family functioning	Behavioural problems	Comments
						*	†	‡	

*Educational problems to be judged present if any of the following have characterized a case during the past 3 years: full or part-time special educational placement, remedial tuition within a regular class, repeating a grade.

†Family functioning to be rated simply as poor or adequate. *Poor* is defined in terms of the presence during the past 3 years of disorganization, deprivation, or significant psychopathology in one or both the parents.

‡Behavioural problems to be judged as present or absent on the basis of whether an individual has received treatment from a mental health agency or professional during the past 3 years (if you want to extend time scale, do so and indicate the age and duration of contact). Recommendations by others, such as school personnel, for involvement with a mental health facility, even if rejected by a proband or his/her family, may also be used to conclude that behaviour problems are present.

under examination. This is because some investigators could not always provide complete data for all their subjects.

47,XXY

Intelligence

Data on intellectual functioning provided by each centre were derived from age-appropriate forms of the Wechsler tests of intelligence. Table II presents a summary of the verbal IQs (VIQ), performance IQs (PIQ), and full-scale IQs (FSIQ) of 73 probands and 60 controls with respective mean ages of 10.4 (SD 2.8) and 10.2 (SD 4.4) years. Thirty-eight of the controls were male. T tests and an analysis of variance of these observations indicated that VIQ but not PIQ was significantly lower in probands than controls. The findings are summarized in Table II.

Educational Progress

Forty-four of 65 (67.6%) 47,XXY cases were judged to have educational problems in contrast to 14 of 63 (22.2%) controls. These results, which are similar to those reported in a previous summary publication [1], are highly significant ($\chi^2 = 27.78$, $P < 0.00001$).

Behavioural Problems

Behavioral problems defined by recommended or actual contact with mental health professionals did not distinguish probands from controls. Nineteen of 65 (29.2%) 47,XXY cases and 13 of 62 (20.9%) controls were judged to have behavioural problems ($\chi^2 = 1.16$, NS).

TABLE II. Verbal IQ (VIQ), Performance IQ (PIQ), and Full Scale IQ (FSIQ) in 47,XXY Probands and Controls

	47,XXY	Controls	t	P
VIQ				
$\bar{X}$	90.11	102.40	4.41	< .01
SD	17.71	14.44		
PIQ				
$\bar{X}$	102.60	104.44	0.84	NS
SD	13.11	12.22		
FSIQ				
$\bar{X}$	97.74	103.62	2.37	< .01
SD	15.74	12.93		
N	73	60		

Relation of Family Functioning to Intellectual, Educational, and Behavioural Development

An examination of the relation of family functioning was undertaken because of an earlier suggestion that 47,XXY males are more susceptible to stress than chromosomally normal individuals [1]. This was performed in relation to intellectual functioning by means of an analysis of variance treating family functioning (poor or adequate) and karyotype (aneuploid or euploid control) as independent variables and the IQ measures VIQ and PIQ as dependent measures. The data are summarized in Table III.

The analysis of variance resulted in significant main effects for karyotype ($F = 9.54$, $P = 0.0026$) and family functioning ($F = 10.96$, $P = 0.0013$) reflecting lower ability levels in extra X subjects and those from dysfunctional homes. There was, however, no significant interaction ($F = 0.71$, NS) between these two variables, indicating that poor family functioning and chromosomal status did not combine to produce intellectual deficits beyond those associated with their independent effects.

This analysis of variance revealed that the interaction of karyotype and VIQ vs PIQ was significant ($F = 10.52$, $P = 0.0016$) reflecting as did the earlier analysis that VIQ was relatively lower than PIQ in the extra X cases than in controls. The interaction of ability type with family function was marginally significant ($F = 3.19$, $P = 0.077$), indicating a tendency for VIQ to be lower than PIQ in dysfunctional families. The three-way interaction of ability type, karyotype, and family functioning was not significant ($F = 0.17$, NS), providing no support for interactive effects beyond the simpler two-way interactions.

The relation of family functioning to the presence of educational problems was examined through three-way log-linear χ^2 tests. The data used in these analyses are presented in Table IV.

TABLE III. Intelligence in 47,XXY Males and Controls in Relation to Family Functioning

	47,XXY		Controls	
	Family problems	No family problems	Family problems	No family problems
VIQ				
$\bar{X}$	81.10	92.95	94.58	103.67
SD	17.38	16.57	14.39	14.01
PIQ				
$\bar{X}$	95.90	103.68	101.00	103.60
SD	11.55	12.47	9.87	12.72
N	21	38	19	30

TABLE IV. Educational Problems in 47,XXY Males and Controls in Relation to Family Functioning

	Family problems	
	Absent	Present
XXY		
Educational problems	22	22
No educational problems	20	1
Controls		
Educational problems	4	10
No educational problems	34	15

TABLE V. Behavioural Problems in 47,XXY Males and Controls in Relation to Family Functioning

	Family problems	
	Absent	Present
XXY		
Behavioural problems	3	16
No behavioural problems	39	7
Controls		
Behavioural problems	0	13
No behavioural problems	38	11

The χ^2 associated with the two-way tests of 1) family functioning and educational problems and 2) karyotype and educational problems were highly significant ($\chi^2 = 14.32$, P = 0.0002; $\chi^2 = 27.78$, $P < 0.00001$, respectively). This indicates that both adverse family functioning and a supernumerary X are associated with educational problems. The three-way analysis of family functioning, karyotype, and presence or absence of educational problems, however, did not indicate that the effect of family functioning on educational status was different in the two subject groups (partial association $\chi^2 = 1.14$, NS).

The relationship of family functioning to the presence or absence of behavioural problems was examined in a similar analysis. The data used for this purpose are presented in Table V. The simple two-way χ^2 test examining the relationship between family functioning and the presence or absence of behavioural problems for the data of the combined aneuploid and control groups was highly significant ($\chi^2 = 55.24$, P = 0.00001). Those relating family functioning to karyotype and karyotype to behavioural functioning were not ($\chi^2 = 0.15$, NS; $\chi^2 = 1.16$, NS, respectively). The three-way relationship between family functioning, karyotype, and behavioural status was not significant (partial association $\chi^2 = 2.11$, NS). These results indicate, therefore, that the presence or absence of behavioural problems was significantly related to family functioning but not to karyotype or to an interaction between karyotype and family characteristics. More simply stated, poor family functioning is associated with behaviour problems to an equal degree in 47,XXY males and controls.

47,XXX

Intelligence

The Wechsler intelligence test results for the collated samples of 32 47,XXX cases and 25 controls are summarized in Table VI. Their respective ages at assessment were 10.5 years (SD = 2.8) and 9.5 years (SD = 2.4), with 15 of the 25 controls females.

The results of simple t tests revealed that the extra X girls had lower IQs on the verbal, performance, and full-scale measures than the controls. An analysis of variance of the VIQ and PIQ data did not yield a significant interaction between karyotype and ability, indicating, unlike the 47,XXY males, that the intellectual limitations of the 47,XXX cases were not confined to verbal areas. However, it should be noted that the VIQ vs PIQ effect was significant (F = 11.72, P = 0.0019), indicating that the latter ability was higher than the former across groups.

Educational Progress

Twenty-five of 35 (71.4%) 47,XXX girls were designated as having educational problems in contrast to nine of 29 (31.0%) controls ($\chi^2 = 10.67$, P = 0.0011).

TABLE VI. Verbal IQ (VIQ), Performance IQ (PIQ), and Full Scale IQ (FSIQ) in 47,XXX Probands and Controls

	47,XXX	Controls	t	P
VIQ				
$\overline{X}$	86.63	105.64	4.72	< .01
SD	16.70	13.71		
PIQ				
$\overline{X}$	95.19	109.60	3.96	< .01
SD	13.86	13.46		
FSIQ				
$\overline{X}$	90.06	108.40	4.89	< .01
SD	14.84	13.39		
N	32	25		

Behavioural Problems

Eleven of the 35 (31.4%) 47, XXX girls and 11 of the 29 (37.9%) controls were judged to have significant behavioural problems ($\chi^2 = 0.30$, NS).

Relation of Family Functioning to Intellectual, Educational, and Behavioural Development

The impact of poor family functioning on intellectual development was examined by an analysis of variance treating quality of family functioning and karyotype as independent variables and VIQ and PIQ as dependent variables. The data used in this analysis are summarized in Table VII. Two of the three significant findings emerging in the analysis of variance were associated with the main effects of karyotype and ability type reflecting the lower IQ of the extra X cases and the lower VIQ than PIQ in both groups. The third significant result was that subjects from poorly functioning homes had lower overall IQs than those from more stable environments ($F = 13.93$, $P = 0.0005$). None of the interactions tested reached significance, indicating that family functioning and chromosomal state did not combine to produce greater deficits than those associated with their independent actions.

The relationships of family functioning to the presence or absence of educational and behavioural problems were examined by means of three-way linear χ^2 tests. The data employed in these analyses are presented in Tables VIII and IX.

The results of χ^2 test of the data in Table VIII revealed that the association between karyotype and educational problems was significant ($\chi^2 = 10.67$, $P = 0.0011$), as was that between family functioning and educational status ($\chi^2 = 10.31$, $P = 0.0013$). The three-way partial association χ^2 involving the variables of karyotype, family functioning, and educational category was not

TABLE VII. Intelligence in 47,XXX Females and Controls in Relation to Family Functioning

	47,XXX		Controls	
	Family problems	No family problems	Family problems	No family problems
VIQ				
$\bar{X}$	76.00	90.17	98.38	109.06
SD	16.18	15.62	15.86	11.54
PIQ				
$\bar{X}$	84.13	98.88	113.59	103.60
SD	8.32	13.46	9.11	13.52
N	8	24	8	17

TABLE VIII. Educational Problems in 47,XXX Females and Controls in Relation to Family Functioning

	Family problems	
	Absent	Present
XXX		
Educational problems	15	10
No educational problems	10	0
Controls		
Educational problems	2	7
No educational problems	16	4

TABLE IX. Behavioural Problems in 47,XXX Females and Controls in Relation to Family Functioning

	Family problems	
	Absent	Present
XXX		
Behavioural problems	3	8
No behavioural problems	22	2
Controls		
Behavioural problems	5	6
No behavioural problems	13	5

(χ^2 = 0.77, NS), indicating that the effects of family functioning on educational progress were not different in the two subject groups.

The corresponding analysis of the data in Table IX produced only one significant result, which was that poor family functioning was associated with behavioural problems (χ^2 = 14.32, P = 0.0002).

47,XYY

Intelligence

The intelligence test results for the combined samples of 28 47,XYY males and eight controls are presented in Table X. Their ages at test were 9.5 (SD = 2.1) and 8.1 (SD = 1.4) years, respectively. Seven of the controls were male.

Both simple t tests and an analysis of variance produced no significant differences between the groups, between VIQ and PIQ, nor an interaction between karyotype and ability type.

TABLE X. Verbal IQ (VIQ), Performance IQ (PIQ), and Full-Scale IQ (FSIQ) in 47,XYY Probands and Controls

	47,XYY	Controls	t	P
VIQ				
$\bar{X}$	100.75	104.50	0.65	NS
SD	14.28	14.48		
PIQ				
$\bar{X}$	108.79	104.62	0.96	NS
SD	16.35	8.58		
FSIQ				
$\bar{X}$	105.00	104.63	0.07	NS
SD	15.22	12.35		
N	28	8		

Educational Problems

Fifteen of the 28 (53.5%) 47,XYY boys had educational problems compared to five of nine (55.5%) controls. Although not significantly different, it should be noted that the number of controls is small, and in both groups educational difficulties were nearly as common as in the intellectually compromised 47,XXY cases. In that the intelligence test results of the 47,XYY males were within average limits, it is unclear why they tended to have poor educational histories.

Behavioural Problems

Ten of 27 (37.0%) 47,XYY cases and five of nine (55.5%) controls were judged to have significant behavioural problems. These proportions are not significantly different ($\chi^2 = 0.94$, NS) and are comparable to those seen in the 47,XXY and 47,XXX series.

Relationship of Family Functioning to Intellectual, Educational, and Behavioural Development

The data employed in the examination of the relationship between family functioning and intelligence are presented in Table XI.

An analysis of variance revealed no significant differences attributable to main effects or interactions associated with karyotype, family functioning, or VIQ vs PIQ.

The observations used in the analysis of 1) family functioning and educational difficulty and 2) family functioning and behavioural disorder are presented in Tables XII and XIII. The only significant findings emerging from three-way log-linear χ^2 analyses of these data were that poor family

TABLE XI. Intelligence in 47,XYY Males and Controls in Relation to Family Functioning

	47,XYY		Controls	
	Family problems	No family problems	Family problems	No family problems
VIQ				
$\bar{X}$	96.14	100.28	99.00	107.80
SD	14.99	13.89	17.44	13.37
PIQ				
$\bar{X}$	108.14	106.06	102.33	106.00
SD	15.74	15.08	9.07	9.03
N	7	18	3	5

TABLE XII. Educational Problems in 47,XYY Males and Controls in Relation to Family Functioning

	Family problems	
	Absent	Present
XYY		
Educational problems	9	6
No educational problems	12	1
Controls		
Educational problems	2	3
No educational problems	4	0

TABLE XIII. Behavioural Problems in 47,XYY Males and Controls in Relation to Family Functioning

	Family problems	
	Absent	Present
XYY		
Behavioural problems	5	5
No behavioural problems	16	1
Controls		
Behavioural problems	2	3
No behavioural problems	4	0

functioning was associated with educational difficulty and behavioural problems. There were no effects attributable to karyotype or interactions between karyotype and family functioning.

XO MOSAICS AND VARIANTS

Intelligence

The intelligence test results for 11 XO/XX mosaics and 11 controls are presented in Table XIV. Their mean ages at examination were 9.3 (SD = 3.6), and 8.5 (SD = 2.7) years, respectively. Six of the controls were female. Simple t tests and an analysis of variance revealed no differences between the groups, between VIQ and PIQ nor an interaction between ability type and group.

The same analyses were carried out on the data for nine XO, partial X subjects and 19 controls. Their mean ages at assessment were 9.0 (SD = 1.4) and 14.1 (SD = 7.0) years, respectively. Eleven controls were female.

The t tests indicated that the mean PIQ of the XO, partial X cases was significantly lower than that of the controls, whereas there was no difference in their VIQs (Table XV). The analysis of variance resulted in one significant difference, an interaction between VIQ/PIQ and group (F = 6.40, P = 0.018) reflecting the relatively low PIQ of aneuploids and the relatively low VIQ of controls.

An analysis of variance contrasting the data for the XO/XX mosaic and XO, partial X cases resulted in one significant difference; the latter group had lower overall scores than the former (F = 5.13, P = 0.036).

TABLE XIV. Verbal IQ (VIQ), Performance IQ (PIQ), and Full-Scale IQ (FSIQ) in XO/XX Mosaics and Controls

	XO/XX Mosaics	Controls	t	P
VIQ				
$\bar{X}$	104.09	104.36	0.05	NS
SD	14.27	12.07		
PIQ				
$\bar{X}$	104.64	106.18	0.26	NS
SD	13.68	13.84		
FSIQ				
$\bar{X}$	104.82	105.82	0.17	NS
SD	13.56	13.65		
N	11	11		

TABLE XV. Verbal IQ (VIQ), Performance IQ (PIQ), and Full-Scale IQ (FSIQ) in XO, Partial X Probands and Controls

	XO, partial X	Controls	t	P
VIQ				
$\bar{X}$	90.44	91.95	0.18	NS
SD	21.72	19.72		
PIQ				
$\bar{X}$	85.78	101.68	2.13	< .05
SD	18.67	17.96		
FSIQ				
$\bar{X}$	87.33	96.36	1.10	NS
SD	20.66	19.44		
N	9	19		

Educational Problems

One of the ten XO/XX mosaics (10%) and four of 11 (36.4%) of their controls had educational problems, an insignificant difference ($\chi^2 = 2.13$, NS). Educational difficulties were reported in six of nine XO, partial X subjects (66.7%) and 14 of 22 of their controls (63.6%), also an insignificant difference ($\chi^2 = 0.16$, NS).

Behavioural Problems

Histories of significant maladjustment were reported in three of nine XO, partial X cases (33.3%) and 13 of 22 (59.1%) of their controls. Behavioural problems were described in one of the ten XO/XX mosaics (10%) and five of the 11 controls (45.5%). Neither difference was significant (respective χ^2 values 0.86 and 3.37).

Family Problems

Poor family functioning was reported in five of nine (55.6%) XO, partial X cases and 15 of 22 controls (68.2%); ($\chi^2 = 0.437$, NS). Three of 11 XO mosaics were said to be from poorly functioning families compared to six of 12 controls. This too was not a significant difference ($\chi^2 = 1.23$). Further analysis of the relation between family functioning and intellectual, educational, and social/emotional development was not carried out because of the limited numbers of cases in these groups.

47,XXY MOSAICS

Intelligence

Table XVI presents intelligence test results for six 47,XXY mosaics and three controls. Their respective mean ages were 11.6 (SD = 2.3) and 9.0

TABLE XVI. Verbal IQ (VIQ), Performance IQ (PIQ), and Full-Scale IQ (FSIQ) in 47,XXY Mosaics and Controls

	47,XXY Mosaics	Controls	t	P
VIQ				
$\overline{X}$	97.33	107.67	1.36	NS
SD	13.16	9.29		
PIQ				
$\overline{X}$	109.33	116.00	0.72	NS
SD	12.99	13.11		
FSIQ				
$\overline{X}$	102.67	113.00	1.29	NS
SD	12.71	10.58		
N	6	3		

(SD = 3.4) years. All the controls were male. T tests and an analysis of variance revealed no group difference, and the only marginally significant result was that VIQ tended to be lower than PIQ in both groups ($F = 5.34$, $P = 0.054$).

The 47,XXY mosaics were also compared to the nonmosaic 47,XXY cases. The results of an analysis of variance revealed no significant difference with the exception that VIQ was significantly lower than PIQ for the data set as a whole ($F = 14.40$, $P = 0.0003$).

Educational Problems

The incidence of educational problems was similar in five 47,XXY mosaics and seven controls. Three of the mosaics and four of the controls had histories of such problems ($\chi^2 = 0.12$, NS).

Behavioural Problems

These were reported in one of five extra X mosaic males and four of seven controls ($\chi^2 = 1.705$, NS).

Family Functioning

One of five 47,XXY mosaics and four of seven controls were from families judged to be dysfunctional, an insignificant difference ($\chi^2 = 1.705$, NS). No other analyses were carried out because of the limited number of cases.

47,XXX MOSAICS

Intelligence

The intellectual findings for the 47,XXX mosaics and their controls are presented in Table XVII. Their respective mean ages at examination were

TABLE XVII. Verbal IQ (VIQ), Performance IQ (PIQ), and Full-Scale IQ (FSIQ) in 47,XXX Mosaics and Controls

	47,XXX Mosaics	Controls	t	P
VIQ				
$\bar{X}$	110.60	109.14	0.18	NS
SD	11.13	16.47		
PIQ				
$\bar{X}$	107.20	111.57	0.49	NS
SD	15.58	14.58		
FSIQ				
$\bar{X}$	110.40	111.43	0.19	NS
SD	14.29	15.63		
N	5	7		

9.1 (SD = 2.1) and 7.9 (SD = 3.1) years. Six of the seven controls were female. Neither t tests nor an analysis of variance resulted in any difference attributable to group, ability, or their interaction. However, an analysis of variance contrasting the 47,XXX mosaics with the 47,XXX nonmosaics resulted in one marginally significant group difference. This involved an interaction between group and ability type (VIQ vs PIQ), suggesting that VIQ was lower than PIQ in 47,XXX females but not 47,XXX mosaics ($F = 3.56$, $P = 0.068$).

Educational Problems

One of five 47,XXX mosaics and three of nine controls had histories of significant educational difficulty ($\chi^2 = 0.244$, NS).

Behavioural Problems

Two of five 47,XXX mosaics and four of nine controls were judged to have behavioural problems ($\chi^2 = 0.011$, NS).

Family Problems

One of the five 47,XXX mosaics and three of nine controls came from dysfunctional families ($\chi^2 = 0.244$, NS).

REFERENCES

1. Stewart DA, Netley CT, Park E: Summary of clinical findings of children with 47,XXY, 47,XYY and 47,XXX karyotypes. In Stewart DA (ed): "Children with Sex Chromosomes Aneuploidy: Follow-Up Studies." New York: Alan R. Liss, Inc., for the March of Dimes Birth Defects Foundation. BD:OAS 18(4), 1982, pp 1–5.

Cellular and Molecular Studies in Human Chromosomal Diseases

Theodore T. Puck, PhD

Eleanor Roosevelt Institute for Cancer Research, Department of Biochemistry, Biophysics and Genetics, and Department of Medicine, University of Colorado Health Sciences Center, Denver, Colorado 80262

INTRODUCTION

Twenty-five years have elapsed since the identification and classification of the human chromosomes [1,2]. During this period, approximately 75 different chromosomal diseases involving complete or partial aneuploidy and translocations have been discovered and characterized in man. These conditions, which include pathologies like Down syndrome, a variety of malignancies, and the group of sex chromosomal diseases, represent an extremely important set of human health-threatening situations, yet in not a single instance have the genes underlying these pathologic conditions been identified or the biochemical nature of the underlying metabolic error been characterized. Furthermore, the causes of these chromosomal defects are still obscure. It appears reasonable to expect that information about these various processes will lead to new and more effective approaches to therapy and prevention of these diseases. Moreover, such advances should also lead to new fundamental understanding of human genetic-biochemical regulatory processes and would illuminate a variety of other disease situations. The purpose of this paper is to describe new methodologies that promise to achieve such needed understanding.

MAPPING OF THE HUMAN GENES

It becomes extremely important to be able to map human genes in any chromosomal location and to achieve the highest possible resolving power.

Birth Defects: Original Article Series, Volume 22, Number 3, pages 307-312

Such data would identify genes that were extra or missing in cases of chromosomal aneuploidy and afford information about the abnormal presence or absence of gene regulatory sequences that might affect specific chromosomal regions within and outside the aneuploid genetic segments. Such studies are particularly important for the sex chromosomal diseases because of the large number of vital genes carried on the X chromosome, the large variety of tissue cells affected by genes on the sex chromosomes, and the phenomenon of lyonization, which appears to act in unique fashion in the sex chromosomes.

Mapping of the human chromosomes can now be carried out in several different fashions. The determination of linkage between a pair of markers can, in a number of cases, furnish an estimate of the distance between the markers involved [3]. The improved procedures of in situ hybridization furnish direct gene mapping data. If the RNA or DNA probe corresponding to a given genetic locus can be prepared, the chromosomal identification and regional mapping position can be determined by molecular hybridization to a preparation of mitotic chromosomes on a microscope slide. The method affords only regional mapping positions, and, since the mapping position obtained is on the mitotic chromosome, one cannot assume a priori that this will conform to mapping position in the true genetic sense. Gene mapping can also be carried out by means of somatic cell hybrids containing specific human chromosomes. Such hybrids can be prepared from the Chinese hamster ovary (CHO) cell and other mammalian cells [4]. Regional mapping of a particular marker can be achieved by subjecting such a cell population to mutation and selection procedures, which yield mutant clones with terminal deletions in the human chromosome. The regional mapping position of the marker of interest is assigned by cytogenetic examination to ascertain the largest terminal deletion retaining the marker of interest.

This method can be modified by application of recombinant DNA techniques so as to achieve a fine-structure mapping that is independent of cytogenetics and that appears to be capable of affording an extremely large resolving power for the locus in question [5]. The method utilizes the principle that repetitive DNA sequences of the human and CHO cell are mutually specific in the Southern blot hybridization. This method affords true genetic mapping and appears to be capable of an increase in resolving power so great that the location of critical genes on the sex, as well as other chromosomes, and determination of regulatory sequences can be secured.

Combination of this technique with that of gene sequencing makes possible, at least in principle, elucidation of the sequence of any genetic locus of the human genome. Understanding of the specific genes and regulatory sequences missing or present in extra fashion in various sex chromosomal diseases should afford important insight into the biochemical basis of the resulting pathology.

PROTEIN ANALYSIS

Important as a change in genomic structure may be, in the last analysis the normal or diseased behavior of a cell depends on the nature and amounts of the different proteins that it contains. Therefore, it becomes necessary to define the changes in protein synthesis caused by specific changes in genetic constitution. So far, no elucidation of such protein differences has been achieved in the sex chromosomal diseases.

An extremely promising approach has been devised by O'Farrell [6], in which two-dimensional gels are prepared and cellular proteins are sorted out on the resulting surface in accordance with isoelectric point values in one dimension and molecular weight in the other. This appears to be a method of extremely great power in that it allows one to examine and ultimately identify large numbers, and possibly even all, of the protein synthesized by any cell population. However, the original methodology suffered from the necessity of making autoradiographs from the resulting gels in which proteins containing labeled amino acids had been deposited at appropriate points. The autoradiographic procedure introduces a time delay that can involve days or weeks before the film is developed; it contributes additional background noise, which can obscure the points of protein deposit, and the amount of blackening of the film is proportional to the actual amount of peptide deposited only over a very narrow range of concentration. For these reasons, quantitative comparison of the amount of different proteins deposited by different kinds of cells differing in genetic composition is difficult.

These difficulties appear now to have been resolved by a new approach devised by Dr. Herbert Anderson of Los Alamos National Laboratory in a collaborative approach with our laboratory. The new device eliminates the need for autoradiography. The gel itself is placed at one end of a cylinder, and at the other end a detector grid of charged wires forming a network consisting of 1 mm squares is placed. An electrostatic field is applied across the length of the cylinder, and a powerful magnetic field collimates the electrons arising from the radioactive peptides deposited on the gel. The electrons traveling down the cylinder produce cascade ionization from the gas existing at low pressure in the cylinder, thus producing a large amount of charge amplification. The radioactive emission from each point on the gel is detected at a corresponding point on the grid, and the data are fed into a computer that delivers a numerical value corresponding to the intensity of peptide deposition at each point on the original gel. The process is rapid and quantitative. Experiments are now underway attempting to determine the qualitative and quantitative differences in protein biosynthesis between fibroblasts of patients with different sex chromosomal conditions.

PROTEIN PROCESSING

The principal problem of the human aneuploid diseases remains essentially as unsolved today as it was more than 2 decades ago: What is the nature of the metabolic abnormality produced in a diploid organism by the presence of extra or missing chromosomal material? Presumably, a full complement of genes is still present, which should suffice to carry out all necessary functions. It has been assumed that some abnormality in regulatory mechanisms may cause pathologic development in the presence of extra or missing chromosome material. Experiments carried out in our laboratory suggest that errors in protein phosphorylation are an important factor in producing the metabolic imbalance of at least some aneuploidies. Experiments were carried out in which CHO cells and their hybrids containing specific human chromosomes were treated with cAMP derivatives; the effect of such treatment on cell survival was measured quantitatively for each of the cell strains employed. It was found that addition of extra human chromosomes greatly changed the growth-inhibitory effect of cAMP derivatives, in some cases inducing great increase in sensitivity and in other cases protecting against this effect. The only known function of cAMP is to increase cellular phosphorylation through activation of particular kinases. Therefore, it seems possible that specific phosphorylating and dephosphorylating actions are associated with the aneuploid condition of specific human chromosomal regions. It is of interest that this phosphorylation hypothesis that invokes changes in protein and other kinds of phosphorylation as a result of chromosomal aneuploidies offers a possible explanation for the fact that pathologies arising from a variety of different chromosomal aneuploidies often involve similar developmental anomalies.

PREVENTION OF HUMAN CHROMOSOMAL DISEASE

Important as is understanding of the detailed biochemistry responsible for the pathology of the aneuploid condition, ultimately one would hope to be able to prevent such diseases. It has been clearly demonstrated that various mutagenic insults can cause virtually every known type of chromosomal mutation. Nevertheless, knowledge adequate to prevent chromosomal disease in man in an effective fashion is still unsatisfactory. What is required is a much more efficient means for detecting and quantitating different kinds of mutagenic actions that result from various environmental exposures. In addition, the biochemistry of various types of mutagenic actions and their associated repair processes must be understood in order to minimize the consequences of exposure to environmental mutagens.

Ames and his collaborators [7] carried out the pioneering studies in measurement of environmental mutagenesis utilizing microorganisms with specific allelic markers whose change could easily be detected in cell populations. The principle was similarly adapted for use with mammalian cells in a variety of different laboratories. However, virtually all these methodologies are designed mainly to detect point mutations, since the marker genes involved are carried on chromosomes that also carry large numbers of other genes essential for cell reproduction. Therefore, deletions and nondisjunctional processes that would lose all or significant parts of the affected chromosome would cause the cell in question to stop multiplication so that its contribution to the mutagenic yield would be lost. This defect was remedied in studies in our own laboratory in which mutagenesis was measured by means of marker genes carried on chromosomes not required for cell multiplication. In this way, it became possible to score point mutations, small deletions, large deletions, and loss of an entire chromosome in a single and convenient experiment [8]. Newer developments with this system also offer promise of quantitative detection of chromosomal translocations.

This methodology appears to have a sensitivity approximately 200 times greater than that of tests that measure predominantly point mutation. It enables one to determine the specific kinds of mutagenic events resulting from different types of exposure and, therefore, offers promise of providing specific and accurate kinds of monitoring of various environmental situations so that conditions harmful to man can be defined and protected against. It promises to provide data that may end the uncertainty and emotion-laden controversy concerning the dangers of human exposure to low-level doses of mutagens. Finally, it should make possible more definitive studies of the biochemistry of mutagenic actions and their repair, which may offer additional means for the prevention of human chromosomal disease [9].

ACKNOWLEDGMENTS

This study was aided by grants from the NIH (PO1 HD0280), the American Cancer Society, the R.J. Reynolds Industries, the Monsanto Company, and the Henry J. Kaiser Family Foundation.

REFERENCES

1. Tjio JH, Puck TT: The somatic chromosomes of Man. Proc Natl Acad Sci USA 44:1229, 1958.
2. Book JA, et al: A proposed standard system of nomenclature of human mitotic chromosomes. JAMA 174:159, 1960.
3. McElheny VK, Abrahamson S (eds): "Assessing Chemical Mutagens: The Risk to Humans" (Banbury Report, Series 1). Cold Spring Harbor, NY: Cold Spring Harbor Laboratory, 1979.
4. Puck TT, Kao FT: Somatic cell genetics and its applications to medicine. Annu Rev Genet 16:225, 1982.
5. Gusella J, et al: Isolation and localization of DNA segments from specific human chromosomes. Proc Natl Acad Sci USA 77:2829, 1980.
6. O'Farrell PH: High resolution, two dimensional electrophoresis of proteins. J Biol Chem 250:4007, 1975.
7. Ames BN, et al: Methods for detecting carcinogens and mutagens with two salmonella/mammalian cell microsome mutagenicity test. Mutat Res 31:347, 1975.
8. Waldren C, et al: Measurement of mutagenesis in mammalian cells. Proc Natl Acad Sci USA 76:1358, 1979.
9. Waldren C, et al: Measurement of low levels of x-ray mutagenesis in relation to human disease. Proc Natl Acad Sci USA 83:4839, 1986.

Index